Automated Deduction – A Basis for Applications

APPLIED LOGIC SERIES

VOLUME 10

SCOPE OF THE SERIES

Logic is applied in an increasingly wide variety of disciplines, from the traditional subjects of philosophy and mathematics to the more recent disciplines of cognitive science, computer science, artificial intelligence, and linguistics, leading to new vigor in this ancient subject. Kluwer, through its Applied Logic Series, seeks to provide a home for outstanding books and research monographs in applied logic, and in doing so demonstrates the underlying unity and applicability of logic.

The titles published in this series are listed at the end of this volume.

Automated Deduction – A Basis for Applications

Volume III

Applications

edited by

WOLFGANG BIBEL

Darmstadt University of Technology, Germany

and

PETER H. SCHMITT

University of Karlsruhe, Germany

KLUWER ACADEMIC PUBLISHERS

DORDRECHT / BOSTON / LONDON

A C.I.P. Catalogue record for this book is available from the Library of Congress.

ISBN 978-90-481-5052-6

Published by Kluwer Academic Publishers,
P.O. Box 17, 3300 AA Dordrecht, The Netherlands.

Sold and distributed in North, Central and South America
by Kluwer Academic Publishers,
101 Philip Drive, Norwell, MA 02061, U.S.A.

In all other countries, sold and distributed
by Kluwer Academic Publishers,
P.O. Box 322, 3300 AH Dordrecht, The Netherlands.

Printed on acid-free paper

TABLE OF CONTENTS

VOLUME III:
APPLICATIONS

VOLUME I:

FOUNDATIONS. CALCULI AND METHODS

VOLUME II: SYSTEMS AND IMPLEMENTATION TECHNIQUES

AFFILIATIONS OF THE AUTHORS

Wolfgang Ahrendt
University of Karlsruhe
Institute for Logic, Complexity
and Deduction Systems
ahrendt@ira.uka.de

Jürgen Brauburger
Darmstadt University of Technology
Department of Computer Science
brauburger@informatik.tu-darmstadt.de

Reinhard Bündgen
IBM Germany Development GmbH
71003 Böblingen, Germany
buendgen@de.ibm.com

Ingo Dahn
Humboldt University Berlin
Institute of Pure Mathematics
dahn@mathematik.hu-berlin.de

Bernd Fischer
Technical University of Braunschweig
Software Technology
38092 Braunschweig, Germany
fisch@ips.cs.tu-bs.de

Jürgen Giesl
Darmstadt University of Technology
Department of Computer Science
giesl@informatik.tu-darmstadt.de

Manfred Kerber
The University of Birmingham
School of Computer Science
Birmingham B15 2TT, United Kingdom
M.Kerber@cs.bham.ac.uk

Jana Köhler
Albert-Ludwigs-University Freiburg
Institute for Informatics
79110 Freiburg, Germany
koehler@informatik.uni-freiburg.de

Michael Kohlhase
University of the Saarland
Department of Computer Science
66041 Saarbrücken, Germany
kohlhase@cs.uni-sb.de

Christoph Kreitz
Cornell University
Department of Computer Science
Ithaca, NY 14853-7501, USA
kreitz@cs.cornell.edu

Hans Jürgen Ohlbach
Imperial College
Department of Computing
London SW7 2BZ, United Kingdom
h.ohlbach@doc.ic.ac.uk

Wolfgang Reif
University of Ulm
Faculty of Computer Science
Department for Software Engineering
and Compiler Construction (PM)
89069 Ulm, Germany
reif@informatik.uni-ulm.de

Gerhard Schellhorn
University of Ulm
Faculty of Computer Science
Department for Software Engineering
and Compiler Construction (PM)
89069 Ulm, Germany
schellhorn@informatik.uni-ulm.de

Johann Schumann
Technical University Munich
Department of Informatics
schumann@informatik.tu-muenchen.de

Gregor Snelting
Technical University of Braunschweig
Software Technology
38092 Braunschweig, Germany
snelting@ips.cs.tu-bs.de

Frieder Stolzenburg
University of Koblenz-Landau
Institute for Informatics
56075 Koblenz, Germany
stolzen@uni-koblenz.de

Jürgen Stuber
Max-Planck-Institute for Informatics
66123 Saarbrücken, Germany
juergen@mpi-sb.mpg.de

Bernd Thomas
University of Koblenz
Institute for Informatics
56075 Koblenz, Germany
bthomas@informatik.uni-koblenz.de

Christoph Walther
Darmstadt University of Technology
Department of Computer Science
Chr.Walther@informatik.tu-darmstadt.de

Part 1

Automated theorem proving in Mathematics

Editor: Michael Kohlhase

INTRODUCTION

We are invited to deal with mathematical activity in a systematic way [...] one does expect and look for pleasant surprises in this requirement of a novel combination of psychology, logic, mathematics and technology.
Hao Wang, 1970, quoted from(Wang, 1970).

The field of mathematics has been a key application area for automated theorem proving from the start, in fact the very first automatically found theorem was that the sum of two even numbers is even (Davis, 1983).

The field of automated deduction has witnessed considerable progress and in the last decade, automated deduction methods have made their way into many areas of research and product development in computer science. For instance, deduction systems are increasingly used in software and hardware verification to ensure the correctness of computer hardware and computer programs with respect to a given specification. Logic programming, while still falling somewhat short of its expectations, is now widely used, deductive databases are well-developed and logic-based description and analysis of hard- and software is commonplace today.

Oddly enough, this development has not carried over to mathematics. It would be plausible that mathematical conferences and journals could demand machine-checkable proofs that verify the correctness of the claimed theorems and complement the informal proofs whose sole purpose is that of communication among mathematicians. It is surely not conservativity on part of the mathematicians that has prevented this vision to become reality. For instance computer algebra systems are nowadays used as standard tools in many areas of mathematics and physics. The main reasons for the lack of widespread computer-supported mathematical reasoning are the insufficient reasoning power for mathematically well-established theories and the lack of user-friendlyness and the lack of support for "theorem proving in the large", i.e. using knowledge bases containing the mathematical knowledge of whole mathematical subfields.

Currently, the reasoning power of automated theorem provers is unevenly distributed: on the one hand, systems such as OTTER (McCune and Wos, 1997) or SPASS (Weidenbach, 1997) have reached a considerable strength,

W. Bibel, P. H. Schmitt (eds.), Automated Deduction. A basis for applications. Vol. III

and have even been used to prove open mathematical problems (McCune, 1997). On the other hand they are more often than not very weak on everyday mathematical tasks, that pose no problem to an average undergraduate student in mathematics. Since these systems are based on universal refutation procedures and only utilize syntactic heuristics in their proof search, they usually surpass humans on problems, where mathematicians have little or no intuition to guide the search for proofs. However, in well-established mathematical subfields, such as e.g. group theory, the battery of mathematical knowledge, proof techniques and intuitions is so vast, that current systems can offer almost no assistance to the working mathematician. It is not likely that this situation will change unless it becomes possible to utilize search heuristics that take the semantics of the mathematical domain into account.

One line of attack on these problems has been to study highly specialized inference procedures that are more tractable and to incorporate them into the universal procedures. Chapter I.2 describes deduction procedures for special theories. The integration of computer algebra systems provides another example, where the integration of external inference engines has extended the scope of deduction systems for mathematics considerably (Suppes and Takahashi, 1989; Kerber et al., 1998).

The other response is to give up on the goal of complete automatization and build interactive deduction systems (see chapter II.1), where the user can directly guide the proof according to his intuition. These systems do not necessarily have the problems with the combinatorics of the search space, but their use still requires too much prior knowledge in logic and is generally too tedious to be a viable alternative to "pen-and-paper" mathematics.

A broad view of the situation has been taken by Hao Wang (one of the founding fathers of the machine-oriented branch of the field of automated reasoning in the 1960s) whose advice from the 1970s is to consider theorem proving for mathematics as a "combination of psychology, logic, mathematics, and technology", i.e. as an integrated effort of various inference components working together. Indeed it seems plausible that the field will only make significant progress in sub-fields of mathematics, where humans have strong guiding intuitions, if we develop inference procedures that are specific for these sub-fields and find system architectures that allow the seamless integration of these procedures as well as a more natural interaction with the user.

Chapters III.1.1 and III.1.4 attack the question of finding alternative architectures that are more suitable for the communication with the user. Chapters III.1.1 "*Lattice-ordered Groups in Deduction*" follows a more bottom-up approach that makes use of the mathematical structure of the problem domain, while chapter III.1.4 "*Proof Planning: A Practical Approach to Mech-*

anized Reasoning in Mathematics" discusses the proof planning paradigm as a cognitively motivated architecture for a mathematical assistant.

Chapters III.1.2 and III.1.3 try to enhance the reasoning power of automated theorem provers by proposing highly specialized and therefore efficient methods for mathematical sub-fields. Chapter III.1.2 "*Superposition Theorem Proving for Commutative Rings*" optimizes an efficient general treatment of equational theories for a particular (but important) mathematical structure, while Chapter III.1.3 "*How to Augment a Formal System with a Boolean Algebra Component*" gives a whole methodology of integrating reasoning about certain properties of sets into a wide range of deduction systems.

Lattice-ordered Groups in Deduction Chapter III.1.1 presents a proof tutor for lattice-ordered groups. This tutor is an interactive theorem prover supported by a set of external reasoning components that include classical automated theorem provers and reasoners specialized to lattice-ordered groups. Ingo Dahn shows that the theory of lattice-ordered groups is particularly well-suited for classical automated theorem provers, since the theory admits an equational formulation as well as a formalization in terms of first-order Horn clauses. Another interesting feature of this theory is that there is a typical interpretation that can be used to test for satisfiability of intermediate results and subgoals; for Horn clauses that do not mention the group operation, validity in this interpretation even entails general validity.

These properties of lattice-ordered groups are used to build a specialized deductive system for this domain using general-purpose automated theorem provers developed within the DFG-Schwerpunkt with only minimal changes. The system goes a long way towards improved usability by providing means of user interaction that are not logic-oriented, but oriented towards mathematical vernacular.

Superposition Theorem Proving for Commutative Rings Chapter III.1.2 presents an instance of the superposition method that promises to lead to automated theorem provers that deal with the theory of commutative rings more efficiently. For practical mathematics it is essential to have specialized inference procedures for the concrete algebraic structures, such as groups, rings, or fields that form the basis of modern algebra and calculus. Jürgen Stuber's work uses a detailed analysis of the properties of equational reasoning in commutative rings to specialize the general superposition inference method (see I.2.11) and to remove redundancy.

The method carries over to theories that can be presented by convergent term rewriting systems, such as Abelian groups, modules and polynomial rings. It may lead to automated theorem provers that deal with these algebraic structures far more efficiently.

How to Augment a Formal System with a Boolean Algebra Component Chapter III.1.3 gives a general method for incorporating reasoning with numerical and other features of sets into certain kinds of formal systems. The basic system may be a logic, but it may for example also be a mathematical programming system for solving equations and inequations. Depending on the type of existentially quantified variables, the combined system can treat different properties of sets. Variables of Boolean type lead to Boolean features of sets, number variables allow to reason about numerical features of sets, e.g. cardinalities.

The method developed by Hans Jürgen Ohlbach and Jana Köhler allows for combining reasoning of axiomatically specified sets with computation and reasoning in the basic system without changing its algorithms. A problem specification can be formulated in a mixed language involving a Boolean part for the sets and a basic part, for example arithmetic equations in the basic system. By means of the *atomic decomposition technique*, the Boolean part is eliminated completely and the whole problem is turned into a pure problem for the basic system.

General reasoning about properties of sets requires higher-order- or set-theoretic methods that are intractable for automated theorem provers except for nearly trivial problems. With the method presented in this chapter, one can make use of efficient special-purpose algorithms instead. Moreover, one can solve problems which are usually not in the realm of logic, for example, like in operations research, finding optimal solutions.

Proof Planning: A Practical Approach to Mechanized Reasoning in Mathematics Chapter III.1.4 presents proof planning as a human-oriented method for mechanized reasoning. In contrast to classical automated theorem proving, the proof planning paradigm does not interpret the process of theorem proving as a search process at the calculus level, i.e. in the search space generated by the inference rules of a (machine-oriented) calculus, but as a hierarchical planning process where the plan operators correspond to human proof methods at various levels of abstraction. Manfred Kerber presents proof planning experiments with the ΩMEGA system and shows that this alternative computational model has several conceptual advantages over the classical paradigm: Proof plans enjoy a close correspondence to human proofs and therefore facilitate human interaction with the machine, the proofs are a means to easily express domain-specific problem solving knowledge and in certain cases they can even be automatically adapted to novel tasks, extending the reasoner's problem solving competence, and finally, the proof planning paradigm can serve as a uniform computational framework for the integration of specialized reasoning procedures and other decision procedures like computer algebra systems at a higher level of abstraction.

References

Davis, M.: 1983, 'A Computer Program for Presburger's Algorithm'. In: J. Siekmann and G. Wrightson (eds.): *Automation of Reasoning*, Vol. 1 Classical Papers on Computational Logic 1957–1966 of *Symbolic Computation*. Springer Verlag, pp. 41–48.

Kerber, M., M. Kohlhase, and V. Sorge: 1998, 'Integrating Computer Algebra Into Proof Planning'. *Journal of Automated Reasoning*. Special Issue on the Integration of Computer Algebra and Automated Deduction; forthcoming.

McCune, W.: 1997, 'Solution of the Robbins Problem'. *Journal of Automated Reasoning* **19**(3), 263–276.

McCune, W. and L. Wos: 1997, 'Otter CADE-13 Competition Incarnations'. *Journal of Automated Reasoning* **18**(2), 211–220. Speacial Issue on the CADE-13 Automated Theorem Proving System Competition.

Suppes, P. and S. Takahashi: 1989, 'An Interactive Calculus Theorem-prover for Continuity Properties'. *Journal of Symbolic Computation* **7**(1), 573–590.

Wang, H.: 1970, 'On the Long-range Prospects of Automated Theorem Proving'. In: *Symposion on Automatic Demonstration*. Springer Verlag.

Weidenbach, C.: 1997, 'SPASS: Version 0.49'. *Journal of Automated Reasoning* **18**(2), 247–252. Speacial Issue on the CADE-13 Automated Theorem Proving System Competition.

INGO DAHN

CHAPTER 1

LATTICE-ORDERED GROUPS IN DEDUCTION

1. INTRODUCTION

Groups and lattices are standard examples of theories used to test deductive systems. In this chapter we investigate how automated theorem provers can support interactive theorem proving in the field of lattice-ordered groups.

The development of a tool that could help mathematicians in the development of proofs has always been a leading motivation in the development of deductive systems. Considerable progress has been achieved towards this aim by the use of interactive theorem provers (Constable *et al.* (1985), Nipkow and Paulson (1992), Gordon and Melham (1993), Farmer *et al.* (1993)). These systems have only very limited possibilities to search for proofs. Instead, they give the user a broad set of proof tactics from which he can chose commands that perform modifications of partial proofs, hopefully leading to a complete proof. Therefore, the user must have a good acquaintance with the available commands to edit a proof.

Many proof tactics require references to axioms or formulas in the proof as parameters. E.g. in order to apply a rule of inference, as the most simple proof tactic, the assumptions that the rule has to use must be selected. Here, the user has to make quite specific choices. So he has to know, what are the available formulas, how they can be used by the tactic and where appropriate formulas can be found in the knowledge base.

On the other side, automated theorem provers have been designed to search for proofs. In each situation they try to find applicable rules and formulas automatically and they have sophisticated methods to avoid redundancies. Though most automated theorem provers can be tuned by setting numerous flags and parameters, they provide default modes that can be applied by users without any knowledge on deduction.

Chapter II.4.14 describes principles that can be applied in order to organize the cooperation of different theorem provers, including the cooperation of automated and interactive provers. While that paper considers the subject from the point of view of deduction technology, we discuss it in this paper rather through the eyes of a mathematician as a potential user.

W. Bibel, P. H. Schmitt (eds.), Automated Deduction. A basis for applications. Vol. III

We shall demonstrate that a combination of both theorem proving paradigms can lead to a system that is easy to use and can assist in the proof of theorems of mathematical significance.

However, we must be aware of a serious gap between the two worlds that we intend to combine. Most automated theorem provers find proofs in a calculus for first order logic. But the vast majority of interesting mathematical problems is formulated in a higher order language. Therefore, interactive theorem provers provide support for formulas in such a language. These formulas cannot be handled immediately by first order provers. Hence, to study the benefits of automated theorem provers for interactive theorem proving, we chose a field of mathematics that poses mathematically interesting first order proof problems.

The theory of lattice ordered groups is well established in mathematics. Glass and Holland (1989) gives a broad survey on the mathematical theory of lattice-ordered groups. This theory can be axiomatized by a purely equational theory. But there is also an axiomatization using a Horn theory of inequalities. Since both theories are logically equivalent, any of them can be used in order to prove advanced theorems. Concerning the possible support by automated theorem provers, this has the consequence that purely equational provers can be used as well as general purpose provers.

Another interesting feature of the theory of lattice-ordered groups is its natural structure as a combination of the theories of lattices and groups with a few connecting axioms.

The first order theory of lattice-ordered groups has become an acknowledged test bed for automated theorem provers. A collection of problems for lattice ordered groups is part of the TPTP library of problems for theorem provers. A report on the behaviour of varying configurations of the DISCOUNT system in this field is in (Fuchs, 1994).

In Section 2 lattice-ordered groups are introduced. We study some of their basic mathematical properties. They will turn out to be relevant to the domain specific configuration of a deductive system. Before we describe such a system, we discuss shortly what a mathematician might expect from a computerized assistant in the field of lattice ordered groups.

Section 4 presents the architecture of an interactive deductive system supported by automated theorem provers and a model checker, specially designed for the domain of lattice-ordered groups.

Section 8 exemplifies the power of the system. The proof of Busulini's theorem is used as a challenge problem. This theorem states that a lattice-ordered group is Abelian if and only if it satisfies the triangle inequality. We shall explain, how the different deductive systems have supported the user editing a proof of the difficult direction of this equivalence.

The proof of Busulini's theorem uses the fact that each element of a lattice ordered group can be presented as a product of its positive and negative part. This Lemma has been proved automatically by the provers DISCOUNT and OTTER. In Section 9 we compare both proofs with each other and with the proof usually found in text books.

The last section gives a critical review of the support that theorem provers can provide actually in the field of lattice-ordered groups.

2. Basics of Lattice Ordered Groups

We shall explain the basic properties of lattice-ordered groups. Details can be found in (Glass and Holland, 1989). Most of the facts discussed here will be applied in Section 4 in order to tune the ILF system for proofs on lattice-ordered groups. We assume that the reader knows the basic concepts of groups and lattices.

DEFINITION 2.1. *A* lattice-ordered group *is a structure*

$$\left(A, \circ, ^{-1}, e, \cap, \cup\right)$$

such that $\left(A, \circ, ^{-1}, e\right)$ *is a group,* $(A, \cap, \cup)$ *is a lattice and the group operation is monotonic increasing in both arguments, i.e. for all* $x_1, x_2, y_1, y_2 \in A$ *we have that*

$$x_1 \leq x_2, y_1 \leq y_2 \rightarrow x_1 \circ y_1 \leq x_2 \circ y_2$$

where $\leq$ *is the usual partial order of the lattice defined by*

$$x \leq y \leftrightarrow x \cap y = x.$$

EXAMPLE 2.1. *The null group with universe* $A = \{e\}$ *and with the trivial order is a lattice-ordered group. It is the only finite lattice-ordered group.*

EXAMPLE 2.2. *The set of integers with addition, additive inverse,* 0, min *and* max *operation is a lattice-ordered group. We denote it by* $\mathbb{Z}$. *If* G *is any lattice-ordered group and a is an element of* G *which is not the neutral element of the group, then the lattice-ordered subgroup of* G *generated by a is isomorphic to* $\mathbb{Z}$.

EXAMPLE 2.3. *Let* $(I, \preceq)$ *be any linear order. Let* $\left(A, \circ, ^{-1}, e\right)$ *denote the group of all permutations of I which are strictly monotonic increasing. For all such permutations* $f_1, f_2 \in A$ *let*

$$(f_1 \cap f_2)(i) = \min\{f_1(i), f_2(i)\},$$
$$(f_1 \cup f_2)(i) = \max\{f_1(i), f_2(i)\}$$

where the minimum and maximum are taken with respect to $\preceq$. Then

$$(A, \circ, ^{-1}, e, \cap, \cup)$$

is a lattice-ordered group. Hence, lattice-ordered groups are in general non-Abelian. Each lattice-ordered group can be presented as a lattice ordered subgroup of a lattice-ordered group constructed this way.

EXAMPLE 2.4. *Each direct product of lattice-ordered groups is a lattice-ordered group.*

It is well known that lattices can be equivalently axiomatized by equations and by Horn clauses stating that each two-element set has a least upper bound and a greatest lower bound with respect to the order $\leq$. Also the Horn clauses stating the monotony of the group operation can be equivalently replaced by the equations

$$\begin{aligned} x \circ (y \cap z) &= x \circ y \cap x \circ z \text{ and} \\ (y \cap z) \circ x &= y \circ x \cap z \circ x \end{aligned}$$

or their duals

$$\begin{aligned} x \circ (y \cup z) &= x \circ y \cup x \circ z \text{ and} \\ (y \cup z) \circ x &= y \circ x \cup z \circ x. \end{aligned}$$

It can be shown that the lattice of each lattice-ordered group is distributive. Subsequently we will always assume, that the distributivity of the lattice is part of each axiom system for lattice-ordered groups that we use.

We generalize some concepts well known for $\mathbb{Z}$ to arbitrary lattice-ordered groups.

DEFINITION 2.2. *An element a is* positive (negative) *if $e \leq a$ ($a \leq e$).*

$a^+ \stackrel{Df}{=} a \cup e$ *and* $a^- \stackrel{Df}{=} a \cap e$ *are called the* positive *and* negative part *of a respectively.*

$|a| \stackrel{Df}{=} a \cup a^{-1}$ *is the* absolute value *of a.*

A lattice-ordered group satisfies the triangle inequality if for all a, b

$$|a \circ b| \leq |a| \circ |b|.$$

THEOREM 2.1. (Busulini). *A lattice-ordered group satisfies the triangle inequality if and only if it is Abelian.*

It is relatively easy to show that each Abelian lattice-ordered group satisfies the triangle inequality. In fact,

$$\begin{aligned} |a| \circ |b| &= \left(a \cup a^{-1}\right) \circ \left(b \cup b^{-1}\right) \\ &= (a \circ b) \cup \left(a \circ b^{-1}\right) \cup \left(a^{-1} \circ b\right) \cup \left(a^{-1} \circ b^{-1}\right) \\ |a| \circ |b| &= (a \circ b) \cup \left(b \circ a^{-1}\right)^{-1} \cup \left(a^{-1} \circ b\right) \cup (b \circ a)^{-1}. \end{aligned}$$

Hence, if the group is Abelian,

$$\begin{aligned} |a| \circ |b| &= (a \circ b) \cup \left(a^{-1} \circ b\right)^{-1} \cup \left(a^{-1} \circ b\right) \cup (a \circ b)^{-1} \\ &= |a \circ b| \cup \left|a^{-1} \circ b\right| \\ |a| \circ |b| &\geq |a \circ b|. \end{aligned}$$

The reverse direction of Busulini's Theorem requires a more complex proof. We shall sketch it in Section 8. This will also require a more detailed discussion of the elementary concepts introduced above.

Now we introduce some theorems from first order model theory. Their application will support the efficient use of automated theorem provers. Semantic methods have been used in several provers in order to guide the search process or to find counterexamples. We mention Gelernter (1959), Reiter (1973), Ballantyne and Bledsoe (1982), and Plaisted and Zhu (1997) as a selection of these approaches. In contrast with these, we shall use semantic information *outside* and *instead* of automated provers.

DEFINITION 2.3. *Let* $L_3 = (\{-1,0,1\}, -, 0, \min, \max)$.

Terms of the language of lattice-ordered groups which do not contain the group operation $\circ$ have values in L_3 when $^{-1}, e, \cap, \cup$ are interpreted as $-, 0, \min, \max$ respectively. Since L_3 is a substructure of the lattice-ordered group $\mathbb{Z}$, each universal sentence without $\circ$ which holds in all lattice-ordered groups must hold in L_3.

Moreover, a partial converse of this statement can be proved. This is based on the following algebraic theorem.

THEOREM 2.2. *If* $(A, \circ, ^{-1}, e, \cap, \cup)$ *is a lattice-ordered group, then*

$$(A, ^{-1}, e, \cap, \cup)$$

is a subdirect product of copies of L_3.

The validity of universal Horn formulas is preserved under subdirect products. Hence we have the following

THEOREM 2.3. *For each universal Horn sentence H in the language of lattice-ordered groups which does not contain the group operation symbol the following conditions are equivalent.*

- *H holds in* L_3,
- *H holds in all Abelian lattice-ordered groups,*
- *H holds in all lattice-ordered groups.*

When a model checker verifies that a sentence of this type holds in L_3, we can conclude from a completeness theorem for a calculus C that there must be a formal proof of this sentence in C from the theory of lattice-ordered groups. Such a formal proof would involve only concepts which are already present in the theory of lattice-ordered groups. However, we don't have any hint, how such a proof could be obtained.

Also the analysis of the proof of Corollary 2.3 does not help. This proof uses groups and formulas as objects and cannot be reformulated in terms of members of a lattice ordered group. However, the use of external tools is quite common in mathematics. E.g. geometric problems are translated into analytic problems through the use of coordinates and arithmetic problems are solved studying functions on the field of complex numbers. We shall return to this point in the next section.

Theorem 2.3 holds actually for a somewhat larger class of formulae. This is based on the observation that positivity and negativity, a^+, a^- and $|a|$ all have explicit definitions without reference to the group operation. Hence Corollary 2.3 holds also for universal Horn formulas that contain additionally these operations. E.g. we can define x^- in L_3 as in each lattice-ordered group by the equation $x^- = x \cap e$. Similarly, 0 and 1 can be considered as the positive elements of L_3, i. e. those elements x satisfying $x \cup e = e$. Hence, when we have checked that

(1.1) x^- is positive $\rightarrow x = e$

holds for all x in L_3, we have also verified that

(1.2) $(x \cap e) \cup e = e \rightarrow x = e$

is true in L_3. Corollary 2.3 can be immediately apply to this formula. Hence (1.2) holds in all lattice-ordered groups. But (1.1) is just a reformulation of (1.2), hence it holds there as well.

We note another theorem that can be used to reduce proof problems for *Abelian* lattice ordered groups.

THEOREM 2.4. *For each universal Horn sentence H in the language of lattice ordered groups the following conditions are equivalent.*

– *H holds in* $\mathbb{Z}$,
– *H can be proved from the theory of all Abelian lattice-ordered groups,*
– *H can be proved from the theory of all linearly ordered groups.*

This theorem is a consequence of the fact that each Abelian lattice-ordered group is a subdirect product of linearly ordered groups. This was discovered by Lorenzen (Lorenzen, 1949) and Sik (Sik, 1960).

3. A User Profile

In the last section we have introduced basic concepts for lattice-ordered groups as they are introduced in any introductory lecture in this field. Students taking such lectures gain some ideas on these concepts and have to derive further theorems as exercises. A major concern of the tutor who has to evaluate these exercises is to verify the correctness of the inferences that the student makes. We are going to describe an environment that can do this work for him. In this sense we are developing a proof checker.

However, to be realistic, we do not assume that the student has any acquaintance with formal logic. He might be neither willing nor able to memorize a large set of powerful proof tactics or to develope its proof in boring detail in a low-level calculus with a few inference rules. So he cannot deliver a proof in a formal calculus that can be checked by simply verifying that certain explicitly given rules of inference are applied in a formally correct way.

Instead, the student will give a rather informal argument by saying that a formula is a consequence of some other formulas. Sometimes he might give too many reasons for an inference, the assumptions might not come in the order in which they have to be applied and he even might forget some necessary assumptions which are obvious to him.

E.g. he might argue, that $a \circ a \circ a$ is positive if a is positive since $\circ$ is a monotonic operation. This argument assumes tacitly the definition of positivity, the fact, that $e \circ e = e$ and the transitivity of the ordering. Moreover, the student can rely on the tutor's ability to see that, given these assumptions, $e \leq a \rightarrow e \leq a \circ a \circ a$ is a valid formula.

The student is not concerned *why* the tutor accepts his argument - whether by intuition, by constructing a proof in a logical calculus or after feeding it to an automated theorem prover. It is only important, that the tutor accepts correct arguments and refutes incorrect inferences.

In a similar way, the user of an editor is not interested in the work of the built-in spell checker as long as it works correctly. Of course, the tutor may not be satisfied with the student's argument and may ask him to provide further evidence or to explain his inference in a more detailed way. But the tutor must take care that the discussion on lattice-ordered groups is not turned into a discussion on formal proofs of trivial details.

For this application which we have in mind, it is neither necessary nor important that a formal proof is obtained (though it is desirable as we explain below). We note that the MIZAR system takes a similar approach. It checks the arguments given by the author of a mathematical article but does not present formal proofs to justify them.

It is a drawback of this approach that the correctness of the argument depends on the correctness of all verification components and their integration. Chapter II.4.14 gives a method to ensure correctness of the integrated system provided that its components are correct.

In order to maximize the evidence of correctness, the verification component must be as simple as possible. Hence arguments must be easy to check. Formal proofs serve exactly this purpose. Moreover, the axioms on which the arguments are based, should be simple and clearly stated.

This simplicity implies, that formal proofs representing complex arguments will become extremely lengthy. Certainly, they will be too lengthy for a human being to work with.

Systems like KIV (Reif, 1992) and OMEGA (Benzmüller et al., 1997) enforce correctness by developing a proof in a single formal calculus. They will translate proofs found by other systems into proofs in their native calculus. This applies also to arguments formulated by the user in some high level language. It is feasible as long as these proofs are small. Hopefully, sometimes, it will be possible to prove the correctness of this translation procedure so that it will not be necessary to perform the translation in reality.

Translation into a single calculus with a simple axiom system is desirable for the sake of safety. But it must be hidden from the working mathematician. When there are other powerful tools (like Corollary 2.3, model checkers or the MIZAR verifier), we assume it is legal to make them available — provided it is made clear that the correctness of explicitly mentioned parts of the argument depends on them. This gives the possibility to replace these parts later, if required, by native formal proofs.

Translation into a single calculus is necessary, if formal proofs are further processed. The extraction of programs from constructive proofs or the verification of proofs from one system by another system are such applications.

4. A TUTOR FOR LATTICE ORDERED GROUPS

We are going to describe a system that can act as a tutor for developing proofs of elementary facts on lattice-ordered groups. In this section we describe the overall concept as it appears to the user. The next section explains the architecture and Section 7 mentions some features of the user interface which turned out to be important for unexperienced users. An example will demonstrate the usability of the system.

The system, called PROOFPAD, is started with a concrete formula to prove. It knows a set of basic axioms on lattice-ordered groups. These are the axioms for lattices in equational form and as conditions on upper and lower bounds, a set of axioms for group theory, the monotonicity conditions and their equational counterparts. Distributivity of the lattice is given in its usual equational form and in the equivalent form

$$x \cap z = y \cap z \wedge x \cup z = y \cup z \rightarrow x = y.$$

This basic theory is structured as a tree of subtheories. A special theory, which is initially empty, collects intermediate results.

The user can prove a theorem by editing a partial proof. Such proofs consist of sequences of lines. Initially, there is only one line. It contains the goal to be proved. The PROOFPAD tries to verify automatically that each line is a logical consequence of the axioms and the preceding lines.

Each line has the status *untried* when it is generated. This status is changed to *tried* when PROOFPAD tries to verify this line. When this attempt is finished, the status is changed automatically to either *proved* or *unproved* or *unprovable*. Thus, PROOFPAD appears to the user to be similar to an on-line spell checker built into a conventional text editor.

The user can edit his partial proof by inserting new formulas as intermediate steps in the PROOFPAD. Unless instructed otherwise, the system will immediately try to prove this formula from the theory of lattice-ordered groups, all lemmas derived so far and a certain number of lines before this new line. For technical reasons explained in Chapter II.4.14 this number of lines is rather small, usually it is 3, i. e. the tutor has a small short-term memory. Hence the user must demand explicitely that more distant lines are taken into account.

When a proof attempt for a certain line fails, the user has to provide further evidence for this line. He can insert more intermediate steps or he can try to give better support to the failed proof attempt. Sometimes a line cannot be proved because a useful preceding line is too far away. It can be also helpful, to restrict the theory from which the line is to be proved, i.e. not to use all axioms for lattice-ordered groups. This method can be error prone because the human user tends to forget axioms which are obvious to him. Therefore, the axioms saying that $\leq$ is a partial ordering are always used, even if the user does not specify them.

Each field of mathematics has its specific proof techniques which must be made available in a domain specific theorem prover. These techniques transform a partial proof in an algorithmic way without explicit construction of a subproof that justifies this transformation. Paramodulation can be seen as such a technique. A specific proof technique relevant to our field is multiplying out, i.e. using distributivity in order to replace subterms of the form $x \cap (y \cup z), x \cup y \cap z$ by $x \cap y \cup x \cap z, (x \cup y) \cap (x \cup z)$ as long as possible. Another technique moves a member of a sequence of terms connected by a commutative and associative operator towards another member of this sequence.

It is important to note that none of the user actions described above requires knowledge of a logical calculus. But all of them might be used by a student explaining his argument to a human tutor.

5. System Architecture

The system described in the last section has been implemented within the ILF system. The services offered by ILF are described in (Dahn *et al.*, 1997). Here we confine ourselves to the ILF configuration which has been developed in order to support proofs in the field of lattice-ordered groups. Figure 1 shows its global architecture.

Conceptually, the ProofPad is an interactive theorem prover which communicates with other systems and with the user through services provided by the ILF foreground system. It takes care of the proof edited by the user. According to user commands it inserts unproved formulas into the proof and annotates hints given by the user. Moreover, the ProofPad extracts particular proof problems.

These proof problems contain the information which formula should be proved and which axioms or proof lines should be used in the proof. Through ILF all proof problems are passed to the problem analyzer running asynchronously within the ILF background system.

There, hidden from the user, a sequence of tests and transformations is performed. Conceptually, the problem analyzer acts like an automated theorem prover which itself can call other provers in order to establish the solvability of proof problems. It is a specific feature of this component to handle several proof jobs simultaneously.

The first subsystem which takes the problem is ThreeLat. It checks, whether the formula to be proved is true in $\mathbb{Z}$ for all instances of the vari-

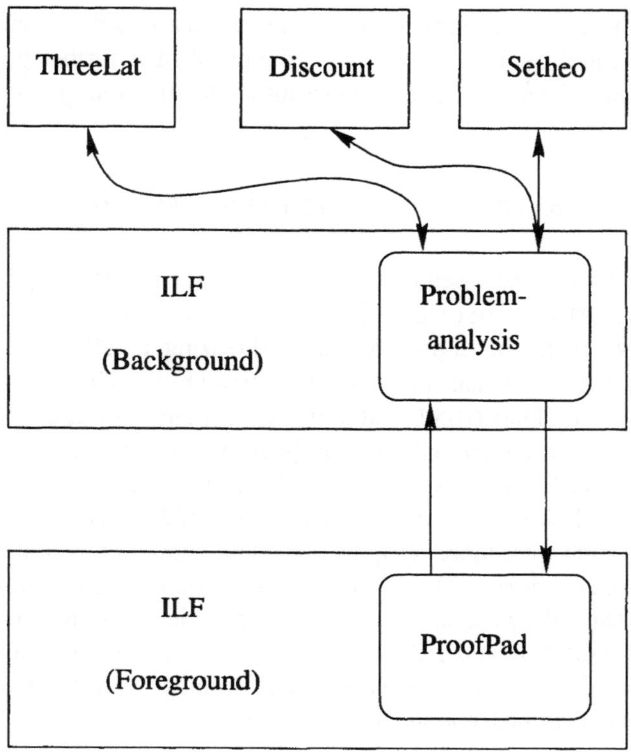

Figure 1. Architecture of a Theorem Prover for Lattice-ordered Groups

ables with $-1,0$ and 1. If this test fails, no attempt to prove it is made and the status of the corresponding line in the foreground is changed to *unprovable*. This may happen when the user entered a formula which is syntactically correct but unprovable from the given assumptions — because of a typing error or simply because the user forgot some necessary assumptions. In this way, THREELAT is a model checker that prevents ILF from starting theorem provers for junk theorems.

During the checks in THREELAT, a flag is set if the formula is not universal Horn or if a relation or operation is found, that is defined using the group operation $\circ$. When this flag is still cleared after the test, we know from Corollary 2.3 that the formula under consideration can be proved from the theory of lattice-ordered groups - though we don't have a proof. Also in this case no further proof attempt is made. and the status of the corresponding line is changed to *proved*. Moreover, a control information is attached to the line which informs the user, that THREELAT has verified the provability.

Now assume that the provability of the formula could be neither established nor refuted by THREELAT. Then the problem is further processed in order to prepare the start of one or more automated theorem provers.

6. DOMAIN-SPECIFIC PREPROCESSING

In the configuration discussed here, ILF contains the automated provers DISCOUNT and SETHEO. DISCOUNT (Denzinger *et al.*, 1997) is a rewrite based theorem prover using the teamwork method (Denzinger, 1995). SETHEO (Letz *et al.*, 1992) is a prover that finds model elimination proofs.

The version of DISCOUNT that was used required an equational axiom system as well as an equational proof problem. When the assertion of the goal was not a positive literal, DISCOUNT could not be used to prove it. Otherwise, whenever a condition was found in a goal, it was added as a new axiom.

Axioms which were more complex than facts, especially conditional equations, could not be given to DISCOUNT. Since lattice-ordered groups are completely axiomatized by the equational part of the axiom system, the completeness of DISCOUNT would still theoretically be able to verify all provable equations. For DISCOUNT, non-equational formulas were transformed into equations whenever possible using the definition

$$x \leq y \leftrightarrow x \cap y = x.$$

Note that replacing $\leq$ by $\cap$ this way can increase the complexity of the formula considerably since the term on the left side of $\leq$ now occurs twice in the equation.

The power of term rewriting methods depends on the appropriate choice of an ordering on the symbols of the language. For a successful application of DISCOUNT it was important that ILF choses an ordering that preferred symbols from the goal as reduction target. DISCOUNT was used in team mode with 2 experts and referees.

Matthias Fuchs (Fuchs, 1994) made numerous experiments to find a configuration of DISCOUNT that was optimal for proofs in the field of lattice-ordered groups. He found that best results were obtained when a goal-oriented heuristic was applied by one of DISCOUNT's experts, working in a team with two other experts. Henceforth we refer to this configuration only.

The goal-oriented expert of DISCOUNT prefers the generation of rewrite rules that are similar to the goal. This similarity was measured in terms of the set of occuring function symbols and their nesting. Also matching subterms in the rule and the goal were taken into account.

SETHEO is a general purpose theorem prover for clausal theories. Therefore, it could treat all proof problems discussed here. Since the axiom systems of these problems consisted of universal Horn formulas, no special transformations of the axioms was necessary. In a more general context it would have been useful to modify the axioms automatically in order to ensure that SETHEO starts its model elimination proof from the goal clause. However, SETHEO has no special tools for treating equalities. Therefore standard axioms for equality where added automatically to each axiom system given to SETHEO.

Usually there are several unproved lines. Each line determines a proof problem and can lead to the configuration of two automated theorem provers. When the user inserts a new line, this creates not just one new proof problem but modifies also subsequent proof problems. Hence, there can be a considerable number of pending proof problems. In order to obtain acceptable response time and sufficient resources for the automated provers, it is necessary to assign the resources of the local network to the provers in an appropriate way. This is one of the services provided by the ILF background system. It is discussed in Chapter II.4.14.

Whenever one of the provers finishes its work successfully, the ILF background system analyzes its output file. As soon as ILF realizes that a proof problem was solved, the problem analyzer is informed and sends a message to the PROOFPAD to change the status of the corresponding line of the proof to *proved*. When none of the provers could prove the goal within a given time limit, the status is changed to *unproved*.

When the user deletes a line, all succeeding lines which had been proved using this line, get the status *untried* again. In order to enable this, provers must be configured to print proofs. In case of the DISCOUNT prover this requires extensive postprocessing. Though THREELAT does not generate a proof, it can verify whether all assumptions in a provable universal Horn formula are needed.

No attempt is made to incorporate the automatically generated subproofs into the user's proof. However, ILF has procedures which can do this later when the user has completed his proof. It is also possible, to obtain LaTeX presentations of the DISCOUNT and SETHEO proofs.

7. THE USER INTERFACE

The PROOFPAD runs under the graphical user interface provided by the ILF system. Its main window is used to issue commands and to get immediate replies from the system. Also interactive proof editing is performed in this window. Two more windows are controlled by the TREEVIEWER — a separate system used within ILF — that can present labelled directed acyclic graphs. One of these windows displays the graph while the other displays the labels of the nodes. Both windows are synchronized. The TREEVIEWER is used to display incomplete proofs, hierarchies of theories and single formulas. Moreover it has menus with some commands for proof editing. It can be also used as an input device for selecting proof lines, subtheories, subformulas or terms. The fourth window is used by other subsystems (especially automated theorem provers) to inform the user about their results.

On each stage of the development of a proof the user can obtain a presentation with the usual mathematical symbols in LaTeX.

Axioms are named by ground terms. Hence it is possible to refer to groups of axioms either by calling subtheories or by giving name templates. E.g. there are three associativity axioms — `asso(*)`, `asso(n)`, `asso(u)` — asserting the associativity of $\circ, \cap, \cup$ respectively. Using the Prolog variable _, `asso(_)` selects the group of all associativity axioms and lets the system decide, which of them is to be used in a concrete case.

Some domain specific proof tactics mentioned in Section 4 manipulate subterms deep inside a proof line. The choice of these subterms is facilitated by the graphical user interface. E.g. if the user indicates that he wants to move a term, the structure of the formula in the current line is presented by the TREEVIEWER and the user can select source and destination of the move with the mouse. Of course, ILF checks that the destination can be really reached from the source passing only through positions with a specific operator which is associative and commutative.

8. A Proof of Busulini's Theorem

In this section we shall demonstrate what can be achieved with the techniques described in the previous sections. We shall explain, how a non-trivial proof for lattice-ordered groups can be interactively generated with the PROOFPAD.

Busulini proved in (Busulini, 1958) that a lattice-ordered group is Abelian if and only if it satisfies the triangle inequality $|x \circ y| \leq |x| \circ |y|$. The easy part of this equivalence was demonstrated in Section 2. Now we sketch the proof from PROOFPAD that the triangle inequality implies the commutativity of the group operation. We indicate, which parts were generated by automated provers. In each of these cases the provers confirmed the corresponding part within 30 seconds. DISCOUNT required 90 seconds more for each problem to generate a proof. ILF was running in a local network consisting of 2 Sparc 10, 4 Sparc 2 and one Sparc 1+.

Henceforth $a, b \ldots$ range over elements of a lattice-ordered group that satisfies the triangle inequality. All formulas in this section had to be entered by the user. However, unless stated otherwise, the axioms and lemmata that justify a particular formula were found by the provers automatically.

The idea of the proof is, to prove the assertion for positive elements first and then to extend it to arbitrary elements through the use of a representation theorem.

LEMMA 8.1. *a^+ is positive for each a.*

This was confirmed immediately by THREELAT.

LEMMA 8.2. *The product of positive elements is positive.*

This was proved by SETHEO.

LEMMA 8.3. *If a is positive, then $|a^{-1}| = a$.*

This was confirmed by THREELAT.

LEMMA 8.4. *If a and b are positive, then $a \circ b \leq b \circ a$.*

Proof SETHEO confirmed that $\left|(a \circ b)^{-1}\right| \leq |b^{-1}| \circ |a^{-1}|$. It needed the hint, to use group theory and the triangle inequality. Then SETHEO proved from Lemma 8.3 and Lemma 8.2 $\left|(a \circ b)^{-1}\right| = a \circ b$ and the assertion.

LEMMA 8.5. *Positive elements commute.*

This was obtained from Lemma 8.4 by SETHEO.

LEMMA 8.6. *$a^+ \circ b^+ = b^+ \circ a^+$ for all a, b.*

SETHEO realized, that this is a consequence of Lemma 8.1 and Lemma 8.5.

LEMMA 8.7. *a^- is negative for each a.*

This was confirmed immediately by THREELAT.

In the proof of the following Lemma all automated provers in ILF cooperated successfully.

LEMMA 8.8. *If a is negative and b is positive, then a and b commute.*

Proof THREELAT confirmed that a^{-1} is positive. Then SETHEO proved from Lemma 8.5 that $b \circ a^{-1} = a^{-1} \circ b$. From this equation DISCOUNT could justify the assertion. The intermediate steps had to be given by the user.

LEMMA 8.9. *$a^{+} \circ b^{-} = b^{-} \circ a^{+}$ for all a and b.*

This was derived by SETHEO from Lemma 8.1, Lemma 8.7 and Lemma 8.8.

LEMMA 8.10. *Negative elements commute.*

Proof If a and b are negative, then THREELAT confirmed that a^{-1} is positive. Hence SETHEO obtained from Lemma 8.5 that $b \circ a^{-1} = a^{-1} \circ b$. This was used by DISCOUNT to prove the assertion.

Now SETHEO proved from Lemma 8.7 and Lemma 8.10

LEMMA 8.11. *$a^{-} \circ b^{-} = b^{-} \circ a^{-}$ for all a, b.*

The following Lemma contains the representation theorem mentioned above.

LEMMA 8.12. *$a^{+} \circ a^{-} = a$.*

This was proved by DISCOUNT from the equational theory of lattice ordered groups without further hints.The proof and a proof of the same formula by OTTER are discussed in the next section.

Now the final assertion can be proved as follows.

DISCOUNT can prove from Lemma 8.6, Lemma 8.9 and Lemma 8.11 that $a^{+} \circ a^{-} \circ (b^{+} \circ b^{-}) = b^{+} \circ b^{-} \circ (a^{+} \circ a^{-})$ and hence that $a \circ b = b \circ a$ by Lemma 8.12.

This proof of Busulini's theorem shows, that all essentials of the proof, especially the Lemmas, had to be given by the user. With the exception of Lemma 8.12 the subproofs found by the automated provers are relatively small. E.g. the maximal subproof found by SETHEO had 15 formulas arranged in a tree of depth 4.

Nevertheless, the use of automated theorem provers has considerably simplified the verification of Busulini's Theorem. The user had not to know any rule of inference and the level of detail that he had to feed into PROOFPAD is comparable to that found in mathematical textbooks. Important details like the selection of the specific assumptions which are necessary to justify the particular formulas of the proof were augmented automatically by the provers. Through the use of ILF, the automated theorem provers could be used without any knowledge of their specifics.

Finally, when the proof has been generated interactively in ILF, it can be presented immediately in LaTeX with a notation used by mathematicians. The automatically generated subproofs can be presented in the same way with any desired level of detail. This presentation is based on a transformation of the subproofs into block structured proofs. The interactively generated proof has also a block structure. An integration of all these proofs into a single block structured proof with a coherent semantics is theoretically possible (see Chapter II.4.14 and (Dahn and Wolf, 1994)), though this has not been implemented.

The representation theorem Lemma 8.12 deserves special attention. Its standard proof uses a particular transformation of the problem. The idea is to prove $a^+ = a \circ (a^-)^{-1}$ instead of $a^+ \circ a^- = a$. Given this trick, the rest of the proof can be found by standard techniques:

$$a \circ (a \cap 1)^{-1} = a \circ \left(a^{-1} \cup 1\right) = 1 \cup a.$$

Note that this proof uses tacitly some other equations, especially $(x \cap y)^{-1} = x^{-1} \cup y^{-1}$, which also have to be proved. This was not necessary for our proof of Busulini's theorem because DISCOUNT found within 30 seconds another proof without further hints. We shall discuss it in the next section.

9. Two Automated Subproofs

The core technology of DISCOUNT is term rewriting. Therefore, DISCOUNT tries to prove an equation by reducing each of its sides to the same normal form. When this is not possible, it tries to find intermediate terms that can be proved to be equal to both sides of the equation. Observe that the problem transformation used in the standard proof of the representation theorem is not accessible with this technique. Hence, DISCOUNT had to find an essentially different proof. It cannot be expected that this proof will be the most simple one.

DISCOUNT analyses which equations are of special importance for a proof and extracts them as lemmata for the final proof presentation (Denzinger and Schulz, 1996). Since simple terms are naturally prefered, these lemmata are sometimes also of mathematical interest for other proofs.

The idea of DISCOUNT's proof of the representation theorem can be expressed by the following chain of equations.

$$a^+ \circ a^- = a^- \circ a^+ = a \circ \left(a^+\right)^- = a$$

In order to justify this, DISCOUNT discovered, that in an arbitrary lattice-ordered group

$$\begin{aligned} x \circ x^- &= x^- \circ x, \\ x^+ \circ x^- &= x^- \circ x^+, \\ \left(x^+\right)^- &= e, \\ \left(x^+\right)^+ &= x^+. \end{aligned}$$

The proof could only be found within the given time limits since the problem analyzer of ILF generated for DISCOUNT in the preprocessing phase an ordering of the symbols that lets DISCOUNT prefer symbols in the goal. Therefore, DISCOUNT did not simplify $(a^+)^-$ to e whenever possible.

DISCOUNT realised within 30 seconds on one Sparc 10 and two Sparc 2 that the assertion could be proved. It took 90 seconds more to work out the proof. No problem specific tuning by the user was made. The proof consisted of 92 equations organized in 19 chains.

OTTER is a saturating general purpose theorem prover with special support for handling equations. Larry Wos (Wos, 1997) has tuned it carefully for proving the same theorem. Then OTTER displayed a proof after 3 seconds on a Sparc 10. OTTER's proof had 33 lines. However, it must be taken into account that OTTER's paramodulation steps can be quite complex compared with the rewrite steps of DISCOUNT. OTTER produces also a very detailed explanation as a so-called proof object. The proof object of this proof had 208 lines (not counting input lines).

OTTER proved essentially that

$$a^+ \circ a^- = (a \circ a \cup a) \cap (a \cup e) = a$$

The formulas in OTTER's proof are more complex than those in the proof from DISCOUNT. Where DISCOUNT tends to introduce derived concepts like the positive and the negative part of an element, OTTER tries to eliminate them. This can hide intermediate results which are of independent mathematical interest. E.g. OTTER's proof contains the formulas

$$\begin{aligned} (x \cap e)^{-1} &= e \cup x^{-1}, \\ (x \cup e) \cap \left(e \cup x^{-1}\right) &= e. \end{aligned}$$

The first of these formulas is (almost) $\left(x^-\right)^{-1} = \left(x^{-1}\right)^+$, while the second, transformed into $x^+ \cap \left(x^{-1}\right)^+ = e$, says that the positive parts of an element and its inverse are mutually orthogonal.

ILF has automatically generated a human readable presentation of the proofs found by DISCOUNT and OTTER. This included a combination of OTTER's proof with the proof object and an elimination of trivial details. For both proofs the presentation in LaTeX, including the axiom system, was 7 pages long.

10. Conclusion and Open Problems

It has been demonstrated that automated theorem provers can give substantial support to the user of an interactive proof editor. Especially, it is possible to edit proofs without explicite reference to rules of inference or logical calculi. The amount of detail that the user has to supply can be reduced to a level approaching the level of usual mathematical discourse. Thus the development of a mathematical tutoring system becomes feasible.

Though the *use* of a system like PROOFPAD does not require knowledge of logic or theorem provers, the *configuration* of the system is based on extensive expertise in theorem proving as well as in the specific field of application.

The combination of theorem provers with different abilities and domain specific methods has led to considerable improvements. Though these subsystems did not find deep mathematical proofs in the present context of lattice-ordered groups, they were able to bridge the gap between mathematical arguments and formal logic. In a few cases, the analysis of subproofs found automatically, has also revealed some mathematically interesting facts. Such an analysis has to be done by the expert who knows about the important concepts of the field. This requires that these proofs can be presented in a language close to the language of the user.

In the example discussed here it was sufficient to use universal Horn formulas as axioms and assertions . In other contexts, when existential quantifiers are involved, general purpose automated theorem provers working with clauses cannot exploit specific structural properties of formulas. The proofs that they find after transforming the proof problem to clausal form will be often hard to understand by the untrained user. This may not be important, as long as the user is only interested in the *confirmation* of his argument and not in its *justification*. A more important restriction is, that many automated theorem provers for first order logic do not document the transformation of a problem into clauses. Hence, when they found a proof they are not able to tell the user which of the axioms they found to be essential.

The setup discussed here is still restricted. Most mathematical problems are formulated in a many sorted language. In the theory of lattice-ordered groups they concern structures and mappings between structures. Some theorems deal with relations between classes of structures. Problems of this kind cannot be even formulated in the input language of most automated theorem provers, notably of those used in ILF. So the use of sorts appears as an urgent need.

In a mathematically meaningful domain certain terms play a designated role. This should be taken into account in the theorem proving process. E. g. rewrite rules may be used to simplify terms. DISCOUNT could discover such rules and SETHEO might benefit from the reduced term complexity obtained by their use. This would require a communication between both automated systems at run time. Chapter II.4.14 describes a way to achieve this.

The axiom system for lattice-ordered groups is rather small compared with the bulk of mathematical knowledge used by a working mathematician. The presentation of this knowledge in a computer in a form easily accessible to this user as well as to automated theorem provers is still a widely unsolved problem.

It has been shown that the use of semantic information, like model checking in L_3, can be very powerful. Still the full benefits of these methods are not well understood. This concerns the conditions under which they are applicable as well as the consequences which they might have for the guidance of the search of conventional proofs. Algebraic and model theoretic properties of the domain under consideration, like Theorem 2.2 and Corollary 2.3, should be investigated as a starting point.

In this way, domain specific tools in combination with general-purpose automated theorem provers can be integrated to build mathematically useful deductive tools.

REFERENCES

Ballantyne, A. M.; Bledsoe, W. W.: On Generating and Using Examples in Proof Discovery, Machine Intelligence 1982, pp. 3–39

Benzmüller, C. and Cheikhrouhou *et al.*: ΩMEGA: Towards a Mathematical Assistant, Proc. CADE-14, Townsville, LNAI 1249, 1997, pp. 252–255.

Busulini, B.: Sulla Relazione triangolare in un l-grupo; Rend. Sem. Math. Univ. Padova, 1958, vol. 28, pp. 68–70.

Constable, R. L. *et al.*: Implementing Mathematics with the Nuprl Proof Development System, Prentice-Hall Inc., 1985

Dahn, I., Gehne, J. *et al.*: Integration of Automated and Interactive Theorem Proving in ILF, Proc. CADE-14, Townsville, LNAI 1249, 1997, pp. 57–60

Dahn, B.I. and Wolf, A. (1994): A Calculus Supporting Structured Proofs, *J. Inf. Proc. Cyb.*, pp. 261–276.

Denzinger, J.: Knowledge-Based Distributed Search Using Teamwork, Proc. ICMAS-95, San Francisco, AAAI-Press, 1995, pp. 81-88.

Denzinger, J.; Schulz, S.: Recording and Analyzing Knowledge-Based Distributed Deduction Processes, Journal of Symbolic Computation Vol. 21, 1996, pp. 523-541.

Denzinger, J. ; Kronenburg, M. ; Schulz, S.: DISCOUNT. A Distributed and Learning Equational Prover, Journal of Automated Reasoning 18(2), 1997, pp. 189-198.

Farmer, William M. and Guttman *et al.*: IMPS: An Interactive Mathematical Proof System, The MIITRE Corporation, 1993.

Fuchs, M.: The Application of Goal-oriented Heuristics for Proving Equational Theorems via the Unfailing Knuth-Bendix Completion procedure. A Case Study: Lattice Ordered Groups, SEKI Report SR-94-02.

Gelernter, H.: Realization of a Geometry Theorem-Proving Machine, In Proc. IFIP, Paris UNESCO House, 1959, pp. 273–282

Holland, W; Charles; (eds.): Lattice-ordered groups, Kluwer Academic Publ., Mathematics and it's Applications, 1989.

Gordon, M.J.C.; Melham, T.F. (eds.): Introduction to HOL: A Theorem Proving Environment for Higher Order Logic, Cambridge Univ. Press, 1993.

Letz, J., Schumann, J. *et al.*: SETHEO: A High-Performance Theorem Prover, JAR, 8(2), pp. 183–212

P. Lorenzen. Über halbgeordnete Gruppen. Math. Z., 52:483-526, 1949.

Nipkow, T. and Paulson, L. C.: Isabelle-91, Proc. CADE-11, Saratoga Springs, LNAI 607, 1992, pp. 673–676.

Plaisted, D.; Zhu, Y.: Ordered Semantic Hyper Linking, Proc. 14th Nat. Conf. on Artificial Intelligence, Providence, Rhode Island 1997

Reif, W.: The KIV-System: Systematic Construction of Verified Software, Proc. CADE-11, Saratoga Springs, LNCS 607, 1992.

Reiter, R.: A Semantically Guided Deductive System for Automatic Theorem-Proving, In Proc. of the 3rd IJCAI, Stanford 1973, pp. 41–46

F. Sik. Über subdirekte Summen geordneter Gruppen. Czechosl. Maih. J., 10:400-424, 1960.

Wos, L.: The Hot List Strategy ANL/MCS-P684-0897, 1997.

JÜRGEN STUBER

CHAPTER 2
SUPERPOSITION THEOREM PROVING FOR COMMUTATIVE RINGS

1. Introduction

Commutative rings are very common both in mathematics and in computer science, as they encompass the basic algebraic properties of addition, subtraction and multiplication for numbers like the integers or the reals. The problem with commutative rings and similar algebraic theories is that their axioms create large search spaces for theorem provers. We address this problem by systematically developing a superposition calculus for first-order problems that contain the theory of commutative rings. The calculus combines certain uses of the axioms into macro inferences, and it restricts inferences to the maximal term within a sum. Thereby it promises to make theorem proving with respect to commutative rings more efficient.

We represent the theory of commutative rings by a convergent term rewriting system CR modulo AC (section 4). We call function symbols not occurring in CR *free* and terms with a free function symbol at their root *atomic*. Then normal forms with respect to CR can be viewed as generalized polynomials. That is, the normal form is a sum of products of atomic terms, which may contain nested polynomials.

Throughout the chapter we use a well-founded ordering $\succ$ on terms and clauses. This ordering is used to prove termination of rewriting, for ordering restrictions on inferences, in the definition of redundancy, and as the induction ordering in the completeness proof. In section 5 we state the requirements for the ordering and show that such an ordering exists.

In section 6 we introduce *symmetrization.* Symmetrization is a weakening of Knuth-Bendix completion that considers only critical pairs with one rule from the theory. We derive a function that map any equation in a certain *commutative ring normal form* directly to a symmetrized set of rules, which contains the original rule and some *extensions*. This avoids the expense of running a completion-like process for every rule.

Using a function for symmetrization also allows to exploit the structure of symmetrized sets. In section 7 we do this by proving a strong critical pair criterion for superpositions between extensions. We show that for any two rules at most one critical pair between extensions needs to be considered.

W. Bibel, P. H. Schmitt (eds.), Automated Deduction. A basis for applications. Vol. III

This is motivated by Gröbner base algorithms, where any two polynomials result in a single critical pair.

In section 8 we present the inference system. It contains no inferences within the theory. Theory superposition inferences, where theory clauses superpose into non-theory clauses, so-called *query clauses*, are restricted to the maximal terms of the top-level sum. Isolation inferences put the maximal monomial on the left-hand side and smaller terms on the right-hand side. Logically this is a combination of several ring axioms. These two inference rules arise from the simplification to commutative ring normal form of an equation. There are no superpositions of query clauses into theory clauses. Instead, superpositions of a query clause into another query clause are done via symmetrization. That is, the superposing equation is restricted to normal form and implicitly represents its symmetrization. Any rule from the symmetrization may then be used to superpose into the other clause. Next, there are Extension Superposition inferences, which are needed to handle critical pairs between extensions. By the redundancy criterion it suffices to compute a single inference for any two ground clauses. Finally, the calculus contains Reflexivity Resolution and Equality Factoring inferences. All inferences except Theory Superpositiona and Isolation are restricted to premises in commutative ring normal form.

In section 9 we prove the refutational completeness of our calculus. We adapt the reduction of counterexamples technique of Bachmair and Ganzinger in chapter I.2.11. Let us first outline their technique and then our modifications. Given a set of ground clauses N that is closed under the inference system, they build a *candidate model* that satisfies the equality axioms and a subset of N. A *counterexample* is a clause in N which is false in the candidate model. The inference system has the *reduction property* for counterexamples. That is, it allows to reduce a minimal non-empty counterexample to another counterexample according to a well-founded ordering, where the original counterexample is a premise of some inference, and the reduced counterexample is the conclusion. By well-founded induction, the candidate model is a model of N if and only if N does not contain the empty clause. The candidate models are represented by convergent ground term rewriting systems.

We modify this construction in several respects. We include the rewriting system for commutative rings in the term rewriting system representing a candidate model, and we keep the term rewriting system symmetrized. The resulting candidate models are reflexive, symmetric, and satisfy the monotonicity axioms and the axioms of commutative rings. However, in general they do not satisfy all instances of transitivity, as extensions from symmetrization may form critical pairs that do not converge. Therefore, we extend the notion of a counterexample so that it can be either a clause in N or a ground instance of transitivity. As before, inferences reduce counterexamples to smaller counterexamples, but only the clauses in N are explicit in the inferences, while the transitivity counterexamples remain implicit. More specifically, Extension Superposition inferences reduce transitivity counterexamples, while the other inferences reduce counterexamples from N. The reduction needs certain instances of transitivity, which must be smaller than the original counterexample. One of the main technical problems in this work is to show that these transitivity instances are indeed small enough.

Finally, in section 10 we discuss lifting. Certain variables at the top induce inferences below variable positions. For some interesting special cases of commutative rings these free variables can be eliminated.

2. Related Work

Our technique extends the technique of Bachmair and Ganzinger in chapter I.2.11. The technique to handle extensions of rules by relaxing transitivity is adapted from their work on AC-superposition (Bachmair and Ganzinger, 1994). Abelian groups or integer modules can be handled analogously, but without the problems with transitivity (Stuber, 1996). For the case of cancellative abelian monoids, the notion of reduction has to be extended such that equations are reduced as a whole (Ganzinger and Waldmann, 1996).

Our work also draws some inspiration from the theory of Gröbner bases, which are convergent reduction systems for polynomials. Theorem 6 is motivated by the fact that for any two polynomials only a single critical pair, a so-called S-polynomial, needs to be considered. The relation between completion for term rewriting systems, which is the basis of our calculus, and Gröbner basis algorithms has already been noted by Buchberger and Loos (1983) and formalized by Bündgen (1996) and Marché (1996). Bündgen introduces the notion of semi-compatible, and implicitly uses the notion of strong symmetrization and normalized rewriting to prove critical pair criteria.

Marché (1996) builds a range of theories from AC to commutative rings into equational completion. This is implemented in the Cime system (Contejean and Marché, 1996). Experiments with Cime have shown a large speedup for the incorporation of abelian groups compared to using the group axioms directly. The terminology in Marché's work differs slightly from ours. What he calls symmetrization corresponds to our notion of commutative ring normal form, while the first component of his normalizing pair corresponds to our notion of symmetrization.

Symmetrization originates from group theory. Greendlinger (1960) is the first to explicitly define the concept of symmetrized set of relations for (non-abelian) groups. Le Chenadec (1986) uses it also for theories like abelian groups, rings and modules and establishes the connection to term rewriting. He does not formally define *symmetrized* for a general setting, but from his examples it becomes clear that the purpose of symmetrization is to let the critical pairs between the built-in theory and other rules converge.

A more general approach to build theories into automated theorem provers is theory resolution (Stickel, 1985). However, because of its generality, theory resolution does not exploit redundancies introduced by specific theories.

In the context of theorem proving, theory reasoning must be integrated into the calculus to be effective. The study of Boyer and Moore (1988) shows that a hierarchical approach, where a black-box decision method is used whenever a problem falls entirely into the domain of the theory, achieves no significant speedup.

3. PRELIMINARIES

We assume that the reader is familiar with superposition theorem proving, term rewriting and reduction orderings, as presented in chapter I.2.11 by Bachmair and Ganzinger. However, since transitivity does not always hold in our interpretations, we do not assume that $\models$ includes equality. Instead we will explicitly write the congruence axioms where they are used. We distinguish the *reflexivity axiom*

$$\mathsf{R} = \{x \approx x\},$$

the *symmetry axiom*

$$\mathsf{S} = \{x \not\approx y \vee y \approx x\},$$

the *transitivity axiom*

$$\mathsf{T} = \{x \not\approx y \vee y \not\approx z \vee x \approx z\},$$

and the *monotonicity axioms*

$$\mathsf{M} = \{x_1 \not\approx y_1 \vee \ldots \vee x_n \not\approx y_n \vee f(x_1, \ldots, x_n) \approx f(y_1, \ldots, y_n) \mid f \in F\}.$$

Then $\mathsf{C} = \mathsf{R} \cup \mathsf{S} \cup \mathsf{T} \cup \mathsf{M}$ is the set of *congruence axioms*. If N is a set of clauses then $g(N)$ denotes the set of ground instances of clauses in N. Furthermore, we have the *associativity axiom*

$$\mathsf{A}(f) = \{f(x, f(y,z)) \approx f(f(x,y),z)\}$$

and the *commutativity axiom*

$$\mathsf{C}'(f) = \{f(x,y) \approx f(y,x)\}$$

for a binary function symbol f. We let $\mathsf{AC} = \mathsf{A}(+) \cup \mathsf{C}'(+) \cup \mathsf{A}(\cdot) \cup \mathsf{C}'(\cdot)$ where $+$ and $\cdot$ are two distinguished binary function symbols. Two terms s and t are AC-*equivalent*, written $s =_{\mathsf{AC}} t$, if $\mathsf{AC} \cup \mathsf{C} \models s \approx t$. An ordering $\succ$ is called AC-*compatible* if $s =_{\mathsf{AC}} s' \succ t' =_{\mathsf{AC}} t$ implies $s \succ t$ for all terms s, t, s' and t'. An ordering is called *total up to* AC *on ground terms* if $s \succ t$, $t \succ s$ or $s =_{\mathsf{AC}} t$ for any two ground terms s and t. A *rewrite rule* $l \Rightarrow r$ (with respect to the ordering $\succ$) is an equation $l \approx r$ such that $l \succ r$. Let R be a set of rewrite rules. Then $t[l']$ *rewrites to* $t[r\sigma]$ *with* AC-*matching*, written $t[l'] \Rightarrow_{\mathsf{AC}\backslash R} t[r\sigma]$, if there exists a rule $l \Rightarrow r$ in R and a substitution σ such that $l' =_{\mathsf{AC}} l\sigma$. We write $s \Downarrow_{\mathsf{AC}\backslash R} t$ for a *rewrite proof* $s \overset{*}{\Rightarrow}_{\mathsf{AC}\backslash R} \cdot \overset{*}{-}_{\mathsf{AC}} \cdot \overset{*}{\Leftarrow}_{\mathsf{AC}\backslash R} t$. By $R^{\Downarrow}$ we denote the interpretation $\{s \approx t \mid s \Downarrow_{\mathsf{AC}\backslash R} t\}$. The system R is *Church-Rosser (modulo* AC*)* if $s \overset{*}{-}_{\mathsf{AC} \cup R} t$ implies $s \Downarrow_{\mathsf{AC}\backslash R} t$, and *terminating (modulo* AC*)* if there exists an AC-compatible reduction ordering that contains R. If R is both Church-Rosser and terminating modulo AC then it is called *convergent modulo* AC or *complete modulo* AC. Two terms t_1 and t_2 are said to *converge* if $t_1 \Downarrow_{\mathsf{AC}\backslash R} t_2$. A *peak* $t_1 \Leftarrow_{R_1} s \Rightarrow_{R_2} t_2$ or a *cliff* $t_1 -_{\mathsf{AC}} s \Rightarrow_R t_2$ converges if s and t converge. Given termination, it suffices to test convergence of all peaks $s \Leftarrow_R \cdot \Rightarrow_{\mathsf{AC}\backslash R} t$ and cliffs $s -_{\mathsf{AC}} \cdot \Rightarrow_{\mathsf{AC}\backslash R} t$, in order to obtain convergence of R modulo AC (Jouannaud and Kirchner, 1986). Convergence of cliffs is ensured by adding AC-extensions: For a rule $l \Rightarrow r$ in R with $l = s + t$ or $l = s \cdot t$ its AC-*extension* is $x + l \Rightarrow x + r$ or $x \cdot l \Rightarrow x \cdot r$, respectively, where x is a new variable (Peterson and Stickel, 1981).

4. Commutative Rings

We present commutative rings with a unit element by the well-known convergent term rewriting system CR modulo AC of Peterson and Stickel (1981) in Figure 1. AC-extensions besides (2.2e) are not necessary. Where appropriate we will also consider CR as a set of equations, which together with AC axiomatize commutative rings. We let $F_{\mathsf{CR}} = \{+, \cdot, -, 0, 1\}$ be the set of function symbols occurring in the theory axioms. Other function symbols are called *free*. We will use the following notational conventions: p, q, r, s, t, u, v and w are used for arbitrary terms. For $n > 0$ we let nt denote a sum $t_1 + \cdots + t_n$ where $t_i =_{\mathsf{AC}} t$ for $i = 1, \ldots, n$. Also, $(-n)t$ denotes $n(-t)$ and $0t$ denotes 0. Terms with a free function symbol at the root will be called *atomic*. ϕ and ψ denote terms of the form $t_1 \cdots t_n$ where t_i is atomic for $i = 1, \ldots, n$. We will

$$x + 0 \Rightarrow x \tag{2.1}$$
$$x + (-x) \Rightarrow 0 \tag{2.2}$$
$$x + y + (-y) \Rightarrow x \tag{2.2e}$$
$$-0 \Rightarrow 0 \tag{2.3}$$
$$-(-x) \Rightarrow x \tag{2.4}$$
$$-(x+y) \Rightarrow (-x) + (-y) \tag{2.5}$$
$$x \cdot 0 \Rightarrow 0 \tag{2.6}$$
$$x \cdot 1 \Rightarrow x \tag{2.7}$$
$$x \cdot (y+z) \Rightarrow (x \cdot y) + (x \cdot z) \tag{2.8}$$
$$x \cdot (-y) \Rightarrow -(x \cdot y) \tag{2.9}$$

Figure 1. The term rewriting system CR

call such a term a ***product***. x, y and z denote variables. To simplify the presentation, our meta-level notation will be modulo ACU for $+$ and $\cdot$. That is, when we write $p = n\phi + p'$ then $n\phi$ denotes n products which are AC-equivalent to ϕ and which occur somewhere in the sum, not necessarily in front. Moreover, neither $n\phi$ nor p' need to be present. Thus p may also be of the form $n\phi$, p' or 0, depending on whether $p = 0$ and/or $n = 0$. Similarly, when we write $\phi = \phi_1\phi_2$ then ϕ_1 or ϕ_2 may be missing in which case $\phi_i = 1$.

5. Term and Clause Orderings

For an introduction to termination orderings for term rewriting we refer the reader to the survey by Dershowitz (1987).

The model construction and in turn the ordering restrictions on inferences are based on a term ordering $\succ$. We call an ordering $\succ$ ***admissible*** if it satisfies the following requirements:

1. It is ***well-founded***, since we want to use it as the basis for inductively building an interpretation. Also, we need results from the theory of term rewriting, which require that the reduction relation is terminating.
2. It has the ***subterm property***. In our construction we assume that all rules which could reduce a term have a left-hand side that is smaller or equal to that term.
3. It has to be ***monotonic*** because equality is monotonic, which is used to rewrite in proper subterms of a term.

4. It is *total* on ground terms up to AC. This is necessary to achieve completeness.
5. It has to be AC-*compatible* because AC-matching is used in rewriting steps.
6. It orients the rewrite rules in CR and rewrite rules of the forms $n\phi \Rightarrow r$ and $-\phi \Rightarrow (n-1)\phi + (-r)$, where $n \geq 2$ and $\phi \succ r$, from left to right. This property is needed for the termination of the term rewriting system.
7. Finally, $\succ$ has the *multiset property* with respect to CR. That is, $t \succ t_i$ for $i = 1, \ldots, n$ implies $t \succ u[t_1, \ldots, t_n]$ where u is a context that contains only F_{CR} symbols and $t, t_1, \ldots, t_n$ are atomic terms. The multiset property implies that for every equation there exists a CR-equivalent equation with a single monomial on the left-hand side which is greater than the sum of the other monomials on the right-hand side.

$\succ$ is extended to equations, literals and clauses in the following way. To an equation $s \approx t$ we assign as its complexity

$$c(s \approx t) = \{s, t\}$$

and order equations by the multiset extension of the term ordering $\succ$. To a literal $[\neg](s \approx t)$ we assign the complexity

$$c(s \approx t) = \{\{s\}, \{t\}\}$$
$$c(\neg s \approx t) = \{\{s, t\}\}$$

and order literals by the twofold multiset extension of the term ordering. This has the effect that literals are ordered according to the lexicographic combination of (i) their maximal term, (ii) their polarity, where negative literals are larger than positive literals, and (iii) their minimal term. Query clauses get as their complexity the multiset of their literals. Finally, to a ground instance

$$t_1 \not\approx s \vee s \not\approx t_2 \vee t_1 \approx t_2$$

of transitivity we assign the complexity $\{\{\{s\}\}\}$ if $s \succ t_1$ and $s \succ t_2$ and $\{\}$ otherwise. We call s the *middle term* of this instance of transitivity. An instance of transitivity whose middle term is not strictly maximal can never become a counterexample, as it is always true in candidate models. The ordering need not distinguish instances of transitivity with the same middle term, since transitivity counterexamples are always reduced to counterexamples with a strictly smaller maximal term. Clauses are ordered by the threefold multiset extension of the term ordering. By this definition instances of transitivity with maximal term s are smaller than any query clause with maximal term s. We denote all orderings by $\succ$. Note that the ordering is not only compatible with the built-in theory, but that it also respects certain properties of

$\approx$ and $\vee$. Namely, all orderings are monotonic, respecting the congruence law for equality; the extension to equations and literals is symmetric; and the extension to clauses respects the associativity and commutativity of $\vee$.

ASSUMPTION 1. *From now on we assume that $\succ$ is an admissible ordering on terms or its extension to equations, literals or clauses as defined above.*

To see that such an ordering exists, consider the lexicographic combination of the following three components. As the first component we use the modified associative path ordering (MAPO) of Delor and Puel (1993) over the precedence

$$f_1 \succ \ldots \succ f_n \succ 1 \succ (\cdot) \succ (-) \succ (+) \succ 0.$$

That is, terms are normalized with the rewrite system D modulo AC, consisting of (2.5), (2.8) and (2.9), flattened, and then compared by the recursive path ordering with respect to the given precedence, where $+$ and $\cdot$ have multiset status and other function symbols have lexicographic status. The MAPO is an AC-compatible simplification ordering that orients the rules in $\mathsf{CR} \setminus D$ and rules of the forms $n\phi \Rightarrow r$ and $-\phi \Rightarrow (n-1)\phi + (-r)$, where $n \geq 2$ and $\phi \succ r$, from left to right, and that has the multiset property. The second component is some AC-compatible simplification ordering that orients the rewrite rules in D from left to right. We may for instance use the ordering for CR of Peterson and Stickel (1981). The third component is the ACRPO of Rubio and Nieuwenhuis (1993) with respect to some arbitrary precedence. This is an AC-compatible simplification ordering that is total up to AC on ground terms. The lexicographic combination of these orderings inherits their properties; hence it is admissible.

Any term t has the form $t_1 + \cdots + t_k$ for $k \geq 1$ where the symbol at the root of t_i is not $+$ for $1 \leq i \leq k$. We say that t_i is the *maximal term* of t if $t_i \succeq t_j$ for $j = 1, \ldots, k$. If the maximal term t_i occurs n times, we say that nt_i is the *maximal monomial* of t. We say that an equation, a literal, a clause or an equational proof has the maximal monomial s if s is the maximal monomial occurring in it. The multiset property ensures that the maximal monomial dominates the size of a term with respect to $\succ$, hence the maximal monomial of a clause occurs in the maximal term of a clause.

6. Symmetrization

A set of rewrite rules S is called *symmetrized* with respect to CR modulo AC if for all peaks $t_1 \Leftarrow_{\mathsf{AC}\setminus\mathsf{CR}} t \Rightarrow_{\mathsf{AC}\setminus S} t_2$ and for all cliffs $t_1 -_{\mathsf{AC}} t \Rightarrow_{\mathsf{AC}\setminus S} t_2$ we have $t_1 \Downarrow_{\mathsf{AC}\backslash(\mathsf{CR}\cup S)} t_2$. The set S is called *strongly symmetrized* with respect to CR modulo AC if it can be partitioned into sets $S_i,\ i \in I$, such that $\mathsf{CR} \cup S_i$ is convergent modulo AC for all $i \in I$.

PROPOSITION 2. *If a set of rewrite rules S is strongly symmetrized with respect to* CR *modulo* AC *then S is symmetrized with respect to* CR *modulo* AC.

Proof. Consider some peak $t_1 \Leftarrow_{\mathsf{AC}\backslash\mathsf{CR}} t \Rightarrow_{\mathsf{AC}\backslash S} t_2$. The rule from S is in some S_i, and by convergence of $\mathsf{CR} \cup S_i$ we get the desired rewrite proof. The same reasoning applies to cliffs. □

Symmetrization is related to Knuth-Bendix completion, and results in a term rewriting system for a single equation combined with CR. Strong symmetrization is weaker than Knuth-Bendix completion, as it derives a complete system only for inputs of the form $\{l \approx r\} \cup \mathsf{CR}$. Peaks between rules that are derived from different input equations need not converge in general. Symmetrization considers only peaks and cliffs where at least one of the rules is from the theory, peaks where both rules are not from CR need not converge in general.

We show below that strong symmetrization allows to normalize proofs with respect to CR, which is not possible with symmetrization alone. However, strong symmetrization is not as widely applicable as symmetrization. For instance, strong symmetrization is not suitable for non-abelian groups, where an equation may have nontrivial overlaps with its own extensions.

We say that an equation is in *commutative ring normal form*, or short in CR-*normal form*, either if it has the form (i) $0 \approx 0$; or if it has the form (ii) $n\phi \approx r$ where $n \geq 1,\ \phi = t_1 \ldots t_k,\ k \geq 0,\ t_1, \ldots, t_k$ are atomic, and $\phi \succ r$. The set of equations in CR-normal form is denoted by $\mathrm{Norm}_{\mathsf{CR}}$. A characteristic property of CR-normal forms that translates to other theories is that they are *left-minimal*. That is, among all CR-equivalent equations they have the minimal left-hand side, where equations are always oriented such that the left-hand side is greater or equal than the right-hand side. As a result, a given equation may have several CR-normal forms with the same left-hand side but different right-hand sides. Since the left-hand side is minimal, it is irreducible with respect to CR. This is in general not the case for the right-hand side. We do not normalize the right-hand side as this would lead to unwanted theory superposition inferences into the smaller side of equations.

To obtain a strong symmetrization of an equation in $\mathrm{Norm}_{\mathsf{CR}}$ we will use the function $\mathcal{S}_{\mathsf{CR}}$ defined in Figure 2. The set of rules $\mathcal{S}_{\mathsf{CR}}(l \approx r)$ is strongly symmetrized and equivalent to $\mathsf{C} \cup \mathsf{AC} \cup \mathsf{CR} \cup \{l \approx r\}$ for any $l \approx r$ in $\mathrm{Norm}_{\mathsf{CR}}$. We call a rule $l' \Rightarrow r'$ in $\mathcal{S}_{\mathsf{CR}}(l \Rightarrow r) \setminus \{l \Rightarrow r\}$ an *extension* (of $l \Rightarrow r$). We

$$
\begin{aligned}
S_{\mathsf{CR}}(0 \approx 0) &= \emptyset && (2.10)\\
S_{\mathsf{CR}}(t \approx r) &= \{t \Rightarrow r\} && (2.11)\\
S_{\mathsf{CR}}(\phi \approx r) &= \{\phi \Rightarrow r\} && (2.12)\\
&\quad \cup g(\{y\phi \Rightarrow yr\}) && (2.13)\\
S_{\mathsf{CR}}(n\phi \approx r) &= \{n\phi \Rightarrow r\} && (2.14)\\
&\quad \cup g(\{n(y\phi) \Rightarrow yr\}) && (2.15)\\
&\quad \cup g(\{x + n\phi \Rightarrow x + r\}) && (2.16)\\
&\quad \cup g(\{x + n(y\phi) \Rightarrow x + yr\}) && (2.17)\\
&\quad \cup \{-\phi \Rightarrow (n-1)\phi + (-r)\} && (2.18)\\
&\quad \cup g(\{-(y\phi) \Rightarrow (n-1)(y\phi) + y(-r)\}) && (2.19)
\end{aligned}
$$

Figure 2. The symmetrization function for commutative rings

extend S_{CR} to sets of equations in the standard way by

$$S_{\mathsf{CR}}(R) = \bigcup_{l \approx r \in R} S_{\mathsf{CR}}(l \approx r)$$

for $R \subseteq \mathrm{Norm}_{\mathsf{CR}}$. Note that $S_{\mathsf{CR}}(R)$ is strongly symmetrized, too.

Such a function S_{CR} can be determined by hand, starting with an equation in normal form[1] and adding critical pairs: For instance, overlapping a rule $n\phi \approx r$ in $\mathrm{Norm}_{\mathsf{CR}}$ where $\phi = t_1 \cdots t_k$ and $n \geq 2$, $k \geq 1$ with distributivity produces the critical pair $x \cdot (n_1\phi + n_2\phi) \approx x \cdot r$, which after simplification results in the rule $n(x \cdot \phi) \Rightarrow x \cdot r$. Critical pairs between rules not in CR are trivial. After exhausting all critical pairs, we get the strongly symmetrized set of rules in the definition of S_{CR}. The technique used is essentially Knuth-Bendix completion, but over term schemas like $n_1\phi + n_2\phi$. Therefore we currently see no way to automate this. To test our results we have run some representative examples with the Cime system (Contejean and Marché, 1996).

Using S_{CR} instead of running a completion-like process has the advantage that the structure of the symmetrized sets is known beforehand. This allows to use it as a theoretical device in the definition of the inference system $\mathcal{CR}$, instead of applying it explicitly to the set of clauses.

We now show that strong symmetrization allows to normalize equational proofs with respect to CR. That is, every term in the proof is normalized with respect to CR, and each rewrite step is replaced by a valley proof. We start by

[1] In general it is not feasible to determine a symmetrized set of equations from an equation that is not in CR-normal form.

showing that the Church-Rosser property does not depend on the choice of general rewriting vs. normalized rewriting. Normalized rewriting gives rules from CR priority over rules from S (Marché, 1996).

For a rewrite relation $\Rightarrow_{\mathsf{AC}\backslash S}$ we define CR-*normalized rewriting with S (modulo* AC*)*, written $\Rightarrow_{\mathsf{AC}\backslash(\mathsf{CR}!S)}$, by $s \Rightarrow_{\mathsf{AC}\backslash(\mathsf{CR}!S)} t$ if and only if $s \overset{*}{\Rightarrow}_{\mathsf{AC}\backslash\mathsf{CR}} u$, u is irreducible with respect to CR and $u \Rightarrow_{\mathsf{AC}\backslash S} t$.

LEMMA 3. *Let* $\mathsf{CR} \cup S$ *be terminating modulo* AC. *Then* $\mathsf{AC}\backslash(\mathsf{CR} \cup S)$ *is Church-Rosser modulo* AC *if and only if* $\mathsf{AC}\backslash(\mathsf{CR}!S)$ *is Church-Rosser modulo* AC.

Proof. Since $(\Downarrow_{\mathsf{AC}\backslash(\mathsf{CR}!S)}) \subseteq (\Downarrow_{\mathsf{AC}\backslash(\mathsf{CR}\cup S)})$ the if-direction is trivial.

For the only-if-direction the proof is by induction on the following proof ordering. Let $\succ$ be $\overset{*}{\Rightarrow}_{(\mathsf{CR}\cup S)/\mathsf{AC}}$ extended by a new minimal element $\bot$. We order proof steps with respect to $\succ$ according to the complexity measure

$$c(s \Rightarrow_{\mathsf{AC}\backslash S} t) = c(t \Leftarrow_{\mathsf{AC}\backslash S} s) = s$$
$$c(s \Rightarrow_{\mathsf{AC}\backslash\mathsf{CR}} t) = c(t \Leftarrow_{\mathsf{AC}\backslash\mathsf{CR}} s) = \bot$$
$$c(s -_{\mathsf{AC}} t) = \bot,$$

i.e., only the larger sides of S-steps count. As the proof ordering we use the multiset extension of the ordering on proof steps. Consider some proof $s \overset{*}{-}_{\mathsf{AC}\cup\mathsf{CR}\cup S} t$. Then $s \Downarrow_{\mathsf{AC}\backslash(\mathsf{CR}\cup S)} t$. Suppose $s \not\Downarrow_{\mathsf{AC}\backslash(\mathsf{CR}!S)} t$, then there exists an S-step $u \Rightarrow_{\mathsf{AC}\backslash S} v$ (or $v \Leftarrow_{\mathsf{AC}\backslash S} u$) in $s \Downarrow_{\mathsf{AC}\backslash(\mathsf{CR}\cup S)} t$ whose larger side u is reducible by CR. For $u' \Leftarrow_{\mathsf{AC}\backslash\mathsf{CR}} u \Rightarrow_{\mathsf{AC}\backslash S} v$ there exists a proof $u' \Downarrow_{\mathsf{AC}\backslash(\mathsf{CR}\cup S)} v$, and we may replace the subproof $u \Rightarrow_{\mathsf{AC}\backslash S} v$ by the smaller subproof

$$u \Rightarrow_{\mathsf{AC}\backslash\mathsf{CR}} u' \Downarrow_{\mathsf{AC}\backslash(\mathsf{CR}\cup S)} v.$$

All in all we get a smaller proof

$$s \Downarrow_{\mathsf{AC}\backslash(\mathsf{CR}\cup S)} u' \Downarrow_{\mathsf{AC}\backslash(\mathsf{CR}\cup S)} v \Downarrow_{\mathsf{AC}\backslash(\mathsf{CR}\cup S)} t$$

and by induction hypothesis $s \Downarrow_{\mathsf{AC}\backslash(\mathsf{CR}!S)} t$, a contradiction. □

A relation $\Rightarrow_R$ is *semi-compatible* if $s \Rightarrow_R t$ implies $u[s] \Downarrow_R u[t]$ for all terms s and t and contexts u.

LEMMA 4. *Let S be a set of ground rules that is strongly symmetrized with respect to* CR *modulo* AC. *Then* CR-*normalized rewriting with S modulo* AC *is semi-compatible.*

Proof. Consider a rewrite step $s \Rightarrow_{\mathsf{AC}\backslash(\mathsf{CR}!S)} t$, and a context u. Then there exists a not necessarily normalized valley proof of $u[s] \approx u[t]$ by putting the context u around every term in the proof that represents the normalized rewrite step. Since this proof contains only one S-step, all rules used in the proof are in $\mathsf{CR} \cup S_i$ for some $i \in I$, where S_i is one of the partitions in the definition of strong symmetrization, and $\mathsf{CR} \cup S_i$ is convergent modulo AC. By lemma 3 there exists a valley proof of $u[s] \approx u[t]$ by normalized rewriting. □

This lemma allows to normalize a single step of an equational proof. Longer equational proofs can be normalized by transforming single steps and pasting the resulting valley proofs together. We will use this technique to perform algebraic operations on proofs. We put a context around every term in the proof and normalize it. Normalization is necessary, because the proof with a raw context is too large to allow the use of certain induction hypotheses. Semi-compatibility ensures that there exist valley proofs between the normalized terms, which thus bound the terms in the proof.

7. Extended Critical Peaks

From symmetrization we get convergence within a set of rules derived from a single equation. To get global convergence we need to consider interactions between symmetrized sets of rules. On the level of term rewriting, the goal of this section is to achieve convergence of a rewrite system built from CR and a strongly symmetrized set of rules S. This is achieved by considering certain *extended critical peaks*. It is desirable to minimize the number of peaks that need to be considered. To this end introduce a notion of redundancy for peaks, and we prove that for any pair of symmetrizations at most one extended peak is not redundant.

We say that a set of ground terms T is *downward-closed* (with respect to $\succ$), if for every ground term t in T all ground terms smaller than t are also in T. Let T be a downward-closed set of ground terms. A term rewriting system R is *Church-Rosser (modulo* AC*) on* T if for any equational proof

$$t_1 -_{\mathsf{AC} \cup \mathsf{AC}\backslash R} t_2 -_{\mathsf{AC} \cup \mathsf{AC}\backslash R} \cdots -_{\mathsf{AC} \cup \mathsf{AC}\backslash R} t_n$$

with $n \geq 1$ and $t_1, \ldots, t_n \in T$ there exists a valley proof $t_1 \Downarrow_{\mathsf{AC}\backslash R} t_n$. The set of *transitivity instances for* T, written T^T, contains all ground instances of transitivity with a middle term from T. Transitivity of $R^{\Downarrow}$ and the Church-Rosser property of R are equivalent (Bachmair and Ganzinger, 1994):

PROPOSITION 5. *Let T be a set of ground terms that is downward-closed with respect to $\succ$, and let R be a ground term rewriting system with $R \subseteq (\succ)$. R is Church-Rosser on T if and only if $R^{\Downarrow} \models \mathsf{T}^T$.*

Given a ground term s, we say that R is *Church-Rosser on terms below [up to] s* if it is Church-Rosser on the set of ground terms smaller than [smaller than or equal to] s. We say that R is *Church-Rosser on monomials below [up to] s* if it is Church-Rosser on the set of ground terms containing only monomials smaller than [smaller than or equal to] s. For $i = 1,2$ let $l_i \Rightarrow r_i$ be a rule in CR-normal form, $R_i = \{l_i \Rightarrow r_i\}$ and $S_i = S_{\mathsf{CR}}(R_i)$. A *peak* of $l_1 \Rightarrow r_1$ and $l_2 \Rightarrow r_2$ is a rewrite sequence $t_1 \Leftarrow_{\mathsf{AC}\backslash S_1} t \Rightarrow_{\mathsf{AC}\backslash S_2} t_2$. Such a peak is called *redundant* if $t_1 \Downarrow_{\mathsf{AC}\backslash(\mathsf{CR}\cup S)} t_2$ for all ground term rewriting systems $\mathsf{CR}\cup S$ where $S = S_{\mathsf{CR}}(R)$, R a ground term rewriting system with $R_1 \cup R_2 \subseteq R$ and $\mathsf{CR}\cup S$ is Church-Rosser on terms below t. A peak is called *extended* if l_1 is irreducible by $\mathsf{CR}\cup S_2$ and l_2 is irreducible by $\mathsf{CR}\cup S_1$. We now apply these concepts to commutative rings and characterize non-redundant extended peaks.

THEOREM 6. *Let $n_1\phi_1 \Rightarrow r_1$ and $n_2\phi_2 \Rightarrow r_2$ be rewrite rules in* $\mathrm{Norm}_{\mathsf{CR}}$ *and assume without loss of generality $n_1 \geq n_2$. The only non-redundant extended peak of these rules has the form*

$$r_1\psi_1 \Leftarrow n_1\phi \Rightarrow (n_1 - n_2)\phi + r_2\psi_2$$

where $\phi =_{\mathsf{AC}} \phi_1\psi_1 =_{\mathsf{AC}} \phi_2\psi_2$ and ψ_1 and ψ_2 have no common factor.

Proof. Consider a peak $t_1 \Leftarrow t \Rightarrow t_2$ of $n_1\phi_1 \Rightarrow r_1$ and $n_2\phi_2 \Rightarrow r_2$. Suppose R is a ground term rewriting system containing these two rules, $S = S_{\mathsf{CR}}(R)$, and $\mathsf{CR}\cup S$ is Church-Rosser on terms below t. Let $S_i = S_{\mathsf{CR}}(l_i \approx r_i)$ for $i = 1,2$. The rules from S_i can be grouped into the following two most general cases:

$$v_i + n_i(w_i\phi_i) \Rightarrow v_i + w_i r_i \tag{2.20}$$

$$-(w_i\phi_i) \Rightarrow (n_i - 1)(w_i\phi_i) + (-(w_i r_i)) \tag{2.21}$$

Other forms may be seen as special cases, where v_i or w_i are omitted, where $n_i = 1$, or where ϕ_i is missing. The analysis of these cases is essentially the same. Without loss of generality we assume $n_1 \geq n_2$.

(1) We start by considering extended peaks of two ground rules $l_i \Rightarrow r_i$ of the form (2.20). Since AC-contexts can be moved into v_i or w_i, we need to consider only overlaps at the root in which the term at the top of the peak is equal to both left-hand sides.

(a) If the sums $n_i w_i \phi_i$ don't overlap in some product, the rewriting steps are independent of each other and the peak is trivially redundant (Figure 3a).

(b) If at least one product overlaps then $w_1\phi_1 =_{\mathsf{AC}} w_2\phi_2$ (Figure 3b). Let w be the greatest common divisor of w_1 and w_2. Then $w_i = ww'_i$ for $i = 1,2$ (Figure 4b). Next, let $w' = w'_1\phi_1 =_{\mathsf{AC}} w'_2\phi_2$. We get the most general nontrivial overlap $v_1 + n_1 ww' =_{\mathsf{AC}} v_2 + n_2 ww'$ where $v_i = v + (n - n_i)ww'$ for some v and n. Then $v + (n - n_1)ww'$ is the part that v_1 and v_2 have in common.

(a) A trivial overlap: $\begin{array}{l} abcw_1 + abcw_1 + \qquad\qquad v_1 \\ \qquad\quad v_2 \qquad\quad + bcdw_2 + bcdw_2 + bcdw_2 \end{array}$

(b) A nontrivial overlap: $\begin{array}{l} abcw_1 + abcw_1 + \qquad v_1 \\ \quad v_2 \quad + bcdw_2 + bcdw_2 + bcdw_2 \end{array}$

(c) A minimal overlap: $\begin{array}{l} abcw_1 + abcw_1 + \; v_1 \\ bcdw_2 + bcdw_2 + bcdw_2 \end{array}$

Figure 3. Overlaps of sums

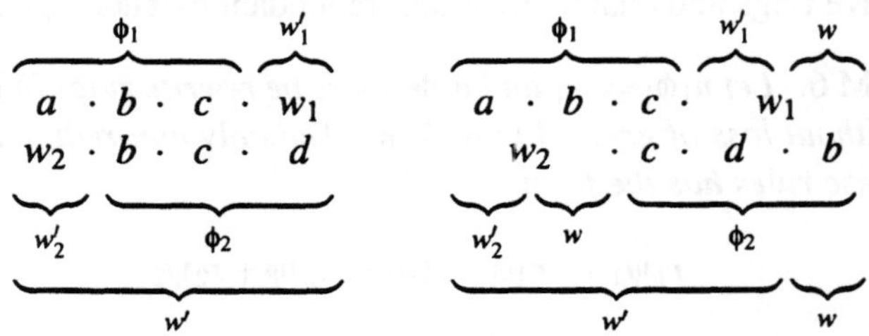

(a) A minimal overlap (b) A non-minimal overlap

Figure 4. Overlaps of products

If v is not empty, we can show redundancy by taking the smaller proof without v and putting the context $v+[]$ around every term in the proof. Hence we may assume that v is empty.

Next we consider the case $w \neq 1$ or $n > n_1$. Figure 5 shows why such a peak is redundant. The non-empty context implies that the peak consisting of the boxed formulas is smaller than $n(ww')$, hence we may assume the existence of the dashed proof. From this we get the dotted proof by putting the context $(n-n_1)ww' + w \cdot []$ around its terms and normalizing it. The terms in the dashed proof have strictly less occurrences of w' than $n_1 w'$, and normalizing transforms them into the same number of occurrences of ww'. Thus, the terms in the dotted proof stay below the bound $n(ww')$. From the Church-Rosser property below $n(ww')$ we obtain that

$$(n-n_1)ww' + r_1 ww'_1 \Downarrow_{\mathrm{AC}\backslash(\mathrm{CR}\cup S)} (n-n_2)ww' + r_2 ww'_2,$$

i.e., the critical peak converges. Note that we get the only non-redundant extended peak if the context is empty.

(c) The only other overlap of two extended rules at the root position occurs between rules of the form (2.21), that is $-(w_i\phi_i) \Rightarrow (n_i-1)(w_i\phi_i)+$

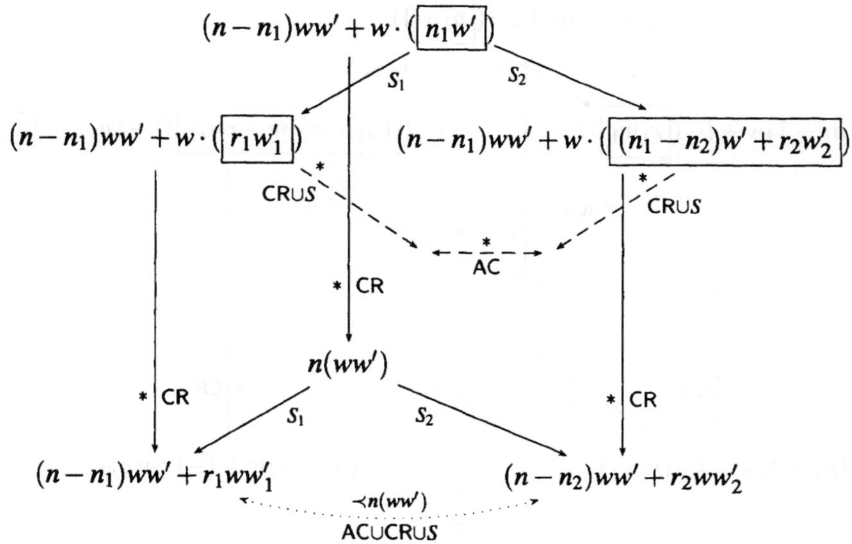

Figure 5. Redundancy of an extended peak

$(-(w_ir_i))$. Then the overlapping term is $-w =_{AC} -(w_1\phi_1) =_{AC} -(w_2\phi_2)$ where we let $w =_{AC} w_1\phi_1 =_{AC} w_2\phi_2$. Figure 6 illustrates why this peak is also redundant. We get the dotted proof from the dashed proof by putting the context $(n_1-1)w + (-([]))$ around the terms in it and using the same construction as in case (b). The dotted proof stays below the bound $-w$, because the normal forms in the transformed proof contain no negated occurrence of the maximal product w, and the ordering is constructed in such a way that any number of positive occurrences is smaller than a single negative occurrence. □

The next lemma combines two results. First, only non-redundant extended peaks need to be considered for the Church-Rosser property. Second, we exploit that only monomials give rise to peaks, in order to extend the Church-Rosser property to terms whose monomials are bounded.

LEMMA 7. *Let $R \subseteq \mathrm{Norm}_{\mathsf{CR}}$ be a ground term rewriting system such that l is irreducible by $\mathcal{S}_{\mathsf{CR}}(R\setminus\{l \Rightarrow r\})$ for each rule $l \Rightarrow r$ in R, let $S = \mathcal{S}_{\mathsf{CR}}(R)$, and let s be a ground monomial. Suppose that $t_1 \Downarrow_{\mathsf{AC}\backslash(\mathsf{CR}\cup S)} t_2$ for every non-redundant extended peak $t_1 \Leftarrow t \Rightarrow t_2$ with $t \prec s$ [$t \preceq s$] that arises from rules $l_1 \Rightarrow r_1$ and $l_2 \Rightarrow r_2$ in R. Then $\mathsf{CR}\cup S$ is Church-Rosser on monomials below [up to] s.*

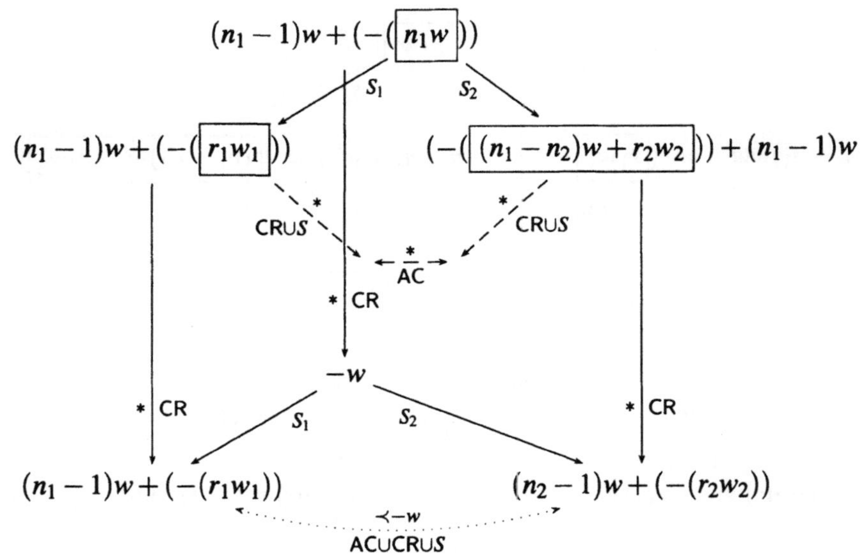

Figure 6. Redundancy of an extended peak

Proof. The proof is by noetherian induction over proofs. Proofs are ordered by the multiset extension of $\succ$ according to of their terms. Consider a proof

$$s_1 -_{\mathsf{AC}\cup\mathsf{AC}\backslash(\mathsf{CR}\cup S)} s_2 -_{\mathsf{AC}\cup\mathsf{AC}\backslash(\mathsf{CR}\cup S)} \cdots -_{\mathsf{AC}\cup\mathsf{AC}\backslash(\mathsf{CR}\cup S)} s_n$$

where $s' \prec s$ [$s' \preceq s$] for the maximal monomial s' of the proof, and assume that for all smaller proofs $s_1' \overset{*}{-}_{\mathsf{AC}\cup\mathsf{AC}\backslash(\mathsf{CR}\cup S)} s_2'$ we have $s_1' \Downarrow_{\mathsf{AC}\backslash(\mathsf{CR}\cup S)} s_2'$.

(1) As CR is convergent and S is symmetrized modulo AC with respect to CR all peaks involving CR and all cliffs with AC converge.

(2) Nonextended peaks do not occur, since left-hand sides of rules are required to be irreducible by other rules.

(3) Consider some extended redundant peak $s_{i-1} \Leftarrow s_i \Rightarrow s_{i+1}$. The induction hypothesis implies that CR $\cup\, S$ is Church-Rosser on terms below s, hence the peak converges by the definition of redundancy.

(4) It remains to consider an extended non-redundant peak

$$r_1\psi_1 \Leftarrow n_1\phi \Rightarrow (n_1-n_2)\phi + r_2\psi_2.$$

This peak converges by assumption, since $n_1\phi$ is a monomial and $n_1\phi \prec s$ [$n_1\phi \preceq s$] from being a term in the proof.

Since all peaks and cliffs on monomials below [up to] s converge, and since CR $\cup\, S$ is terminating modulo AC, we obtain that CR $\cup\, S$ is Church-Rosser on monomials below [up to] s. □

8. The Ground Inference System

In this section we define a ground inference system CR for commutative rings, which we will later show to be refutationally complete. We assume a function on non-empty ground clauses which selects one of the literals in the clause, such that this literal is either some negative literal, or a positive literal that is maximal in the entire clause. For each premise of an inference rule we have the additional implicit restriction that the literal upon which the inference operates is selected.

The inference rules Theory Superposition and Isolation simplify equations to CR-normal form. It is not necessary to use rewriting with CR eagerly and to go all the way to a normal form on both sides of the equation. Only those inferences that involve the maximal terms of the sum are needed, since this maximal terms will ultimately become the left-hand side. The other non-maximal terms will end up in r, hence it is not necessary to reduce them once they have been separated from the maximal terms. Consider an example of simplification where $a \succ b$:

$$\begin{aligned} a+a+b &\approx -((-a)+(b+0)) \\ a+a+b &\approx (-(-a))+(-(b+0)) \\ a+a+b &\approx a+(-(b+0)) \\ a &\approx (-(b+0))+(-b) \end{aligned}$$

The first redex is $-((-a)+(b+0))$, which includes both a and b. It is rewritten to the top-level sum $(-(-a))+(-(b+0))$ where $(-(-a))$ is the maximal term, and this is in turn rewritten to a. The last step isolates a on the left-hand side. Note that $b+0$ is not reduced, since it is not necessary. Of course, if the term $t+0$ were present also on the non-ground level, it would be a good idea to reduce it to t. But in general this is not the case, as there could be a variable instead of 0.

Let us now make this intuition formal. For any term t we have $t =_{\mathsf{AC}} t_1 + \cdots + t_n$ where $n \geq 1$ and t_i has a function symbol other than $+$ at the root. We *select* a redex at position π in t for CR-rewriting if it is comparable to a position π' at the root of a maximal term in t. That is, π' is either below, above or equal to π. By reducing both sides of an equation in this way, we get an equation of the form $nt + p \approx mt + q$ where $n, m \in \mathbb{Z}$, $t \succ p$, $t \succ q$, and t is irreducible with respect to CR. It only remains to isolate the maximal terms on one side.

THEORY SUPERPOSITION $$\frac{[\neg](p[l'] \approx q) \vee C}{[\neg](p[r\sigma] \approx q) \vee C}$$

if (i) $l \Rightarrow r$ is a rule in CR, (ii) $l\sigma =_{AC} l'$, (iii) $p \succeq q$, and (iv) l' is a selected redex in $p[l']$.

ISOLATION $$\frac{[\neg](n\phi + p \approx m\phi + q) \vee C}{[\neg]((n-m)\phi \approx q + (-p)) \vee C}$$

if (i) $n \geq m$, (ii) $\phi = t_1 \cdots t_k,\ k \geq 1,\ t_i$ atomic for $i = 1,\ldots,k$, (iii) ϕ is irreducible with respect to CR, and (iv) $\phi \succ p$ and $\phi \succ q$.

REFLEXIVITY RESOLUTION $$\frac{p \not\approx q \vee C}{C}$$

where (i) $p =_{AC} q$, and (ii) $p \not\approx q$ is in CR-normal form.

SUPERPOSITION $$\frac{l \approx r \vee D \qquad [\neg](s[l''] \approx t) \vee C}{[\neg](s[r'] \approx t) \vee C}$$

where (i) $[\neg](s \approx t)$ is in CR-normal form, (ii) $l \approx r$ is in CR-normal form, (iii) $l' \Rightarrow r' \in \mathcal{S}_{CR}(l \approx r)$, and (iv) $l' =_{AC} l''$.

EQUALITY FACTORING $$\frac{s \approx t \vee s' \approx t' \vee C}{t \not\approx t' \vee s' \approx t' \vee C}$$

where (i) $s =_{AC} s'$, (ii) $s \approx t$ is selected, and (ii) $s \approx t$ and $s' \approx t'$ are in CR-normal form.

EXTENSION SUPERPOSITION $$\frac{l_1 \approx r_1 \vee C_1 \qquad l_2 \approx r_2 \vee C_2}{t_1 \approx t_2 \vee C_1 \vee C_2}$$

where (i) $l_1 \approx r_1$ and $l_2 \approx r_2$ are in CR-normal form, and (ii) $t_1 \Leftarrow t \Rightarrow t_2$ is a non-redundant extended peak of $l_1 \approx r_1$ and $l_2 \approx r_2$.

Although it is not explicit in the inference, we say that this inference has as its *main premise* the transitivity instance $t_1 \not\approx t \vee t \not\approx t_2 \vee t_1 \approx t_2$. This will allow uniform definitions below, as this instance of transitivity is the counterexample that motivates the inference. The explicit premises are *side premises*. For Superposition the right premise is the main premise and the left

$$R_C = \bigcup_{D \prec C} E_D \qquad S_C = S_{\mathsf{CR}}(R_C) \qquad I_C = (\mathsf{CR} \cup S_C)^{\Downarrow}$$

$$E_C = \begin{cases} \{l \Rightarrow r\} & \text{if (i) } C \in N \text{ is false in } I_C, \\ & \text{(ii) } C = l \approx r \vee C', \\ & \text{(iii) } C \text{ is reductive for } l \approx r, \\ & \text{(iv) } l \approx r \text{ is in CR-normal form,} \\ & \text{(v) } l \text{ is irreducible by } \mathsf{CR} \cup S_C, \text{ and} \\ & \text{(vi) } C' \text{ is false in } (\mathsf{CR} \cup S_C \cup S_{\mathsf{CR}}(l \approx r))^{\Downarrow} \\ \emptyset & \text{otherwise.} \end{cases}$$

$$R^C = R_C \cup E_C \qquad S^C = S_{\mathsf{CR}}(R^C) \qquad I^C = (\mathsf{CR} \cup S^C)^{\Downarrow}$$

$$R_N = \bigcup_C E_C \qquad S_N = S_{\mathsf{CR}}(R_N) \qquad I_N = (\mathsf{CR} \cup R_N)^{\Downarrow}$$

Figure 7. The candidate model

premise is a side premise. Unary inferences only have a main premise and no side premises.

9. Refutational Completeness

We present the construction of a candidate model for a given set of ground clauses N, which will be used for the completeness proof.

A ground clause $C \vee s \approx t$ is called *reductive* for $s \Rightarrow t$ if $s \approx t \succ C$ and $s \succ t$. Let N be a set of ground clauses. We define an interpretation I_N inductively, based on the total well-founded ordering $\succ$ on ground clauses. For any ground clause C Figure 7 defines the set E_C of rules produced by C, the rewrite systems R_C and S_C which contain rules and their symmetrizations produced by clauses smaller than C, the rewrite systems R^C and S^C which incorporate E_C, and the corresponding interpretations I_C and I^C. At each step we assume that for every ground clause $D \prec C$ the sets E_D, R_D, R^D, S_D, S^D, I_D and I^D are defined. Finally, R_N, S_N and I_N are the sets of rules and the interpretation for all clauses. If $E_C \neq \emptyset$ we say that C *produces* E_C, or that C is *productive*. I_N is called a *candidate model* for N. A clause C is a *counterexample* for I_N if C is false in I_N.

PROPOSITION 8. *Let $C = C' \vee l \approx r$ be a ground clause that produces $l \Rightarrow r$. Then C' is false in I_N.*

Let $\mathsf{CR}_0 = \mathsf{R} \cup \mathsf{S} \cup \mathsf{M} \cup \mathsf{AC} \cup \mathsf{CR}$.

LEMMA 9. *The axioms in* CR_0 *are true in any candidate model* I_N.

Proof. The convergent term rewriting system CR is included in the rules used to define I_N and rewriting is modulo AC, hence we have $s \Downarrow t$ for all axioms $s \approx t$ in $\mathsf{AC} \cup \mathsf{CR}$. The equality axioms except transitivity follow from the definition of $\Downarrow$. □

Note that transitivity does not hold for every interpretation constructed in this way, as there are cases where two extended rules overlap, and where $\mathsf{CR} \cup S_N$ is not Church-Rosser. The notions of redundancy are designed to allow reduction of counterexamples. If a clause is redundant and it is a counterexample for I_N then it can be reduced to a smaller counterexample for I_N in $N \cup g(\mathsf{T})$. If an inference is redundant and its conclusion is false in I_N then it can be reduced to a smaller counterexample for I_N in $N \cup g(\mathsf{T})$.

Formally, let $N^{\prec C}$ be the set of ground clauses in N which are smaller than C. Similarly, let $\mathsf{T}^{\prec C}$ denote the set of ground instances of transitivity which are smaller than C in the clause ordering. A ground clause C is *redundant* in a set of ground clauses N (with respect to CR) if $N^{\prec C} \cup \mathsf{T}^{\prec C} \cup \mathsf{CR}_0 \models C$. A ground inference with main premise D, side premises $C_1, \ldots, C_n$ and conclusion C is *redundant* in N (with respect to CR) if

$$N^{\prec D} \cup \mathsf{T}^{\prec D} \cup \mathsf{CR}_0 \cup \{C_1, \ldots, C_n\} \models C.$$

A set N of ground clauses is *saturated up to redundancy* with respect to an inference system $\mathcal{C}$ if all inferences in $\mathcal{C}$ from non-redundant premises in N are redundant. We say that an inference system $\mathcal{C}$ has the *reduction property* if whenever a non-empty clause D is the minimal counterexample for I_N in $N \cup g(\mathsf{T})$ then $\mathcal{C}$ contains an inference with main premise D, side premises $C_1, \ldots, C_n$ and conclusion C such that C is false in I_N and $C_1, \ldots, C_n$ are productive.

THEOREM 10. $\mathcal{CR}$ *has the reduction property.*

Proof. Let D be the minimal counterexample for I_N and assume that D is not the empty clause. Then all instances of transitivity in $\mathsf{T}^{\prec D}$ are true in I_N.

(1) Suppose that D is the instance $s_1 \not\approx s \vee s \not\approx s_2 \vee s_1 \approx s_2$ of transitivity. Then $\mathsf{T}^{\prec D} = \mathsf{T}^{\prec s}$ and $\mathsf{CR} \cup S_N$ is Church-Rosser on terms below s by proposition 5. However, since D is false, $\mathsf{CR} \cup S_N$ is not Church-Rosser on terms up to s. By lemma 7 there exists a non-redundant extended peak $t_1 \Leftarrow s \Rightarrow t_2$ which arises from rules $l_1 \Rightarrow r_1$ and $l_2 \Rightarrow r_2$ in R_N such that $t_1 \approx t_2$ is false in I_N. The rules $l_1 \Rightarrow r_1$ and $l_2 \Rightarrow r_2$ are produced by clauses $C_i = C_i' \vee l_i \approx r_i$ for $i = 1,2$. Hence C_1' and C_2' are false in I_N, and the conclusion $C = C_1' \vee C_2' \vee t_1 \approx t_2$ of the Extension Superposition inference with main premise D and side premises C_1 and C_2 is false in I_N.

(2) Otherwise D is a ground clause in N. Then the transitivity instances smaller than D include all instances of transitivity with a middle term smaller than or equal to s, where s is the maximal term of D. Hence $\mathsf{CR} \cup S_N$ is Church-Rosser on terms up to s by proposition 5. D is not redundant, since it is the minimal counterexample for I_N. Let $D = D' \vee L$ where L is the selected literal in D.

(2.1) Suppose $L = [\neg](p \approx q)$ is not in CR-normal form, $p \succeq q$, p is reducible by CR and L is not of the form $[\neg](n\phi + p' \approx m\phi + q')$, where ϕ is irreducible with respect to CR, $\phi \succ p'$, $\phi \succ q'$ and $n \neq 0$. The terms p and q have the form $s_1 + \cdots + s_k$ and $t_1 + \cdots + t_{k'}$ where $k, k' \geq 1$. It suffices to consider rewrite steps that may reduce or eliminate a maximal term s_i or t_j of these sums. This will either lead to an irreducible maximal term ϕ as desired, or another term will become maximal and subsequent reduction steps on that term become applicable. Hence we may assume that l' is a selected redex in p, and that a rule $l \Rightarrow r$ in CR reduces l' to $r\sigma$. Then a Theory Superposition inference with main premise D and conclusion $C = [\neg](p[r\sigma] \approx q) \vee D'$ is in $\mathcal{CR}$. The main premise follows from the conclusion and instances of transitivity whose maximal term is bounded by p. Since the main premise is false in I_N, the conclusion is false in I_N, too.

(2.2) Suppose $L = [\neg](p \approx q)$ is not in CR-normal form, L is of the form $[\neg](n\phi + p' \approx m\phi + q')$, where ϕ is irreducible with respect to CR, $\phi \succ p'$, $\phi \succ q'$ and $n \geq m$. Then consider the Isolation inference with main premise D and conclusion $C = D' \vee L'$ where $L' = [\neg]((n-m)\phi \approx q' + (-p'))$.

(2.2.1) If L is positive let us assume that C is true in I_N. Since D' is false in I_N, the literal L' is true in I_N and there exists a rewrite proof

$$(n-m)\phi \Downarrow_{\mathsf{AC}\backslash(\mathsf{CR}\cup S_N)} q' + (-p').$$

In Figure 8 this is the dashed proof operating on the boxed formulas. We put the context $m\phi + p' + []$ around every term in it and replace single steps of the original proof by normalized valley proofs, using semi-compatibility. The monomials in the resulting dotted proof are bounded by $m\phi$ or $n\phi$, the maximal monomials of p and q. Using lemma 7, we extend the Church-Rosser property from terms up to s to monomials up to s', where s' is the maximal monomial of the clause. Then $n\phi + p' \Downarrow_{\mathsf{AC}\backslash(\mathsf{CR}\cup S_N)} m\phi + q'$, and L is true in I_N, a contradiction. We conclude that C is false in I_N. This is the only case where we need to refine the bound on the Church-Rosser property to monomials.

(2.2.2) If on the other hand L is negative then L is false in I_N, that is, $n\phi + p' \Downarrow_{\mathsf{AC}\backslash(\mathsf{CR}\cup S_N)} m\phi + q'$. To show that L' is false in I_N, we use the same technique as before. We apply the context $(-m)\phi + (-p') + []$ to each term in the proof $n\phi + p' \Downarrow_{\mathsf{AC}\backslash(\mathsf{CR}\cup S_N)} m\phi + q'$ and get a proof $(n-m)\phi \overset{*}{-}_{\mathsf{AC}\cup\mathsf{CR}\cup S_N} q' + (-p')$ that is bounded by s. By the Church-Rosser property of $\mathsf{CR} \cup S_N$

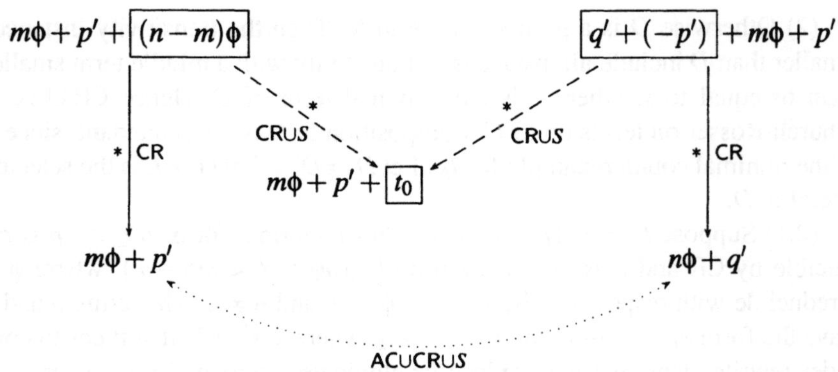

Figure 8. Isolation for positive literals

the terms $(n-m)\phi$ and $q' + (-p')$ converge and L' is false in I_N. Since D' is also false in I_N, the conclusion C is false in I_N.

(2.3) Suppose that $L = [\neg](p \approx q)$ is in CR-normal form, $p \succ q$ and p is reducible by S_C. Then there exists a clause $C_1 \prec D$ in N that produces a rule $l \Rightarrow r$ such that $p = u[l'']$ and $l'' =_{\mathsf{AC}} l'$ for some rule $l' \Rightarrow r'$ in $\mathcal{S}_{\mathsf{CR}}(l \Rightarrow r)$. Consider the Superposition inference with main premise D, side premise C_1 and conclusion $C = D' \vee L'$, where $L' = [\neg](u[r'] \approx q)$. Since C_1 is productive, $l' \approx r' \in \mathcal{S}_{\mathsf{CR}}(l \approx r)$ is true in I_N. Using $l' \approx r'$, transitivity, monotonicity and AC we obtain that L' is false in I_N. The maximal term of the instances of transitivity used is bounded by p, which in turn is bounded by the maximal term of D. Hence these instances of transitivity are contained in $\mathsf{T}^{\prec D}$. We conclude that C is false in I_N.

(2.4) Suppose that $L = [\neg](p \approx q)$ is in CR-normal form, $p \succ q$, and p is irreducible by S_C. Then $p \approx q$ is false in I_N, and since L is false in I_N it is a positive literal that is not productive. This can only be the case because D' is true in $(\mathsf{CR} \cup S_C \cup \mathcal{S}_{\mathsf{CR}}(p \approx q))^{\Downarrow}$. The only way that this can happen is that there is another positive equation with maximal term l in D'. That is, $D' = p \approx q' \vee D''$, such that $q \Downarrow_{\mathsf{AC}\backslash(\mathsf{CR} \cup S_C)} q'$. Consider the Equality Factoring inference with main premise D and conclusion $C = q \not\approx q' \vee p \approx q' \vee D''$. Both $q \not\approx q'$ and D' are false in I_N, hence C is false in I_N. □

This is the main part of the completeness proof. It remains to combine it with saturation.

THEOREM 11. *Let N be a set of ground clauses that is saturated up to redundancy with respect to $\mathcal{CR}$. Then $N \cup \mathsf{CR} \cup \mathsf{AC} \cup \mathsf{C}$ is unsatisfiable if and only if N contains the empty clause.*

Proof. CR is sound and has the reduction property. The rest of the proof is analogous to the proof of theorem 7 in chapter I.2.11. □

We have shown that the inference system CR is refutationally complete. To use this as the basis of a theorem prover, the same notion of fair theorem proving derivations as for the case of standard superposition can be used (see chapter I.2.11).

10. Lifting

It remains to consider how this ground inference system can be lifted to non-ground clauses. Instead of formally stating such an inference system, which is a rather tedious technical exercise, we discuss certain problems that distinguish this case from superposition theorem proving. Notably, one cannot avoid all overlaps below variable positions. The most problematic case is that of variables in top positions, like x in $x+p \approx q$. *Shielded* variables, which occur below a free function symbol somewhere else in the clause, cannot become the maximal term when instantiated and pose no problem. An unshielded variable x can become maximal, and inferences below x, which cannot be avoided in general, arise from Theory Superposition and Isolation. Also, variables below free function symbols and immediately below $\cdot$ are problematic, as for instance x in $f(n(x\cdot\phi))$. Any productive equation $m\psi \approx r$ where $2 \leq m \leq n$ gives rise to a superposition inference with such a subterm. In this case there are also many Theory Superposition inferences, in particular with distributivity, which replaces $x\phi$ by $y\phi + z\phi$.

In the presence of certain additional axioms, problematic variables can be eliminated. Waldmann (1997) has done this for the case of cancellative abelian monoids and Stuber (1996) has discussed it for the case of integer modules. The main idea is to remove top-level variables by simplification with the additional axioms. To treat variables inside atomic subterms one can use abstraction, replacing $C[f(x\cdot\phi)]$ by $C[f(y)] \vee x\cdot\phi \not\approx y$ and applying the same technique to x, which is now in the CR-context at the top.

11. Conclusion

We have presented a calculus for commutative rings that promises to be more efficient than the direct use of the ring axioms in a more general theorem prover. The calculus has not yet been implemented. The implemented system that is closest to our theory is the Cime system for Knuth-Bendix completion

(Contejean and Marché, 1996). It achieves substantial speedups when theory reasoning is used.

The technique is not limited to commutative rings. Given a convergent term rewriting system for some other theory, the notions of ordering, normal form, symmetrization, and inference system can be carried over. Obstacles could be the lack of a suitable ordering or a highly irregular symmetrization.

We also believe that the technique presented in this chapter can be generalized further using notions of reduction other than term rewriting. For instance, by using non-symmetric rewriting it should be possible to treat theories that include an ordering, such as ordered abelian groups, ordered rings or lattice-ordered groups (see chapter III.1.1).

Acknowledgements: The work reported here was supported in part by Deutsche Forschungsgemeinschaft under grant GA 261/7-1.

References

Bachmair, L. and H. Ganzinger: 1994, 'Associative-Commutative Superposition'. In: N. Dershowitz and N. Lindenstrauss (eds.): *Proc. 4th Int. Workshop on Conditional and Typed Rewriting*, Vol. 968 of *Lecture Notes in Computer Science*. Jerusalem, pp. 1–14, Springer. Revised version of TR MPI-I-93-267, 1993.

Boyer, R. S. and J. S. Moore: 1988, 'Integrating Decision Procedures into Heuristic Theorem Provers: A Case Study of Linear Arithmetic'. In: J. E. Hayes, D. Michie, and J. Richards (eds.): *Machine Intelligence 11*. Oxford: Clarendon Press, Chapt. 5, pp. 83–124.

Buchberger, B. and R. Loos: 1983, 'Algebraic Simplification'. In: *Computer Algebra: Symbolic and Algebraic Computation*. Springer, 2nd edition, pp. 11–43.

Bündgen, R.: 1996, 'Buchberger's algorithm: The term rewriter's point of view'. *Theoretical Computer Science* **159**, 143–190.

Contejean, E. and C. Marché: 1996, 'CiME: Completion Modulo E'. In: H. Ganzinger (ed.): *Proc. 7th Int. Conf. on Rewriting Techniques and Applications*, Vol. 1103 of *Lecture Notes in Computer Science*. New Brunswick, NJ, USA, pp. 18–32, Springer.

Delor, C. and L. Puel: 1993, 'Extension of the associative path ordering to a chain of associative commutative symbols'. In: *Proc. 5th Int. Conf. on Rewriting Techniques and Applications*, Vol. 690 of *Lecture Notes in Computer Science*. Montreal, pp. 389–404, Springer.

Dershowitz, N.: 1987, 'Termination of Rewriting'. *Journal of Symbolic Computation* **3**, 69–116.

Ganzinger, H. and U. Waldmann: 1996, 'Theorem Proving in Cancellative Abelian Monoids (Extended Abstract)'. In: M. A. McRobbie and J. K. Slaney (eds.): *13th*

Int. Conf. on Automated Deduction, Vol. 1104 of *Lecture Notes in Artificial Intelligence*. New Brunswick, NJ, USA, pp. 388–402, Springer.

Greendlinger, M.: 1960, 'Dehn's algorithm for the word problem'. *Communications on Pure and Applied Mathematics* **13**, 67–83.

Jouannaud, J.-P. and H. Kirchner: 1986, 'Completion of a set of rules modulo a set of equations'. *SIAM Journal on Computing* **15**(4), 1155–1194.

Le Chenadec, P.: 1986, *Canonical Forms in Finitely Presented Algebras*. London: Pitman.

Marché, C.: 1996, 'Normalised Rewriting: an Alternative to Rewriting modulo a Set of Equations'. *Journal of Symbolic Computation* **21**, 253–288.

Peterson, G. E. and M. E. Stickel: 1981, 'Complete Sets of Reductions for Some Equational Theories'. *Journal of the ACM* **28**(2), 233–264.

Rubio, A. and R. Nieuwenhuis: 1993, 'A precedence-based total AC-compatible ordering'. In: C. Kirchner (ed.): *Proc. 5th Int. Conf. on Rewriting Techniques and Applications*, Vol. 690 of *Lecture Notes in Computer Science*. Montreal, pp. 374–388, Springer.

Stickel, M. E.: 1985, 'Automated Deduction by Theory Resolution'. *Journal of Automated Reasoning* **1**(4), 333–355.

Stuber, J.: 1996, 'Superposition Theorem Proving for Abelian Groups Represented as Integer Modules'. In: H. Ganzinger (ed.): *Proc. 7th Int. Conf. on Rewriting Techniques and Applications*, Vol. 1103 of *Lecture Notes in Computer Science*. New Brunswick, NJ, USA, pp. 33–47, Springer.

Waldmann, U.: 1997, 'A Superposition Calculus for Divisible Torsion-Free Abelian Groups'. In: M. P. Bonacina and U. Furbach (eds.): *Int. Workshop on First-Order Theorem Proving*. Schloß Hagenberg, Linz, Austria, pp. 130–134, Johannes Kepler Universität, Linz, Austria.

HANS JÜRGEN OHLBACH & JANA KÖHLER

CHAPTER 3

HOW TO AUGMENT A FORMAL SYSTEM WITH A BOOLEAN ALGEBRA COMPONENT

1. INTRODUCTION

Reasoning with Boolean Algebras is just propositional reasoning. This is well investigated and a lot of good algorithms have been developed. Other formal systems, for example mathematical programming for reasoning about arithmetical equation systems, are equally well developed. Combining such a system with a Boolean component where the Boolean expressions are interpreted as sets, would allow one to use arithmetical algorithms to reason about numerical features of sets.

In this chapter we introduce the atomic decomposition technique as a means to combine computation and reasoning in a given formal system with reasoning about Boolean Algebras and features of the elements of the Boolean Algebra which make sense in the given basic system. Propositional reasoning is invoked in a kind of compilation phase which eliminates the Boolean Algebra part of the problem completely and shifts the main reasoning problem to the basic system. The decomposition method works for combinations of formal systems with a Boolean Algebra component, where the Boolean terms are embedded in *bridging functions* mapping the Boolean parts to objects the given formal system can understand.

Before we start developing the mathematical part, let us introduce the basic idea with a simple example.

1.1. *An Introductory Example*

If Henry has two sons and three daughters, he has five children. In a formal language we can state this as

$$(3.1) \qquad |sons| = 2 \wedge |daughters| = 3 \rightarrow |children| = 5$$

where $|\ldots|$ denotes the set cardinality function. Checking the validity of this formula involves a combination of Boolean reasoning with arithmetic equation solving. Without further information, sons, daughters and children can be arbitrary sets having arbitrary overlaps with each other. Therefore implication (3.1) is not valid in general.

W. Bibel, P. H. Schmitt (eds.), Automated Deduction. A basis for applications. Vol. III

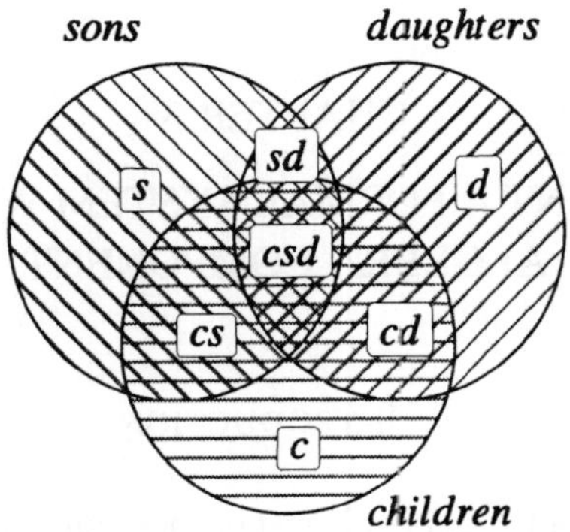

Figure 1. A general set structure

Figure 1 shows the most general way, three different sets can overlap with each other. As one can see, the three sets can be built up from seven mutually disjoint and unseparated areas. We gave these areas names with the following meaning:

c = children, not sons, not daughters.
s = sons, not children, not daughters.
d = daughters, not children, not sons.
cs = children, which are sons, not daughters.
cd = children, which are daughters, not sons.
sd = sons, which are daughters, not children.
csd = children, which are both sons and daughters.

The original sets can now be obtained from their "atomic" components:

$$\begin{aligned} children &= c \cup cs \cup cd \cup csd \\ sons &= s \cup cs \cup sd \cup csd \\ daughters &= d \cup cd \cup sd \cup csd. \end{aligned}$$

Moreover, since this decomposition is mutually disjoint and exhaustive, the cardinalities of the sets just add up:

$$\begin{aligned} |children| &= |c| + |cs| + |cd| + |csd| \\ |sons| &= |s| + |cs| + |sd| + |csd| \\ |daughters| &= |d| + |cd| + |sd| + |csd|. \end{aligned}$$

Formula (3.1) can now be rewritten into

$$(3.2)\qquad \begin{array}{l} |s|+|cs|+|sd|+|csd|=2 \wedge |d|+|cd|+|sd|+|csd|=3 \\ \quad \rightarrow\ |c|+|cs|+|cd|+|csd|=5 \end{array}$$

or by dropping the cardinality function |...|

$$(3.3)\qquad \begin{array}{l} s+cs+sd+csd=2 \wedge d+cd+sd+csd=3 \\ \quad \rightarrow\ c+cs+cd+csd=5. \end{array}$$

In (3.3) we interpret the symbols $c, s, \ldots$ directly as the numbers denoting the cardinality of the corresponding sets. This makes sense because the sets are finite and mutually disjoint. This way, problem (3.1) has been transformed into a pure non-negative linear Diophantine equation problem. Diophantine equations are equations with integer-valued variables. They are called *linear* if no products of different variables occur and *non-negative* if variables are constrained to non-negative integers. Formula (3.3), of course, is still not valid. Further information is necessary.

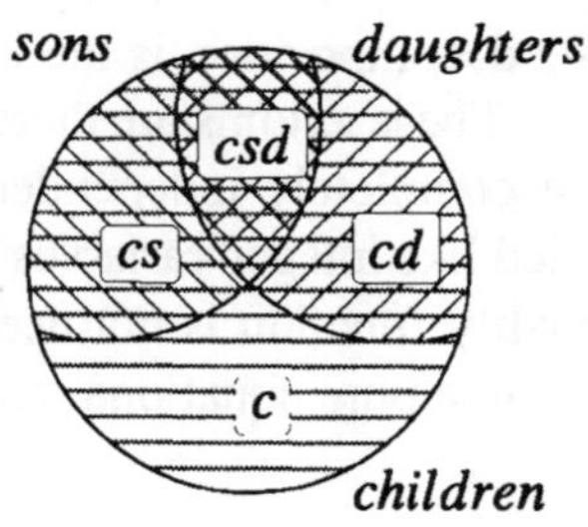

Figure 2. The correct subset relationships.

First of all we add the subset information. Both sons and daughters are children. This means the first picture was a bit too general. We are in a more specific situation which is depicted in figure 2. Here all sets not containing c are empty reflecting the subset information. That means $s=0$, $d=0$, $sd=0$ holds and we can simplify problem (3.3) to

$$(3.4)\qquad cs+csd=2 \wedge cd+csd=3 \quad \rightarrow \quad c+cs+cd+csd=5.$$

which is still not valid. The next piece of information, we need, is that there are no hermaphrodites. This means the intersection of sons and daughters is empty: $csd=0$. We obtain

$$cs=2 \wedge cd=3 \quad \rightarrow \quad c+cs+cd=5.$$

Finally, we exploit the fact that there are no other types of children besides sons and daughters, i.e. $c = 0$. We end up with

$$(3.5)\quad cs = 2 \wedge cd = 3 \quad \rightarrow \quad cs + cd = 5$$

and this is in fact valid.

We get a more abstract view of what we did, if we change the argumentation slightly. The information that *children* are partitioned into *sons* and *daughters* can be represented as propositional formulae:

$$(3.6)\quad \begin{array}{ll} sons \rightarrow children & daughters \rightarrow children \\ children \rightarrow sons \vee daughters & sons \rightarrow \neg daughters. \end{array}$$

There are only three propositional models of these formulae: $\{children, sons, \neg daughters\}$, $\{children, daughters, \neg sons\}$ and $\{\neg children, \neg sons, \neg daughters\}$.

The first one corresponds to the set of sons only, the second one to the set of daughters only and the third one to all other objects which are not children. This triggers the decomposition of $|children|$ into $|cs| + |cd|$, $|sons|$ into $|cs|$ and $|daughters|$ into $|cd|$ (the negated propositions in the models can be ignored). Built into the decomposition is the information contained in the propositional axioms (3.6). These axioms are therefore not needed any longer. The decomposed sets are completely independent of each other. Therefore terms like $|cs|$ can be treated like integer variables with the constraint $|cs| \geq 0$. This way even the cardinality function is not needed any longer. The whole problem has been reduced to a pure equational reasoning problem.

1.2. *Boolean Algebras*

The introductory example illustrates the basic idea of the method. Combining propositional reasoning with arithmetic equation solving, however, is only one application of the atomic decomposition. The method can be described in a much more abstract way leading to a wider range of applications. For an appropriate mathematical treatment of the abstract method we need some basic facts and results about Boolean Algebras which may not be familiar to the average reader. Therefore we summarise the material about Boolean Algebras which is relevant for the purposes of this paper. Details can be found in every modern textbook about lattice theory and Boolean Algebra.

A Boolean Algebra $A = (D_A, \bot_A, \top_A, \sqcap_A, \sqcup_A, {}'^A)$ consist of a non-empty *domain* D_A, two distinguished elements $\bot_A$ and $\top_A$, the two binary functions $\sqcap_A$ (meet) and $\sqcup_A$ (join) and the inverse function ${}'^A$. They observe the familiar laws commutativity, associativity, distributivity for $\sqcap_A$ and $\sqcup_A$, the absorption laws for $\sqcap_A$ and $\sqcup_A$ and the inverse law for ${}'^A$.

A Boolean Algebra A filter in a Boolean Algebra

Figure 2.

Examples for Boolean Algebras are the two element Boolean Algebra **2** consisting of just $(\bot, \top)$, the powerset algebra 2^S of some basic set S, or the equivalence classes of propositional formulae (the Lindenbaum algebra for propositional logic).

A *partial order* $x \sqsubseteq_A y$ iff $x \sqcap_A y = x$ can be introduced. It makes $\bot_A$ the smallest element and $\top_A$ the largest element. A *difference function* $x \setminus_A y \stackrel{\text{def}}{=} x \sqcap_A y'^A$ is also very common.

Figure 1.2 shows a very intuitive picture of a Boolean Algebra with smallest element $\bot$ and largest element $\top$.

A Boolean Algebra is *generated* by a set P of objects if all its elements can be obtained by meets, joins and the inverse of elements of P. It is *finitely generated* if P is finite. The Lindenbaum algebra of propositional logic is generated from the set of Boolean variables.

A smallest element above $\bot$ is called an *atom* in a Boolean Algebra. (There also 'anti-atoms'. These are the biggest elements below $\top$.) A Boolean Algebra is *complete* iff it contains all (*infinite*) meets and joins. It is *atomic* iff all its elements x can be obtained as the join of all the atoms below x, i.e.

$$(3.7) \quad x = \bigsqcup_{a \sqsubseteq_A x,\ a \text{ atomic}} a.$$

Finite Boolean Algebras are always complete and atomic. The powerset algebra is also complete and atomic. Its atoms are the singleton sets.

A *filter* F in a Boolean Algebra A is a *proper* subset which is upwards closed and contains all meets. (The dual concept to a filter is an *ideal*). An *ultrafilter* is a maximal filter. An ultrafilters F is characterized by the following properties:

1. for each element x in A, either $x \in F$ or $x' \in F$.
2. $x \sqcap y \in F$ iff $x \in F$ and $y \in F$.
3. $x \sqcup y \in F$ iff $x \in F$ or $y \in F$.
4. $\top \in F$.

Because of condition 1, an ultrafilter F cuts a Boolean Algebra in two halfs, the set of elements which are in F and the set of elements which are not in F. This actually induces a homomorphism h_F from A into the binary algebra **2**.

THEOREM 1.1. (Stone's Representation Theorem (Stone, 1936)).
For every Boolean Algebra A there is an isomorphism ι *between A and the Boolean Algebra of the set of sets of ultrafilters of A:*

$$\iota(x) \stackrel{\text{def}}{=} \{F \mid F \text{ is an ultrafilter in } A \text{ and } x \in F\}.$$

Let $\iota(A)$ *be the corresponding set algebra' isomorphic to A.* ■

If A is complete and atomic then ι maps A to the full powerset algebra of its ultrafilters, otherwise ι is only an embedding into it. The ultrafilters in A are actually the atoms in $\iota(A)$. If A is complete and atomic, and F is an ultrafilter then $\iota^{-1}(\{F\})$ is an atom in A.

2. Syntax of the Languages Involved

We need three components in the syntax. The first component of our syntax is the language L_E of some basic system E which we want to augment with a Boolean component. In order to cover as many systems as possible, we do not specify E in detail. The only requirement is that L_E has constant and function symbols, or free variables, which can be existentially quantified. The system E may be very concrete in the sense that everything except the free variables has a fixed meaning. Typical examples for E are systems of arithmetical equations and inequations, but E may be any other suitable formal system.

The second component is the Boolean Algebra component. It is specified in the usual way.

Given a set P of Boolean variables, the Boolean term language is:

$$L_B(P) ::= P \mid L_B(P) \sqcap L_B(P) \mid L_B(P) \sqcup L_B(P) \mid L_B(P)'.$$

Other functions, for example set difference $\setminus$, can be defined as abbreviations in the language.

Important: $L_B(P)$ and L_E need not share any symbols!

A *Boolean axiom* is of the form $c = \top$ where $c \in L_B(P)$.

$x \sqsubseteq y$ can be used as an abbreviation for $x' \sqcup y = \top$ and $x = y$ can be used for $x' \sqcup y = \top$ together with $y' \sqcup x = \top$.

EXAMPLE 2.1. *If $P = \{children, sons, daughters\}$ then $\{sons \sqcap daughters, children \sqcup sons'\} \subseteq L_B(P)$, 'sons $\sqsubseteq$ children' and 'daughters $\sqsubseteq$ children' are Boolean axioms.* ■

As a bridge between these two languages L_E and $L_B(P)$ we need a distinguished set F of *bridging functions*, different from all other symbols involved. A typical example for a bridging function is the cardinality function mapping sets to integers. A bridging function symbol may have any finite arity. Each argument position, however, can take either a Boolean term as argument, or an L_E-term. For convenience, we assume that a bridging function of arity $n+k$ reserves the first n arguments for Boolean terms and the remaining k arguments for L_E-terms.

The combined language is defined as follows:

DEFINITION 2.2. (The combined language $L_{BE}(P)$).
If $s[x] \in L_E$ where x is some term occurring at some position in s, $f \in F$ with arity $n+k$, $t_1, \ldots, t_n \in L_B(P)$, $s_1, \ldots, s_m \in L_E$ then $s[x/f(t_1, \ldots, t_n, s_1, \ldots, s_k)] \in L_{BE}(P)$. No other terms are in $L_{BE}(P)$. ■

$L_{BE}(P)$ is essentially like L_E, but L_E-term positions can be occupied by $L_B(P)$-terms embedded in a bridging function. The combined language is such that the Boolean parts and the L_E-parts are separated by the functions in F. The Boolean part is the lower part in the term tree, then there comes some bridging function and the upper part belongs to L_E.

EXAMPLE 2.3. *E = arithmetic, $F = \{|.|, f, a, c\}$ with the informal meaning: $|.|$ is the set cardinality function, f means 'combined fortune', a means 'average income' and the binary function c means 'consumption of'. Well formed $L_{BE}(P)$ axioms are now*

$|sons \sqcup friends| \geq 5$

(there are more than 5 sons and friends.)

$f(children) \geq 100000$

(the combined fortune of the children exceeds 100000.)

$a(daughters) = 10000$

(the average income of the daughters is 10000.)

$c(children \sqcup friends, cigarettes) = 0$

(The children and friends do not smoke cigarettes.) ■

Bridging functions are instances of Dov Gabbay's fibring functions, he introduced for combining different logical systems (Gabbay, 1996; Gabbay, 1998). In Gabbay's fibring approach, these functions are objects in the semantics of the combined language. They link models of the component languages with each other. In our system these functions have a syntactic representation, which means that one can specify their properties and reason with them explicitly.

3. Semantics of the Languages Involved

The language L_E comes with its natural semantics. The only feature we need is that an interpretation $\Im_E$ for E maps the free variables and constant symbols to elements of E's domain and interprets function symbols as functions in the usual way. In an arithmetical language, $\Im_E$ may for example represent a solution of an equation system.

The language $L_B(P)$ is to be interpreted as a complete and atomic Boolean Algebra, usually, but not necessarily as a set algebra.

The interpretation is therefore a homomorphism $\Im_B : L_B(P) \mapsto A$ where A is a complete and atomic Boolean Algebra.[1]

$\Im_B$ satisfies a Boolean axiom $\varphi = \top$, i.e.
$\Im_B \models (\varphi = \top)$ iff $\Im_B(\varphi) = \top_A$.

Since the language $L_B(P)$ and L_E do not share any symbols, we can define a combined interpretation $\Im_{BE}$ as the union of the interpretations $\Im_B$ and $\Im_E$. The interpretation of the bridging function symbols also becomes part of $\Im_{BE}$.

The interpretation of the bridging function symbols in F can, but need not be fixed. Certain structural conditions, however must be satisfied. First of all, bridging functions must be type conform. That means for a bridging function symbol f with n Boolean arguments and m L_E-arguments, a combined interpretation $\Im_{BE}(f)$ must map tuples consisting of n elements of the Boolean Algebra and m elements of E's domain to an element of E's domain.

But there is the extra requirement of *additivity*, which is essential for making the atomic decomposition work. If a bridging function is applied to the join of two disjoint objects x and y of a Boolean Algebra ($x \sqcap_A y = \bot_A$), then it must be possible to apply the bridging function to x and y separately and then combine the result. For the set cardinality function for example, this is a very natural property: $|x \cup y| = |x| + |y|$, provided x and y are disjoint.

To ensure additivity, we require that the *additivity axioms* are satisfied, either by building them into the system, or by making them an explicit part

[1] In many practical applications, A is even finite. We have not investigated the case where A is not complete or not atomic.

of a specification. The additivity axioms are special interaction axioms for fibring functions in the sense of Gabbay (Gabbay, 1996; Gabbay, 1998).

DEFINITION 3.1. (Additivity Axioms).
The additivity axioms for a bridging function $f \in F$ with arity $n+k$ are:
$x \sqcap y = \bot \to f(\ldots, t_{i-1}, x \sqcup y, t_{i+1}, \ldots)$
$* = g_i(f(\ldots, t_{i-1}, x, t_{i+1}, \ldots), f(\ldots, t_{i-1}, y, t_{i+1}, \ldots))$
for each $i \in \{1, \ldots, n\}$ of the Boolean argument positions, where $g_i(x,y)$ is some term in L_E. ■

EXAMPLES 3.2. *If $x \cap y = \emptyset$ then $|x \cup y| = |x| + |y|$. Therefore the bridging function 'cardinality' is additive for set algebras and the function g_1 in the definition above is the addition of numbers.*

The same holds if the function 'combined fortune' f in the example above is interpreted in the intended way. The situation for the bridging function 'average income' 'a' mentioned in the same example is a bit more complicated. To get the intended behaviour, we have to require

$$a(x \sqcup y) = \frac{|x|a(x) + |y|a(y)}{|x| + |y|}.$$

In a different setting we may map the Boolean Algebra terms to, say, sets of axioms of some logic, and the algebra E consists of the set of models of this logic. A bridging function may now be a function M mapping axiom sets to sets of models. This function is additive as well because

$$M(x \sqcup y) = M(x) \cap M(y)$$

holds. Here the function g is set intersection. ■

DEFINITION 3.3. (Problem specification).
A problem specification *now consists of three parts*

1. *a set $\mathcal{A}$ of Boolean axioms,*
2. *a set $\mathcal{E}$ of $L_{BE}(P)$ formulae,*
3. *the bridging function additivity axioms $\mathcal{B}$ for each bridging function (Def. 3.1).*

The satisfiability *problem is to find out whether such a specification is consistent, i.e. it has an interpretation satisfying all three parts.* ■

4. ATOMIC DECOMPOSITION

The Boolean language $L_B(P)$ is isomorphic to the propositional formula language. The correspondences are $\sqcap \sim \wedge, \sqcup \sim \vee, ' \sim \neg, \sqsubseteq \sim \rightarrow, = \sim \leftrightarrow$. Therefore we can also have interpretations into binary truth values. As algebraic objects, these binary truth value interpretations are homomorphisms, $M : L_B(P) \mapsto \mathbf{2}$. Therefore we can summarise the situation in the following picture.

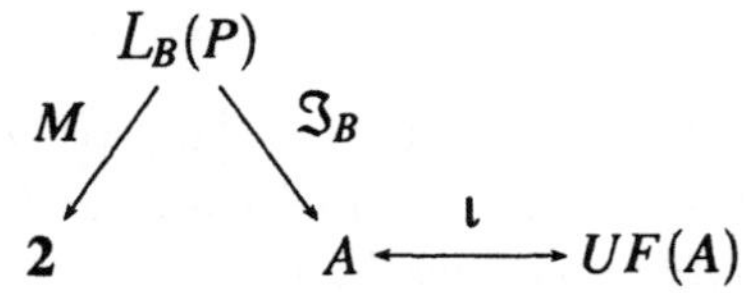

PROPOSITION 4.1. (Models and ultrafilters).
Let A be a consistent set of Boolean axioms and $\Im_B : L_B(P) \mapsto A$ be an interpretation satisfying A. There is a (surjective) on-to mapping μ from the set of propositional models of A to the set of ultrafilters of A.

Proof: A induces an equivalence relation on $L_B(P)$ terms such that $\Im_B$ as a model of A maps equivalent formulae to the same element of A. Each element x in A is therefore the image $\Im_B(c_x)$ of some Boolean A-equivalent terms c_x. For each x in A we therefore choose a Boolean term c_x with $\Im_B(c_x) = x$. $\Im_B(c_x') = x'^A$ is automatically guaranteed.

a) Let m be a propositional model of A. We define

$$\mu(m) \stackrel{\text{def}}{=} \{\Im_B(c) \mid m \models c\}$$

and check the properties 1 – 4 of ultrafilters (page 61).
1.) for $x \in A$, either $m \models c_x$ or $m \models c_x'$. Therefore either $x \in \mu(m)$ or $x'^A \in \mu(m)$.
2.) $x \sqcap_A y \in \mu(m)$ iff $m \models c_x \sqcap c_y$ iff $m \models c_x$ and $m \models c_y$ iff $x \in \mu(m)$ and $y \in \mu(m)$. The proof for the third condition is analogous and the fourth condition is trivial.

b) Let F be an ultrafilter in A. We define $m = \mu^{-1}(F)$ to be a propositional interpretation such that

$$m \models c \text{ iff } \Im_B(c) \in F.$$

Using the conditions 1 – 4 for ultrafilters again, it is straightforward to prove that m is an interpretation. m satisfies A because $\Im_B$ satisfies A and therefore it maps all terms ψ with '$\psi = \top$' $\in A$ to $\top_A$. Thus, $\Im_B(\psi) \in F$ and therefore $m \models \psi$.

The definitions in a) and b) guarantee $\mu(\mu^{-1}(F)) = F$. The mapping is not injective. For example in the extreme case where A is the binary Boolean Algebra **2**, there is only one ultrafilter consisting of $\{\top\}$ only. All propositional models are mapped to this single ultrafilter. ■

As a consequence of Stone's representation theorem we also get an on-to mapping from the propositional models of $\mathcal{A}$ to the atoms in A, provided A is complete and atomic. In this case the isomorphism ι yields a one-to-one mapping between A's atoms and A's ultrafilters, which are the atoms of $\iota(A)$. Therefore $\nu(m) \stackrel{\text{def}}{=} \iota(\mu(m))$ yields for a model m a corresponding atom in A.

COROLLARY 4.2. (Propositional Models and Atoms).
If A is complete and atomic there is an on-to mapping ν from the set of propositional models of $\mathcal{A}$ to the set of atoms in A.

Furthermore we have for each $x = \Im_B(c)$ in A, atom a in A and for each propositional model m with $\nu(m) = a$:
$a \sqsubseteq_A x$ iff $m \models c$. ■

EXAMPLE 4.3. *Let us illustrate this with a fragment of the introductory example. Suppose the Boolean axiomatization is*

$$\mathcal{A} = \{sons \sqsubseteq children, daughters \sqsubseteq children\}.$$

This corresponds to the propositional clause set

$$\{\{\neg sons, children\}, \{\neg daughters, children\}\},$$

which has five models:

1. $\{\neg sons, \neg daughters, \neg children\}$
2. $\{\neg sons, \neg daughters, children\}$
3. $\{\neg sons, daughters, children\}$
4. $\{sons, \neg daughters, children\}$
5. $\{sons, daughters, children\}$.

Each of these models corresponds to an area in Figure 2. The first one corresponds to the area surrounding the circle, no children at all. The second one corresponds to c, the third one to cd, the fourth one to cs and the fifth one to csd. These are the atoms of the Boolean Algebra consisting of the set of sets which can be obtained by forming their set-union. ■

We can now introduce a syntactic representation for the propositional models of $\mathcal{A}$ and extend the interpretation $\Im_B$ to map the syntactic representation of the models, let us call them *syntactic atoms*, to the atoms of A:

$$\text{If } m \models \mathcal{A} \text{ then } \Im_B(m) \stackrel{\text{def}}{=} \nu(m) \in A.$$

The precise structure of the syntactic representation of A's models can be left to an implementation. Quite convenient is the representation we used in the introductory example, as a set or string of Boolean variables.

Corollary 4.2 and (3.7) now yields the basis for our atomic decomposition calculus

$$\Im_B(c) = \bigsqcup_{m \models c, m \models A} \Im_B(m)$$

for each Boolean term c.

This allows us to map Boolean terms to sets of syntactic atoms. A set of such atoms denotes the join of the interpretations of its members. Therefore we define for sets $\{m_1, \ldots, m_n\}$ of atoms

$$\Im_B(\{m_1, \ldots, m_n\}) \stackrel{\text{def}}{=} \Im_B(m_1) \sqcup_A \ldots \sqcup_A \Im_B(m_n).$$

DEFINITION 4.4. (Atomic Decomposition). *We define a decomposition function α_A which maps Boolean terms to sets of syntactic atoms (propositional models of A).*

$$\begin{array}{lcll} \alpha_A(p) & = & \{m_1, \ldots, m_n\} & \textit{where } p \textit{ is a Boolean variable} \\ \alpha_A(x \sqcup y) & = & \alpha_A(x) \cap \alpha_A(y) & \textit{and } m_i \models p \\ \alpha_A(x \sqcap y) & = & \alpha_A(x) \cup \alpha_A(y) & \\ \alpha_A(x') & = & M \setminus \alpha_A(x) & M \textit{ is the set of all models of } A. \end{array}$$

■

Notice that α_A yields sets of syntactic atoms by performing set operations on sets of syntactic atoms. This is no longer an operation within one of the languages, but with datastructures of the corresponding implementation.

Using that $\Im_B(m_1) \sqcap_A \Im_B(m_2) = \bot_A$ for two different atoms m_1 and m_2 it is now straightforward to show with structural induction

(3.8) $\quad \Im_B(c) = \Im_B(\alpha_A(c))$

for all Boolean terms c. For each '$\psi = \top$' $\in A$ this means in particular $\top_A = \Im_B(\psi) = \Im_B(\alpha_A(\psi))$ and therefore

(3.9) $\quad \alpha_A(\psi) =$ set of all atoms in A.

The next proposition confirms that the Boolean axioms A are no longer needed after the decomposition.

LEMMA 4.5. (Elimination of the Boolean axioms).
A specification $\{A, B, E\}$ consisting of the Boolean axioms A, the additivity axioms B and some $L_{BE}(P)$-formulae E is satisfiable if and only if $\{B, \alpha_A(E)\}$ is satisfiable.

Proof: If $\{A,B,E\}$ is satisfied by some interpretation $\Im$, then because of (3.8), the Boolean terms c and $\alpha_A(c)$ are mapped to the same objects. Therefore $\alpha_A(E)$ is satisfied as well.

Now suppose there is some model $\Im$ for $\{B,\alpha_A(E)\}$. $\Im$ maps the syntactic atoms m_i occurring in $\alpha_A(E)$ to some objects in some domain. Let $M \stackrel{\text{def}}{=} \{k_1,\ldots,k_l\} \stackrel{\text{def}}{=} \{\Im(m_1),\ldots,\Im(m_n)\}$ where $m_1,\ldots,m_n$ are all syntactic atoms. We define the Boolean Algebra A to be the powerset algebra of M. A is complete and atomic. The atoms in A are just the objects $\Im(m)$ where the m are the syntactic atoms.

$\Im$ is extended to $\Im_B$ such that $\Im_B$ maps Boolean variables p to $\{\Im(m) \mid m \in \alpha_A(p)\}$. Because of (3.9), it maps each ψ with '$\psi = \top$' $\in A$ to the interpretation of the set of all atoms, which is just the top element in A. Thus, $\Im_B$ satisfies A.

Structural induction using definition 4.4 yields $\Im_B(c) = \Im_B(\alpha_A(c))$. Since $\Im$ satisfies $\alpha_A(E)$, $\Im_B$ now must satisfy E. Thus, the original specification is satisfiable. ■

Having eliminated the Boolean part, we use now the additivity axioms B to eliminate the bridging functions together with the atoms introduced by α_A. This is done in two steps. In the first step we eliminate the additivity axioms.

LEMMA 4.6. (Elimination of Additivity Axioms).
Let $\{B,E'\}$ be a decomposed specification and let $E'' \stackrel{\text{def}}{=} \rho(E')$ be the result of an exhaustive left to right application of the equations in B to E, taking the sets $\{m_1,\ldots,m_k\}$ as joins $m_1 \sqcup \ldots \sqcup m_k$. (The bridging functions in E' then occur only with a single atom m as argument, no longer with a set of atoms).

$\{B,E'\}$ is satisfiable iff E'' is satisfiable.

Proof: The additivity axioms in B are conditioned equations. The conditions $x \sqcap y$ are satisfied for atoms x and y. Thus, the equation itself can be used as rewrite rules. Using them as rewrite rules is an equivalence transformation. Thus, E' and E'' are equivalent, given B. What remains to be shown is that whenever there is a model for E'', there is also one satisfying B. Let $\Im$ satisfy E''. As in the proof of proposition 4.5 we construct the algebra A as the powerset algebra of the interpretations of the m_i. Starting with the interpretations of the m_i as atoms of the powerset algebra, each of its elements can now be constructed as the join of some smaller elements. Therefore we can take the additivity axioms for the bridging functions f themselves as a recursive definition for the values of $f(x_1,\ldots,x_n,s_1,\ldots,s_k)$. The basis of the recursion is the interpretation of the terms $f(m_1,\ldots,m_n,s_1,\ldots,s_k)$ for the atoms m_i, which is provided by $\Im$. This way we get an interpretation satisfying B and E. ■

The last step is now to eliminate the bridging terms $f(m_1,\ldots,m_n,s_1,\ldots,s_k)$ in the transformed specification E''. If $k=0$, each such term can just be replaced by a new (existentially quantified) L_E-variable or constant symbol. If $k>0$, $f(m_1,\ldots,m_n,\ s_1,\ldots,s_k)$ is translated into a term $f'_{m_1,\ldots,m_n}(s_1,\ldots,s_k)$ where $f'_{m_1,\ldots,m_n}$ is a fresh new L_E-function symbol.

LEMMA 4.7. (Elimination of the Bridging Functions).
We define a replacement operation β *which replaces all bridging function symbols* f *with* n *Boolean and* k L_E*-arguments by corresponding* L_E*-terms.* β *introduces for each term* $f(m_1,\ldots,m_n,\ldots)$ *a new* L_E*-function (or constant) symbol* $f'_{m_1,\ldots,m_n}$ *and replaces terms* $f(m_1,\ldots,m_n,\ s_1,\ldots,s_k)$ *with* $f'_{m_1,\ldots,m_n}(s_1,\ldots,s_k)$.

E'' *is satisfiable if and only if* $\beta(E'')$ *is satisfiable.*

Proof: If E'' is satisfied by some interpretation $\mathfrak{I}$ we just choose $\mathfrak{I}(f'_{m_1,\ldots,m_n})(x_1,\ldots,x_k) \stackrel{\text{def}}{=} \mathfrak{I}(f(m_1,\ldots,m_n,x_1,\ldots,x_k))$.

This guarantees that $\beta(E'')$ is satisfied as well.

For the other direction, suppose $\mathfrak{I}$ satisfies $\beta(E'')$. For the interpretation of the terms $f(m_1,\ldots,m_n,s_1,\ldots,s_k)$ we just map the atoms m_i to themselves and define the interpretation of the bridging functions f to be

$$\mathfrak{I}(f)(m_1,\ldots,m_k,x_1,\ldots,x_k) \stackrel{\text{def}}{=} \mathfrak{I}(f'_{m_1,\ldots,m_n})(x_1,\ldots,x_k).$$

This is possible because neither the atoms m_i nor the bridging functions f occur in $\beta(E'')$. The terms $f(m_1,\ldots,m_n,s_1,\ldots,s_k)$ in E'' are now interpreted exactly like the terms $f'_{m_1,\ldots,m_n}(s_1,\ldots,s_k)$ in $\beta(E'')$. Nothing else is changed and therefore E'' is satisfied as well. ■

As a consequence of the lemmas 4.5, 4.6 and 4.7 we get a general soundness and completeness result.

THEOREM 4.8. (Soundness and Completeness).
A problem specification (A,E,B) *is satisfiable iff and only if* A *is satisfiable and the transformed specification* $\beta(\rho(\alpha_A(E)))$ *is satisfiable.* ■

The inference procedure derived from this theorem comprises the following steps: in order to check satisfiability of a combined specification (A,E,B), first compute the syntactic atoms derived from the propositional models of A. If there are no models then the specification is unsatisfiable. If there are models, decompose the Boolean terms occurring in E into sets of syntactic atoms. Use the additivity axioms in E to push the bridging functions down to the level of single atoms. Then replace the resulting 'bridging terms' with variables or composed L_E-terms, and check the result with an E-satisfiability checker.

If satisfiability in E is decidable we get a decision procedure for the combination with the Boolean language. Satisfiability for this combination is then decidable as well. However, if the decision problem in E is exponential, the complexity of the combined procedure does may get double exponential because exponentially many new variables may be introduced. In section 5 below this problem is discussed in more detail.

EXAMPLE 4.9. *Using our method, it is quite straightforward to enrich classical propositional logic with a component for reasoning about sets. The bridging functions are just predicates on sets.*

As an example, consider again the specification (3.6) of children partitioned into sons and daughters. As bridging function we take the predicate $beautiful$. $beautiful(x)$ *is intended to mean 'every element of x is beautiful'. The additivity axiom is now*

$$beautiful(x \sqcup y) = beautiful(x) \wedge beautiful(y).$$

This holds even without the condition of x and y being disjoint.

The formulae in the combined language are for example

1) $beautiful(children)$
2) $beautiful(daughters) \rightarrow happy\text{-}father.$

The decomposition turns 1) into $beautiful(sons) \wedge beautiful(daughters)$ *and leaves 2) essentially the same. Replacing the bridging function terms with new predicate symbols yields the pure propositional formulae*

$beautiful\text{-}sons$
$beautiful\text{-}daughters$
$beautiful\text{-}daughters \rightarrow happy\text{-}father$

from which $happy\text{-}father$ *can be derived with Modus Ponens.*

EXAMPLE 4.10. *If the basic logic E is first-order predicate logic, we can enrich this logic with Boolean terms as arguments of the predicates, provided the additivity axioms hold. Typically they would again look like*

$$p(x \sqcup y, \ldots) = p(x, \ldots) \wedge p(y, \ldots).$$

The first steps of the procedure transforms each literal p with a Boolean term as argument into $p(\{m_1, \ldots, m_k\}, \ldots)$. *Instead of applying the additivity axioms now, which might yield a big conjunction for the positive literals and a big disjunction for the negative literals, one can build them into a new theory resolution rule:*

$$\frac{\begin{array}{l} p(\{m_1,\ldots,m_k\},s),C \\ \neg p(\{u_1,\ldots,u_l\},s'),D \qquad s\sigma = s'\sigma \end{array}}{(p(\{m_1,\ldots,m_k\}\setminus\{u_1,\ldots,u_l\},s),C)\sigma.}$$

5. Optimisations

A formula with l Boolean variables may in the worst case have 2^l models. For all of them one has to generate syntactic atoms. This makes the whole approach questionable. Fortunately there are some optimisations which can reduce the number of syntactic atoms considerably.

5.1. *Relevancy Principle*

A Boolean variable p occurring in the Boolean axioms A of some problem specification (A,B,E), but not in the E-formulae E does not contribute much to the problem solution. Boolean variables are implicitly existentially quantified. That means A is in fact short for $\exists p\, A$ if p does not occur in E. A is a propositional formula, and therefore the existentially quantified p can be eliminated using a quantifier elimination procedure (Gabbay and Ohlbach, 1992). The result is some formula A' which is equivalent to $\exists p\, A$, but does not contain p. In the propositional case, elimination of the existentially quantified p amounts to generating all resolvents with p in the clause form of A. The resolvents represent all consequences of p and therefore p is no longer necessary (Ackermann, 1935).

This way one can have large databases of Boolean axioms, but for the actual problem at hand, the atomic decomposition takes into account only the relevant Boolean variables.

5.2. *Factoring Principle*

A Boolean axiomatization A which can be split into separate parts $A_1,\ldots,A_r$ such that the parts mutually do not share Boolean variables, simplifies the atomic decomposition as well. The set of propositional models for A can be factored into the product $M_1\times,\ldots,\times M_r$ of the set of models for the A_i. The mapping μ from the set of propositional models to the set of ultrafilters of the algebra A, we established in proposition 4.1, implies that the ultrafilters, and thus the whole algebra A can be factored into the product

$A_1 \times \ldots \times A_r$ of its components. The atoms of such a product have the form $(\ldots, \bot_{i-1}, a_i, \bot_{i+1}, \ldots)$ where a_i is an atom of A_i and all other components are the bottom elements of the other algebras.

This can be exploited for the representation of the syntactic atoms. They can have the form $(\ldots, \bot, m_i, \bot, \ldots)$ where m_i is a syntactic atom of the component M_i. A further simplification is possible by just storing m_i and labelling it with the information 'belongs to M_i'.

This reduces the overall number of syntactic atoms from $|M_1| \cdot \ldots \cdot |M_r|$ to $|M_1| + \ldots + |M_r|$, which is an exponential improvement.

The meaning of this simplification makes sense also from an application point of view. As an illustration, consider some A axiomatizing, say family relationships as in the introductory example, and in addition relationships between the makes of cars. If there are no axioms saying something about the union or intersection of people and cars, then the factoring operation implicitly imposes that there is no object which is at the same time a person and a car. Therefore the whole Boolean Algebra is split into the part with sets of people and the part with sets of cars. People and cars together are represented by tuples in the product algebra. On the calculus side we therefore get syntactic atoms which represent either people or cars, but none for the intersection of both.

6. Extensions to Other Algebras

Many other algebras are extensions of Boolean Algebras. Relational Algebras provide composition and inverse of binary relations. Boolean Algebras with operators are the algebraic counterparts of modal logics (Jónsson and Tarski, 1951). Both have useful applications in computer science and the knowledge representation area. In particular Boolean Algebras with operators bring extra dimensions into a logical system, for example time dependency.

The elements of these algebras are still elements of the underlying Boolean Algebra, and all terms in the syntactic representation denote elements of this Boolean Algebra. Therefore in principle the mechanism of atomic decomposition still should work. Unfortunately there are in general infinitely many terms, which means that the syntactic representation of the atoms may become infinite. In many cases, however, the relevancy principle can help. We must make sure that only the terms occurring in the problem at hand, or at least a finite superset of them is needed, and all others can be ignored. In this case the representation of the atoms becomes finite again. For the finitely many relevant terms one can first compute the atoms as ordinary propositional models where all the terms in the extension of the Boolean Algebra are treated as Boolean variables. These atoms are conjunctions of terms in the language of the extension of the Boolean Algebra. For each of these conjunctions one has to check whether they are consistent in the extended algebra, using a decision procedure for this algebra. The inconsistent conjunctions are deleted from the atomic decomposition. If satisfiability in the extended algebra is decidable then there are only finitely many tests to be done, and the decomposition method works.

The problem to ensure that only a finite superset of the terms occurring in the problem specification is needed, is in many algebras related to a finite model property. For example in the algebra corresponding to the modal logic K, the finite model property yields that there is always a finite tree model as deep as the maximal number of nestings of modal operators occurring in the problem (Chellas, 1980). For all modal terms with deeper nestings, the $\Box$ versions become $\top$ and the $\Diamond$ versions become $\bot$.

The finiteness condition has to be analyzed for each class of algebras, but for may cases well known results can be exploited for this purpose.

7. Summary

Sophisticated inference procedures as for example mathematical programming techniques are usually restricted to the language they have been developed for. Any extension of the language is extremely difficult to incorporate into the procedure.

In this contribution we have developed a technique for incorporating a Boolean component into such a language without affecting the calculus. With the atomic decomposition of the Boolean terms we were able to eliminate the Boolean component completely from a mixed problem specification and reduce the given problem to the language of the basic system.

In the Boolean component one can axiomatize sets and set-theoretic relationships between them. The interface between the two languages is provided by a set of 'bridging functions'. They map the sets to objects of the basic system. Different bridging functions can be used simultaneously to deal with different features of the sets. A typical such bridging function is the cardinality function mapping sets to non-negative integers. Any inference system that can deal with arithmetic can therefore be used to reason about cardinalities and other numeric attributes of sets.

An interesting feature of our approach is the fact that more information at the Boolean side, i.e. more Boolean axioms, in general simplifies matters. More Boolean axioms usually means less propositional models, which in turn means less syntactic atoms. Less syntactic atoms eventually means less variables to deal with in the arithmetic part.

We have sketched an extension of the method to more complex algebras such as Relational Algebras or Boolean Algebras with operators. But the details for these extensions still have to be worked out.

REFERENCES

Ackermann, W.: 1935, 'Untersuchung über das Eliminationsproblem der mathematischen Logik'. *Mathematische Annalen* **110**, 390–413.

Chellas, B. F.: 1980, *Modal Logic: An Introduction.* Cambridge: Cambridge University Press.

Gabbay, D. M.: 1996, 'An Overview of Fibred Semantics and The Combination of Logics'. In: F. B. K. Schulz (ed.): *Proceedings of FroCoS'96, Frontiers of Combining Systems*. pp. 1–56, Kluwer.

Gabbay, D. M.: 1998, *Fibring Logics*. Oxford University Press.

Gabbay, D. M. and H. J. Ohlbach: 1992, 'Quantifier elimination in second–order predicate logic'. In: B. Nebel, C. Rich, and W. Swartout (eds.): *Principles of Knowledge Representation and Reasoning (KR92)*. pp. 425–435, Morgan Kaufmann.

Jónsson, B. and A. Tarski: 1951, 'Boolean Algebras with Operators, Part I'. *American Journal of Mathematics* **73**, 891–939.

Stone, M. H.: 1936, 'The Theory of Representations for Boolean Algebras'. *Transactions of American Mathematical Society* **40**, 37–111.

MANFRED KERBER

CHAPTER 4

PROOF PLANNING: A PRACTICAL APPROACH TO MECHANIZED REASONING IN MATHEMATICS

1. INTRODUCTION

The attempt to mechanize mathematical reasoning belongs to the first experiments in artificial intelligence in the 1950 (Newell et al., 1957). However, the idea to automate or to support deduction turned out to be harder than originally expected. This can not at least be seen in the multitude of approaches that were pursued to model different aspects of mathematical reasoning. There are different dimension according to which these systems can be classified: input language (e.g., order-sorted first-order logic), calculus (e.g., resolution), interaction level (e.g., batch mode), proof output (e.g., refutation graph), and the purpose (e.g., automated theorem proving) as well as many more subtle points concerning the fine tuning of the proof search.

In this contribution the proof planning approach will be presented. Since it is not the mainstream approach to mechanized reasoning, it seems to be worth to look at it in a more principled way and to contrast it to other approaches. In order to do so I categorize the systems according to their purpose. The different attempts can be roughly categorized into three classes: machine-oriented automated theorem provers, interactive proof checkers, and human-oriented theorem provers. In the rest of the introduction we shall take a closer look at these different categories.

1.1. *Machine-Oriented Automated Theorem Provers*

It seems as though logicians had worked with the fiction of man as a persistent and unimaginative beast who can only follow rules blindly, and then the fiction found its incarnation in the machine. Hence, the striving for inhuman exactness is not pointless, senseless, but gets direction and justification. One may even claim that a new life is given to the Hilbert program of the Entscheidungsproblem *which von Neumann thought was thoroughly shattered by Gödel's discoveries. Although a universal decision procedure for all mathematical problems is not possible, formalization does seem to promise that machines will do a major portion of the work that takes up the time of research mathematicians today.*

H. Wang, 1960, quoted from (Siekmann and Wrightson, 1983)*, p.260*

W. Bibel, P. H. Schmitt (eds.), Automated Deduction. A basis for applications. Vol. III

By machine-oriented theorem provers, I mean theorem provers based on computational logic theories such as resolution, paramodulation, or the connection method, that is, systems based upon some computer-oriented inference system, and which derive their strength from their ability to maintain and manipulate very large search spaces.

Such systems have been successfully applied in different fields of logic and mathematics (see, e.g., (Wos et al., 1984)). The strength of these systems is truly remarkable and even open mathematical problems, like recently the question whether Robbins algebras are Boolean algebras (McCune, 1997), can be solved with the assistance of an automated theorem prover.

On the other hand, observing the blind search behavior of such a system as it fails to solve a problem that seems trivial to us as humans can be disappointing. The general complexity results clearly demonstrate that no algorithm can be constructed to practically solve arbitrary tasks, even propositional logic is in a class that is generally considered intractable. The mathematical knowledge and semantics that guide human search are clearly missing: for instance, function symbols such as "powerset" are treated in the same way as function symbols such as "successor", even in contexts a mathematician would never contemplate. Furthermore, although many applications of these systems are important and interesting (for instance, a test for the independence of the axioms of a given system, or their use in software and hardware verification), such applications differ substantially from the daily work of a mathematician when proving theorems. In particular these systems give either a perfect answer (a formal proof) or no answer at all. There is no good way of telling how far one is away from a solution.

1.2. *Proof Checkers*

As a kind of dream I played (in 1968) with the idea of a future where every mathematician would have a machine on his desk, all for himself, on which he would write mathematics and which would verify his work. But, by lack of experience in such matters, I expected that such machines would be available in 5 years from then. But now, 23 years later, we are not that far yet.

N.G. de Bruijn (Nederpelt et al., 1994)*, p.210*

Interactive proof checking and proof development systems have been built with the aim of achieving a new standard of rigor in mathematical proof. As pointed out by de Bruijn (see (Nederpelt et al., 1994)), the developer of one of the earliest systems, Automath, only a small part of today's mathematical literature is absolutely flawless. To improve this situation, interactive proof checkers have been developed that carry out the meticulous final checking. In a similar spirit, proof development systems such as Nuprl, (Constable et al.,

1986), were built, where the user essentially develops a proof and the system ensures that every step is correct. In future, this might result in new standards for the acceptance of mathematical papers: for each of his/her theorems an author would have to deliver a machine-readable proof that could be checked by the proof checking system of the editor. For this purpose it is not important whether the author produces the proof himself/herself or it is generated automatically. For simplicity, proof-checking system designers have as a first step offered interactive-only environments, where users must specify all details themselves.

For several reasons, Automath and most other comparable systems have not reached any broad acceptance as a working instrument for mathematicians. One reason is that there is a loss in brevity by a factor of 10 to 20 when using Automath compared to normal mathematical texts. Modern systems (like Nuprl) normally incorporate a very powerful construct, the so-called tactic (Gordon et al., 1979), which dramatically improves their user-friendliness. Tactics are programs that manipulate the current state of the proof not only by the application of a single calculus step, but by a whole series of such steps. In this way one user interaction, namely the call of a single tactic, results in a sequence of steps. Tacticals allow to combine tactics to compound tactics, in particular the availability of conditionals and loops make the construct very powerful. Since all tactics eventually result in the application of calculus level rules, a tactic may fail to be applicable, but if it is it cannot produce faulty proofs. The acceptance of interactive systems is increasing and using interactive techniques, a large set of proofs has been constructed in mathematical applications as well as for program verification. Nevertheless there is one major problem. They incorporate far too little automated deductive support; the proof is found by the user with some help from the machine, rather than vice versa: a lot of machine support with some conceptual help from the user. The machine can be considered as a slave that meticulously checks steps, but not as an intelligent assistant which makes suggestions or produces own ideas how to prove the theorem. Just as machine-oriented provers, these systems do not offer any help how to find a solution of an open sub-goal, unless they can fully settle it. In some sense the "interaction" can be viewed more as a process of "action" (call of a tactic) on the side of the user and "reaction" (execution of a tactic) on the side of the system than as a mutual interaction.

1.3. *Human-Oriented Theorem Provers*

We secure our mathematical knowledge by demonstrative reasoning, *but we support our conjectures by* plausible reasoning ... *Demonstrative reasoning is safe, beyond controversy, and final. Plausible reasoning is hazardous, controversial, and provisional. ... In strict reasoning the principal thing is to distinguish a proof from a guess, a valid demonstration from an invalid attempt. In plausible reasoning the principal thing is to distinguish a guess from a guess, a more reasonable guess form a less reasonable guess. ... [plausible reasoning] is the kind of reasoning on which his [a mathematician's] creative work will depend.* *G. Pólya* (Pólya, 1954), *p. vi*

Human-oriented theorem provers as for instance Gelernter's geometry prover are among the first AI systems (Gelernter, 1959). Although surely machine-oriented and interactive theorem provers are in the focus of interest, human-oriented theorem-proving systems have recently attracted growing attention after the initial enthusiasm for machine-oriented theorem provers died down somewhat and the limitations of these systems became more apparent. By human-oriented theorem provers, I mean systems that model human proof-search behavior. In particular the distinction between demonstrative and plausible reasoning plays a significant rôle, while search in the first two approaches (machine-oriented and interactive) is done in the space of demonstrative reasoning, humans normally search for proofs using plausible reasoning. Ideally the object language in human-oriented theorem proving would be the mathematical colloquial language. Since this would cause many problems from natural language understanding in its train, normally the object language is a formal language that sufficiently approximates the mathematical colloquial language.

Human-oriented theorem provers find proofs by performing steps of plausible reasoning. These steps can normally be translated to low-level formal steps that can be fed into a proof checker (e.g., one based on some variant of the natural deduction calculus). While the strength of traditional theorem provers essentially relies on fixed domain-independent search strategies, the deductive power of human-oriented systems mainly resides in (user-written) domain-specific heuristics.

Both aspects of theorem proving are of extra-ordinary importance and a large body of nowadays mathematics (also that of Newton, Leibniz, and Euler) started by inspirational work that was not logically justified. Indeed "rigorous mathematics as a going concern is a rather rare phenomenon in the 4000-year-history of mathematics" (Kleiner and Movshovitz-Hadar, 1997), p.17. This does not mean that rigorous proofs are not worth striving for, on the contrary an informal proof can be considered as a major step towards a formal proof. Indeed it may be useful to view proofs ranging on a very wide spectrum of vague proof ideas over concrete proof plans to fully expanded formal proofs in a logical calculus.

There are different attempts to model human-oriented theorem proving, e.g., some based on expert systems or learning systems. The most prominent one is based on the idea to consider the process of plausible reasoning as a planning process, consequently this approach is called proof planning. It is first introduced by Bundy (Bundy, 1988). In the next section this approach is described in detail.

2. PROVING AS PLANNING

While logic provides an explanation how the steps of a proof fit together it is inadequate to explain many common observations about mathematical proofs. ... Mathematicians distinguish between understanding each step of a proof and understanding the whole proof. It is possible to understand at one level without understanding at the other – either way round. A step by step understanding is adequately represented by a logical proof. We propose representing an overall understanding by associating a proof plan with each proof.

A. Bundy (Bundy, 1991)

The most prominent attempt to separate levels in proving is the proof planning approach introduced by Bundy (Bundy, 1988). The basic idea is to distinguish two levels: a planning level, which can be seen as the level of plausible reasoning, and a plan execution and verification level that corresponds to demonstrative reasoning. By the integration of abstraction into the planning process a wide spectrum reaching from vague proof ideas over concrete plans to calculus level proofs can be realized.

In proof planning a plan operator is called *method* and consists of a tactic (in the sense of proof checkers, see Section 1.2) *and* a declarative specification. The declarative specification is used to construct plans and the tactic to execute them. An important feature of the plan approach is that the specification may be incomplete and hence the search space for a proof plan often is considerably smaller than the search space on the calculus level. Of course a price has to be paid for this simplification, namely, the execution of proof plans may fail. There is, however, justified hope that in many cases even failed proof plans contain useful information that can either be adapted automatically to an executable proof plan or that can at least be communicated to the user (or another system) to provide useful hints for their proof search. One of the main strengths of the proof planning approach is the possibility to encode domain-specific control knowledge in the methods.

In a first approximation, planning for proofs can be done by applying techniques from the planning field, e.g., hierarchical planning. Methods can be very general, for instance, correspond essentially to one application of a rule at the calculus level. More specific methods comprise proof ideas, for instance, to use a homomorphy property, to apply mathematical induction, or to use diagonalization. The most specific methods consist of full proofs for specific problems. Typically methods can be formulated in terms of proof schemata, that is, partial proofs, in which expressions may be replaced by meta-variables.

A successful realization of the plan-based approach is the OYSTER-CLAM-system (Bundy et al., 1990), where the Nuprl-tactics have been extended to methods (OYSTER is a reimplementation of Nuprl and CLAM is the proof planner built on top of OYSTER). A method there can be viewed as a unit consisting of a *procedural* tactic and a *declarative* specification. CLAM has been applied in particular to problems from the field of mathematical induction and a large body of the heuristic knowledge of the Boyer-Moore prover (Boyer and Moore, 1979) is encoded in CLAM methods. This makes CLAM to a strong theorem prover in the field of mathematical induction.

Since the working language and the proof formalism of proof planning systems are usually human-oriented, proof planners can easily support user interactions in the spirit of a proof checker. The plan-based framework is particularly well-suited to encode *domain-specific* problem-solving knowledge. In the current form proof planning and machine-oriented theorem proving should not be considered as exclusive alternatives but rather as supplementary approaches, the first is particularly good in finding global (and more detailed) proof ideas and the latter to fill remaining gaps.

In order to build a useful proof development environment, we took in the ΩMEGA system the approach to combine both the reasoning power of automated theorem provers as logic engines *and* that of the proof planning mechanism. We think that the combination of these complementary approaches inherits more of the advantages than of the drawbacks, because for most tasks domain-specific as well as domain-independent problem-solving know-how is required. In the rest of this section the proof planning approach taken in ΩMEGA will be described in more detail. For a detailed description of the other features of ΩMEGA see (Huang et al., 1994; Benzmüller et al., 1997).

2.1. *Methods*

The concept of a method is central to the reasoning process, since methods are the basic units which make up proof plans and are carried out to obtain the proof by bridging the gap between premises and conclusions. The set of all available methods constitutes the basic reasoning repertory of a reasoner, it is constantly enlarged and enriched with increasing experience. As a first approximation the process of learning mathematical reasoning by a student can be considered as the acquisition of new methods rather than the adoption of a formal calculus. While certain methods are specific to a certain field (e.g.,

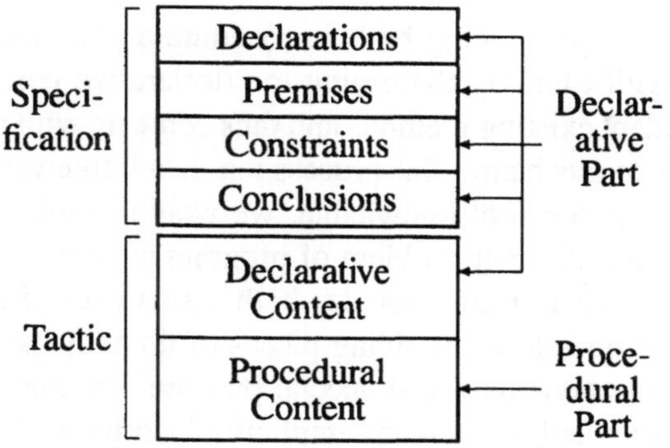

Figure 1. The Structure of Methods

in analysis the methods to prove continuity of real-valued functions by the ε-δ criterion), others are domain-independent (e.g., indirect proof). A possible formalization of the theorem proving process can be considered as a planning process of combining methods to build high-level proofs.

In a more sophisticated creative stage, given methods are not only applied and combined to build more or less complicated proofs, but existing methods are adapted to situations for which no particularly well-suited methods are available. Note that this is fundamentally different from the logicistic view where a sound and complete calculus is assumed and there is no need for augmenting the calculus rules by new ones (where on the other hand the theoretical complexity results for the complete calculi are devastating).

In order to model aspects of this creative process we have adopted a declarative approach to the representation of methods. Since it can be pretty long winded (and hence inadequate) to represent all aspects of methods in a purely declarative form, we formalize a method as a 6-tuple as displayed in figure 1, where only the last one is procedural.

The *Declarations* represent the signature that declares the meta-variables used in the method, *Premises* consist of schemata of proof lines which are used by this method to prove the conclusions, *Conclusions* are schemata of proof lines which this method is designed to prove, the *Constraints* contain additional restrictions on the premises and the conclusions, which cannot be formulated in terms of proof line schemata. The *Declarative Content* consists of declarative knowledge that is interpreted by the procedural content (currently restricted to schemata of partial proofs), and finally the *Procedural Content* is either a standard procedure interpreting the declarative content or a special purpose inference procedure.

Generalizing the tactic part of a method from just a procedure (in Bundy's framework) to a pair containing both a procedure and a piece of declarative knowledge is significant. By discerning the declarative part of tactics, it is much easier to adapt existing methods and thus come up with novel ones. This makes it possible to mechanize this process too. In a framework where a tactic only consisted of procedural knowledge, we would in effect be confronted with the much more difficult problem of program synthesis. Our framework is cast so general that it accommodates both a small set of general purpose procedures which operate by applying pieces of domain-specific declarative knowledge and an open-end set of special purpose reasoning procedures, in which required knowledge is already implicitly incorporated.

Viewed within a planning framework, the method structure can be partitioned into different units as illustrated in figure 1. The *Declarations*, *Premises*, the *Constraints*, and the *Conclusions* slots together specify whether a method is applicable in a particular proof state. While *Premises* and *Conclusion* are proof line schemata, in the *Constraints* information can be expressed that cannot be specified by proof line schemata (for details, see (Huang et al., 1994)). The applicability information will normally be refined by further bits of information, for instance the proof history and the current planning state in order to give a heuristic estimation of the probability that the method can significantly contribute to finding a proof plan.

While the specification contains all information necessary for the successful generation of a proof plan, the *declarative content* and the *procedural content* slots contain the information for the execution of the plan operator. *Together* they form the *tactic*, which is a pure procedure in Bundy's framework. Up to now we only use two special cases: the one, where the declarative content is a proof schema (that is, a proof fragment with meta-variables) and the procedural content is a LISP procedure that takes this schema as an argument, instantiates it and puts it at the right place into the current proof fragment. In the other case the procedure completely ignores the proof schema and constructs new proof fragments purely procedurally. Methods with such purely procedural tactics roughly correspond to CLAM methods.

For an example of a method see figure 2. In the declarations we distinguish meta-variables of different sorts, in particular proof lines (we use a linearized version of natural deduction proofs, see (Andrews, 1980)), justifications, object variables, object constants, and terms (which include formulae in the higher-order framework, we adopted). In figure 2, the premises and conclusions read: use lines L_1, L_2, and L_3 to derive line L_6. The labels $\oplus$ and $\ominus$ tell which lines have to be present, will be constructed, or will be deleted when planning with the method. These labels will be discussed in detail in section 2.2. The constraints mean that Ψ_1 and Ψ_2 are obtained by replacing X in Ψ by C and $F(C)$, respectively. Further constraints, for instance, that the

Method : hom1-1	
Declarations	L_1,L_2,L_3,L_4,L_5,L_6:prln X,Y:var H_1,H_2,H_6:list(prln) J_1,J_2,J_3:just Φ,Ψ,Ψ_1,Ψ_2:term P,F,C:const
Premises	$L_1,L_2,\oplus L_3$
Constraints	$\Psi_1 := \Psi[X/C]$ & $\Psi_2 := \Psi[X/F(C)]$ & ...
Conclusions	$\ominus L_6$
Declarative Content	(L_1) $H_1 \vdash \forall Y\ \Phi$ (J_1) (L_2) $H_2 \vdash \forall X\ P(X) \leftrightarrow \Psi$ (J_2) (L_3) $H_6 \vdash P(C)$ (J_3) (L_4) $H_6 \vdash \Psi_1$ (def-e L_2,L_3) (L_5) $H_6 \vdash \Psi_2$ (OPENL_1,L_4) (L_6) $H_6 \vdash P(F(C))$ (def-i L_2,L_5)
Procedural Content	schema − interpreter

Figure 2. The hom1-1 method.

hypothesis sets H_1 and H_2 must be subsets of H_6 and that the constant function symbol F must not occur in the formula Φ are not displayed in figure 2 in order not to complicate the presentation.

The idea of hom1-1 can informally be described as follows: *If f is a given function, p a defined predicate and the goal is to prove $p(f(c))$, then show $p(c)$ and use this to show $p(f(c))$.* The very idea is that f is a homomorphism for the property p and that f can be "rippled out" (compare (Bundy, 1988; Hutter, 1990)). Note that line L_5 is an open line that does not occur in the specification and therefore does not enter the planning state. This leads to an abstraction in the planning process: since line L_5 is not considered by the planner, after the completion of the plan it will be inserted into the proof tree as an open line by the application of the tactic of hom1-1. This will result in a recursive call of the planner in the following verification phase.

2.2. *Proof Planning with Methods*

To give an account of the proof planning process itself, we should first note that the goal of proof planning is to fill gaps in a given partial proof tree by forward and backward reasoning. In our framework we follow a STRIPS-like planning paradigm (Fikes and Nilsson, 1971), where the planning operators correspond to the methods. Thus from an abstract point of view the planning process is the process of exploring the search space of *planning states* that is

Method `def-i` (Definition Introduction)		
Premises	`Definition:`	$\forall\bar{x}\, P(\bar{x}) \leftrightarrow \Psi_{\bar{x}}$
	$\oplus$ `expanded-line:`	$\Psi_{\bar{t}}$
Conclusions	$\ominus$ `defined-line:`	$P(\bar{t})$

Method `def-e` (Definition Elimination)		
Premises	`Definition:`	$\forall\bar{x}\, P(\bar{x}) \leftrightarrow \Psi_{\bar{x}}$
	$\ominus$ `defined-line:`	$P(\bar{t})$
Conclusions	$\oplus$ `expanded-line:`	$\Psi_{\bar{t}}$

Figure 3. The specification of the methods `def-i` and `def-e`

generated by the *planning operators* in order to find a complete *plan* (that is a sequence of instantiated planning operators) from a given *initial state* to a *terminal state*.

Concretely a *planning state* is a subset of lines in the current partial proof that correspond to the boundaries of a gap in the proof. This subset can be divided into *open lines* (that must be proved to bridge the gap) and *support lines* (that can be used to bridge it). The *terminal state* is reached when there are no more open lines. In the planning process new open lines enter the planning state as subgoals by backward reasoning from existing open lines and new support lines by forward reasoning from existing support lines. In order to achieve this with a uni-directional planning mechanism like STRIPS, the planning direction must be independent of the reasoning direction.

A key feature of our approach is that the planning operators are directly derived from the specifications of the methods. However, the specification only gives a static description (viewed from the completed proof) of the method which is inadequate for the dynamic behavior needed in proof planning. Statically a method derives its *conclusions* from its *premises*. Dynamically, it is important to declare which lines in the specification have to be present in the planning state for the method to be applicable (we will call them *required* lines), and which are constructed by the method. We will do this by labelling the latter lines which will be inserted into the planning state by the label "$\oplus$". Additionally it is useful to specify that some of the required lines will not be used again by the planner. We will mark such lines with a "$\ominus$". Note that the required lines consist of the unmarked ones and those that are marked with "$\ominus$". This labelling in effect determines the direction (forward vs. backward) of reasoning of the method. In order to illustrate these labels, figure 3 shows the specification of two simple methods. These methods are only a simplified

Planning State 1

(!)	1.	1;	$\vdash \forall\sigma\,\mathsf{Sym}(\sigma) \leftrightarrow \forall x,y\,(\langle x,y\rangle \in \sigma \rightarrow \langle y,x\rangle \in \sigma)$
(!)	2.	2;	$\vdash \mathsf{Sym}(\rho)$
(?)	9.	1,2;	$\vdash \mathsf{Sym}(\mathsf{conv}(\rho))$

`def-i 1 9`
`def-e 1 2`

Planning State 2

(!)	1.	1;	$\vdash \forall\sigma\,\mathsf{Sym}(\sigma) \leftrightarrow \forall x,y\,(\langle x,y\rangle \in \sigma \rightarrow \langle y,x\rangle \in \sigma)$
(!)	3.	1,2;	$\vdash \forall x,y\,\langle x,y\rangle \in \rho \rightarrow \langle y,x\rangle \in \rho$
(?)	8.	1,2;	$\vdash \forall x,y\,\langle x,y\rangle \in \mathsf{conv}(\rho) \rightarrow \langle y,x\rangle \in \mathsf{conv}(\rho)$

Figure 4. Using `def-i` and `def-e` in the planning process

version of a more general class of methods applying definitions and theorems. For a study of this class which approximates basic proof steps encountered in informal mathematical practice see (Huang, 1994).

Figure 4 shows an example of the effect of the methods `def-e` and `def-i` on the Planning State 1, in which $\mathsf{Sym}(\mathsf{conv}(\rho))$ is to be proved using the assumptions in lines 1 and 2. In planning states we mark open lines by "?" and support lines by "!". The method `def-i` applies a definition for a predicate P (i.e., $\forall\overline{x}\; P(\overline{x}) \leftrightarrow \Psi_{\overline{x}}$) to an open line and `def-e` applies the same definition to a supporting line. It is obvious that in both methods the line of the definition is required when applying these methods because it is not sensible to "guess" a definition[1]; furthermore this line must not be deleted since it might be used more often than once. Therefore the proof line `Definition` has no label. Clearly in both methods `defined-line` must be a required line and `expanded-line` can be constructed by the methods (and therefore is labelled with "$\oplus$"). Furthermore `defined-line` in `def-i` (and analogously in `def-e`) is useless after the application because an open line must be proved only once. Hence "Planning State 1" is transformed by the application of `def-i` and `def-e` to "Planning State 2", in which the definition of symmetric is unfolded.

Planning works by heuristically selecting the best method from all applicable ones, applying the planning operator to the planning state (that is, inserting lines labelled by $\oplus$ and deleting those labelled by $\ominus$) until there are no more gaps (this is of course a search process and may result in backtracking). Once a complete proof plan is found, all methods in the proof plan are

[1] This could be sensible at a more sophisticated level of proof planning. However, this "guessing" should be implemented by a different method in order to clearly separate it from simply applying a definition.

1.	1;	$\vdash$	$\forall\sigma\,\forall x,y\,\langle x,y\rangle \in \mathsf{conv}(\sigma) \leftrightarrow \langle y,x\rangle \in \sigma$	(Hyp)
2.	2;	$\vdash$	$\forall\sigma\,\mathsf{Sym}(\sigma) \leftrightarrow \forall x,y\,(\langle x,y\rangle \in \sigma \to \langle y,x\rangle \in \sigma)$	(Hyp)
3.	1,2;	$\vdash$	$\mathsf{Sym}(\rho)$	(J3)
4.	1,2;	$\vdash$	$\forall x,y\,\langle x,y\rangle \in \rho \to \langle y,x\rangle \in \rho$	(`def-e` 2 3)
5.	1,2;	$\vdash$	$\forall x,y\,\langle x,y\rangle \in \mathsf{conv}(\rho) \to \langle y,x\rangle \in \mathsf{conv}(\rho)$	(OPEN 1 4)
6.	1,2;	$\vdash$	$\mathsf{Sym}(\mathsf{conv}(\rho))$	(`def-i` 2 5)

Figure 5. Proof fragment resulting from the application of the tactic `homl-1`

successively applied. Note that such a method application phase need not lead to a complete proof of the problem at hand, since we do not require methods to be sound or complete with respect to their specifications (e.g, the plan may be not executable, since not all meta-variables are bound). Furthermore the proof segments inserted by the methods may still contain open lines (see, e.g., the `homl-1` method) that define further gaps which still have to be closed by the proof planner. Therefore the verification phase may result in a recursive call of the planner or in backtracking. While the first possibility calls for a refinement of the plan found and can be used to model a simple form of hierarchical planning, the latter rejects the plan and calls the proof planner in order to find a different plan.

In general in a certain planning state there will be different methods that are applicable. For instance, in "Planning State 1" (figure 4) the method `homl-1` (figure 2) is applicable as well if we assume that the definition of conv is given too. In the application conv, Sym, and ρ are substituted for the meta-Variables F, P, and C, respectively. While in figure 4 we filled the gap between $\mathsf{Sym}(\rho)$ and $\mathsf{Sym}(\mathsf{conv}(\rho))$ which were both existing lines, the method `homl-1` can propose $\mathsf{Sym}(\rho)$ as a new line for proving $\mathsf{Sym}(\mathsf{conv}(\rho))$ together with the definitions of Sym and conv. The proof fragment resulting from the application of the tactic `homl-1` is displayed in figure 5.

Let us have a look at the justifications in this proof fragment. Lines 1 and 2 are justified as hypothesized definitions (and are matched against the meta-variables J_1 and J_2 in `homl-1`). J3 has to be instantiated in the rest of the current proof planning phase. The justifications of lines 4 and 6 stand for the subproofs generated by the applications of the tactics of these methods, whereas the justification of line 5 defines a new gap with support lines containing lines 1 and 4.

2.3. *Mechanical Adaptation of Methods*

The reasoning power of human-oriented systems is not derived from a complete underlying reasoning calculus, but relies on domain-specific problem solving knowledge. Such knowledge is, however, practically inevitably incomplete and this leads to a limited reasoning power of the corresponding systems. While this holds for plan based systems and human mathematicians alike, the latter can go beyond their specialized knowledge. They have the ability to adapt existing methods to novel situations, which is one of the main features contributing to our problem solving competence, see (Pólya, 1945). By adopting a declarative approach for formulating methods, we propose a way to mechanize some aspects of this adaptation process. The construct used to perform this adaptation is called *meta-method*. It is a procedure which takes as input a method, and possibly further parameters (in particular the current state of proof planning) and generates a new method with the same procedural part.

Let us take a closer look at this process by the adaptation of the `homl-1` method (cf. figure 2). `homl-1` simplifies a problem by generating an intermediate goal, where a *unary* function symbol, represented by the meta-variable F, is eliminated in an open line that starts with a unary predicate symbol (represented by the meta-variable P). Suppose we are facing the similar problem, where a *binary* function symbol is involved, as in the proof that the intersection of symmetric relations is itself a symmetric relation. In this case, a variant of `homl-1` is needed, which can handle the case with a binary function symbol (i.e. "$\cap$") in a similar way.

In the following we illustrate how the meta-method `add-argument` generates a binary version `homl-2` (cf. figure 6) from the unary version `homl-1` (cf. figure 2). While `homl-1` is applicable to situations with a unary predicate constant P and a unary function constant F, `homl-2` handles situations with a unary predicate constant P and a binary function constant F'. Note that P, F, and F' are meta-variables standing for object constants.

The meta-method `add-argument` takes as input a method `M` and a unary function or predicate constant `F` used in `M`. This meta-method is supposed to add an argument to the key constant symbol `F`, the modified constant is called $\mathtt{F}'$. `add-argument` creates a method $\mathtt{M}'$ by carrying out the following modification on the declarative part of `M`: firstly replace $\mathtt{F}(x)$ by $\mathtt{F}'(x,y)$ and augment the corresponding quantifications, secondly replace $\mathtt{F}(a)$ by $\mathtt{F}'(a,b)$ (b has to be a new meta-variable), and thirdly if a occurs in a proof line, but not in a term $\mathtt{F}(a)$, a copy of this line will be inserted into the proof schema, replacing a by b (in the example line L_4 is copied from L_3). The specification of the method has to be adapted correspondingly. The procedural part of `M`,

Method : hom1-2	
Declarations	$L_1,L_2,L_3,L_4,L_5,L_6,L_7,L_8$:prln X,Y:var H_1,H_2,H_8:list(prln) J_1,J_2,J_3,J_4:just $\Phi,\Psi,\Psi_1,\Psi'_1,\Psi_2$:term P,F',C,D:const
Premises	$L_1,L_2,\oplus L_3,\oplus L_4$
Constraints	$\Psi_1 := \Psi[X/C]$ & $\Psi'_1 := \Psi[X/D]$ & $\Psi_2 := \Psi[X/F'(C,D)]$ & ...
Conclusions	$\ominus L_8$
Declarative Content	(L_1) $H_1 \vdash \forall Y\ \Phi$ (J_1) (L_2) $H_2 \vdash \forall X\ P(X) \leftrightarrow \Psi$ (J_2) (L_3) $H_8 \vdash P(C)$ (J_3) (L_4) $H_8 \vdash P(D)$ (J_4) (L_5) $H_8 \vdash \Psi_1$ (def-e L_2,L_3) (L_6) $H_8 \vdash \Psi'_1$ (def-e L_2,L_4) (L_7) $H_8 \vdash \Psi_2$ (OPEN L_1,L_5,L_6) (L_8) $H_8 \vdash P(F'(C,D))$ (def-i L_2,L_7)
Procedural Content	schema-interpreter

Figure 6. The hom1-2 method.

however, is taken over unchanged for the new method.

In figure 7 it is shown how the hom1-2 method simplifies the problem of showing that the intersection of two symmetric relations is symmetric too. Analogously a method hom2-1 (for handling a unary function symbol and a binary predicate symbol) can be obtained by applying add-argument with the arguments hom1-1 and P.

2.4. *Proof Planning with Meta-Methods*

It is possible to create new methods by applying meta-methods to existing methods. However, up to now the work on meta-methods is still in an experimental state and is not part of the standard distribution of ΩMEGA (see http://www.ags.uni-sb.de/projects/deduktion/), in particular only first steps were made towards the integration of the automatic creation of new methods into the planning algorithm. In order to automatically call a meta-method different questions have to be heuristically answered. The first question is, when should the search for a proof plan be interrupted in order to create a new method. The second question is, how is it possible to find a method from the knowledge base that is well-suited for adaptation.

1.	1;	$\vdash \forall \rho,\sigma \forall x,y\, \langle x,y\rangle \in (\rho \cap \sigma) \leftrightarrow \langle x,y\rangle \in \rho \wedge \langle x,y\rangle \in \sigma$	(J1)
2.	2;	$\vdash \forall \sigma\, \mathsf{Sym}(\sigma) \leftrightarrow \forall x,y\, (\langle x,y\rangle \in \sigma \rightarrow \langle y,x\rangle \in \sigma)$	(J2)
3.	1,2;	$\vdash \mathsf{Sym}(\rho)$	(J3)
4.	1,2;	$\vdash \mathsf{Sym}(\sigma)$	(J4)
5.	1,2;	$\vdash \forall x,y\, \langle x,y\rangle \in \rho \rightarrow \langle y,x\rangle \in \rho$	(`def-e` 2 3)
6.	1,2;	$\vdash \forall x,y\, \langle x,y\rangle \in \sigma \rightarrow \langle y,x\rangle \in \sigma$	(`def-e` 2 4)
7.	1,2;	$\vdash \forall x,y\, \langle x,y\rangle \in (\rho \cap \sigma) \rightarrow \langle y,x\rangle \in (\rho \cap \sigma)$	(OPEN 1 5 6)
8.	1,2;	$\vdash \mathsf{Sym}((\rho \cap \sigma))$	(`def-i` 2 7)

Figure 7. The proof resulting from the application of the tactic `hom1-2`

A heuristic answer to the first question is: When there is no or no sufficiently promising method for simplifying the problem create a new one. Note that in the setting where the calculus level rules are encoded as methods too, there will in general be applicable methods as long as the problem is not yet solved. Two general heuristics for choosing a method to adapt are: first, organize methods in a hierarchy of mathematical theories and prefer methods that belong to the same theory as the current problem or whose theory is close to that of the problem. Second, use general conflict solving strategies (as in OPS5) that favor the methods and meta-methods with the most specific specification. A way to reduce the computational cost of the application of meta-methods is to create only the specification of the methods initially, go on with planning, and create the tactic part by need only (that is, in the execution phase).

3. Further Experiments

In this section I shall give a short account of further experiments we have done with the declarative framework of proof planning. The framework itself was developed by looking at a class of diagonalization methods. Historically diagonalization arguments were adapted from Cantor's proof that the powerset of a set has bigger cardinality than the set itself. In (Huang et al.,) we present a detailed trace protocol of an evolution of diagonalization methods: firstly an initial method is manually extracted from a concrete proof for Cantor's theorem (Further experiments on diagonalization are described in (Melis, 1994; Cheikhrouhou, 1997)). This initial method is then adapted for subsequent problems, namely the problem that the continuum has bigger cardinality than the natural numbers and the halting problem. By abstracting existing methods we come up with a fairly abstract method capable of dealing with all the three problems, since it captures the very key idea of diagonalization. The modifications used are, however, sensitive to the representation of the problems. If the formulations are more different than the chosen ones, the reformulation efforts will increase. It is worthwhile noting that the meta-methods used do not rely on any particular properties of diagonalization and most of them have been employed in other contexts too. In addition we investigated the mechanical modification of methods via a particular meta-method `connective-to-quantifier` which modifies methods by transforming connectives to quantifiers, that is, $\wedge$ to $\forall$ and $\vee$ to $\exists$ as well as adapting corresponding justifications like $\wedge$-elimination to $\forall$-elimination.

In (Kerber and Sehn, 1997) we have applied the proof planning framework to the problem of proving the ground completeness of resolution and have described meta-methods that can transfer the proof plan to a proof plan for proving ground completeness of lock resolution.

In recent experiments (Kerber et al., 1996; Kerber et al.,) we demonstrate how computer algebra systems can be integrated into a proof plan based proof development environment. The very idea is to consider the computer algebra system as a particular method that can be called during the proof planning process. If the computer algebra system is enhanced by a verbose mode which generates communicable and checkable proofs in form of methods the protocol information can be represented in form of a high-level hierarchical proof plan. This makes it possible to achieve two goals simultaneously, namely to produce high-level descriptions of the calculations of the computer algebra system for communicating them to human users *and* to produce low-level ones for mechanical checking. Of course it is a major piece of work to add such a verbose mode to a non-trivial computer algebra system and we have demonstrated the principle for a toy computer algebra system only. In our approach the proof planning framework offers a clean framework for integrating computer algebra systems into a theorem development environment without weakening the correctness requirements.

4. Conclusions and Future Work

Up to now human-oriented proof planning and machine-oriented techniques have different (only partly overlapping) niches in which they can be successfully applied. The results of machine-oriented approaches are particularly impressive in areas where we as humans have no good intuition and face essentially the same search space (even more impressive when the resulting proof is short and hence step-by step understandable as in the case of the Robbins problem). This, however, is neither the way human mathematicians normally work and nor the way that is taken in human-oriented approaches like proof planning. If possible, human mathematicians rather try to find reformulations of the problem that can more easily be tackled (e.g., by introducing new concepts and/or generalizing the theorem). While with our approach of meta-methods we make first steps towards the answer of a similar question, the question of how to efficiently reformulate problems is at least of the same importance. Interactive experiments clearly demonstrate that reformulations can also play a significant rôle in automated theorem proving, see (Kerber and Präcklein, 1996). However, current theorem proving environments can only insufficiently support this process.

Nowadays proof planners have reached a strength that enables them to solve – without interaction – typical problems of high school mathematics and partly undergraduate mathematics, given a representation of the methods and control knowledge involved. This is very encouraging since by its hierarchical nature the approach has the potential to scale up. In other words proof planning seems not only to offer a framework for formulating and communicating high-level ideas on an interactive level, but there is also justified hope that the heuristic knowledge of whole mathematical fields can be modeled in this approach. This shows a way out of the problems with the combinatorial explosion, machine-oriented theorem provers are normally faced with in the attempt to prove deep mathematical theorems. However, before proof planners form a useful tool to working mathematicians, another complexity problem must be solved: A proof planner can work successfully only when it is filled with the multitude of useful methods that human mathematicians use in that field. To build up such a knowledge base is a non-trivial task. As in other knowledge-based systems knowledge elicitation will be the main bottleneck. In order to avoid this bottleneck it would be extremely helpful if we were able to simulate the process how a student can learn new methods (either by imitating his/her teacher or by solving a collection of well chosen exercises). I consider this question as one of greatest research challenges in the field. Its answer would not only give us new insights in human learning, but also constitute a major step forward towards an artificial mathematical assistant.

ACKNOWLEDGEMENTS

Most of the work described in this contribution was done while I was working in the Ω group at the Universität des Saarlandes in Saarbrücken. The work benefits from many discussions with my former colleagues, in particular

L. Cheikhrouhou, X. Huang, M. Kohlhase, E. Melis, A. Präcklein, J. Richts, A. Sehn, J. Siekmann, and V. Sorge. My view on this joint work has been sharpened by many inspiring discussions with my colleagues here in Birmingham. The Ω project is supported by the Deutsche Forschungsgemeinschaft and benefitted a lot from the KEIM project in the Schwerpunkt Deduktion and the work of D. Fehrer and D. Nesmith.

References

Andrews, P.: 1980, 'Transforming Matings into Natural Deduction Proofs'. In: W. Bibel and R. Kowalski (eds.): *Proc. of the 5th CADE*. Springer, LNCS 87.

Benzmüller et al., C.: 1997, 'ΩMEGA: Towards a mathematical assistant'. In: W. McCune (ed.): *Proc. of the 14th CADE*. pp. 252–255, Springer, LNAI 1249.

Boyer, R. and J. S. Moore: 1979, *A Computational Logic*. Academic Press.

Bundy, A.: 1988, 'The Use of Explicit Plans to Guide Inductive Proofs'. In: E. Lusk and R. Overbeek (eds.): *Proc. of the 9th CADE*. pp. 111–120, Springer, LNCS 310.

Bundy, A.: 1991, 'A Science of Reasoning'. In: *Computational Logic: Essays in Honor of Alan Robinson*. MIT Press.

Bundy, A., F. van Harmelen, C. Horn, and A. Smaill: 1990, 'The OYSTER-CLAM system'. In: M. Stickel (ed.): *Proc. of the 10th CADE*. pp. 647–648, Springer, LNAI 449.

Cheikhrouhou, L.: 1997, 'Planning Diagonalization Proofs'. In: G. Brewka, C. Habel, and B. Nebel (eds.): *Proc. of the 21th KI, Annual German Conference on Artificial Intelligence*. pp. 377–380, Springer, LNAI 1303.

Constable et al., R.: 1986, *Implementing Mathematics with the Nuprl Proof Development System*. Prentice Hall.

Fikes, R. and N. Nilsson: 1971, 'STRIPS: A New Approach to the Application of Theorem Proving to Problem Solving'. *Artificial Intelligence* **2**, 189–208.

Gelernter, H.: 1959, 'Realization of a geometry theorem-proving machine'. In: *Proc. of the International Conference on Information Processing, UNESCO*.

Gordon, M., R. Milner, and C. Wadsworth: 1979, *Edinburgh LCF: A Mechanized Logic of Computation*, LNCS 78. Springer.

Huang, X.: 1994, 'Reconstructing Proofs at the Assertion Level'. In: A. Bundy (ed.): *Proc. of the 12th CADE*. pp. 738–752, Springer, LNAI 814.

Huang, X., M. Kerber, and L. Cheikhrouhou, 'Adaptation of Declaratively Represented Methods in Proof Planning'. *Annals of Mathematics and Artificial Intelligence* **forthcoming**.

Huang, X., M. Kerber, J. Richts, and A. Sehn: 1994, 'Planning Mathematical Proofs with Methods'. *Journal of Information Processing and Cybernetics, EIK* **30**, 277–291.

Huang et al., X.: 1994, 'Ω-MKRP: A Proof Development Environment'. In: A. Bundy (ed.): *Proc. of the 12th CADE*. pp. 788–792, Springer, LNAI 814.

Hutter, D.: 1990, 'Guiding Induction Proofs'. In: M. Stickel (ed.): *Proc. of the 10th CADE*. pp. 147–161, Springer, LNAI 449.

Kerber, M., M. Kohlhase, and V. Sorge, 'Integrating Computer Algebra into Proof Planning'. *Journal of Automated Reasoning* **forthcoming.**

Kerber, M., M. Kohlhase, and V. Sorge: 1996, 'Integrating Computer Algebra with Proof Planning'. In: J. Calmet and C. Limogelli (eds.): *DISCO-96 – Proc. of Design and Implementation of Symbolic Computation Systems*. pp. 204–215, Springer, LNCS 1128.

Kerber, M. and A. Präcklein: 1996, 'Using Tactics to Reformulate Formulae for Resolution Theorem Proving.'. *Annals of Mathematics and Artificial Intelligence* **18**, 221–241.

Kerber, M. and A. Sehn: 1997, 'Proving Ground Completeness of Resolution by Proof Planning'. In: D. Dankel (ed.): *FLAIRS-97 – Proc. of the Tenth International Florida Artificial Intelligence Research Symposium*. pp. 372–376, Florida AI Research Society, St. Petersburg, Florida, USA.

Kleiner, I. and N. Movshovitz-Hadar: 1997, 'Proof: A Many-Splendored Thing'. *Mathematical Intelligencer* **19**, 16–26.

McCune, W.: 1997, 'Solution of the Robbins Problem'. *Journal of Automated Reasoning* **19**, 263–276.

Melis, E.: 1994, 'Representing and Reformulating Diagonalization Methods'. Technical Report CMU-CS-94-174, School of Computer Science, Carnegie Mellon University, Pittsburgh, Pennsylvania, USA.

Nederpelt, R., H. Geuvers, and R. de Vrijer (eds.): 1994, *Selected Papers on Automath*. North-Holland.

Newell, A., C. Shaw, and H. Simon: 1957, 'Empirical Explorations with the Logic Theory Machine: A Case Study in Heuristics'. In: *Proc. of the 1957 Western Joint Computer Conference*. McGraw-Hill.

Pólya, G.: 1945, *How to Solve It*. Princeton University Press.

Pólya, G.: 1954, *Mathematics and Plausible Reasoning*. Princeton University Press.

Siekmann, J. and G. Wrightson: 1983, *Automation of Reasoning – 1. Classical Papers on Computational Logic 1957–1966*. Springer.

Wos, L., R. Overbeek, E. Lusk, and J. Boyle: 1984, *Automated Reasoning – Introduction and Applications*. Prentice Hall.

Part 2

Automated Deduction in Software Engineering and Hardware Design

Editor: Johann Schumann

JOHANN SCHUMANN

INTRODUCTION

1. INTRODUCTION

Computers have gained a tremendous importance in everyday life. The processors have by far outnumbered human beings (Storey, 1996). However, only few processors are found in "classical computers" like PCs, workstations, or supercomputers. Many more are used within embedded systems to control applications and machinery from car brakes and washing machines to nuclear plants. The software controlling the processors has reached many billions of lines of code.

Software and hardware are expected to work error-free and reliable. Many applications exist where a high reliability must be ensured, because failures are costly (with respect to human lives, environmental issues, or money). Such applications can be found in nearly all areas, e.g., aviation, nuclear power plants, medicine, transportation, process control, banking. A good collection of incidents in connection with such systems is given in P. Neumann's book (1995) where also an assessment of computer related risks is given.

For the area of hardware design and software engineering numerous approaches and methods to increase safety and reliability of computer systems have been developed. One approach is to use *formal methods*. Formal methods employ techniques from mathematical logic and discrete mathematics for the description, specification, and construction of computer systems. Formal methods allow to write down the behavior of the system under consideration (in a very broad sense) and to express properties about the system. This requires to explicitly address all assumptions and to draw all conclusions within this logic framework. Reasoning is done with a usually small number of formally defined inference rules.

Much theoretical and practical work has been done in this respect, too much to be mentioned here in any detail. The WWW-site (Formal Methods Archive, 1998) contains numerous pointers to important information about formal methods and their applications (e.g., books, papers, systems, case studies). Furthermore, a variety of good textbooks and conferences (e.g., "Formal Methods Europe", "Computer Aided Verification", or "Automated Software Engineering") cover these topics. The degree of formalization can

W. Bibel, P. H. Schmitt (eds.), Automated Deduction. A basis for applications. Vol. III

range from mathematical notation with informal analysis to full formal specification languages which allow (or even require) formal proof checkers and theorem provers.

Regardless of the degree of formalization, however, formal methods are successful in industrial real-world applications only, if there exists *tool support* for them. Tools must be *powerful*, *easy to use* (without a profound background in logic and formal methods), and *automatic* (i.e., no avoidable user interactions are required). This imposes strong requirements on the deductive components underlying these tools.

Traditionally, only interactive theorem provers or provers with a small degree of automatic processing (e.g., NQTHM (Boyer and Moore, 1988; Boyer and Moore, 1990), HOL (Gordon, 1988), Isabelle (Paulson, 1994), or PVS (Crow et al., 1995)) have been used for reasoning in this area. These systems, however, require detailed knowledge about the prover and the underlying logic and calculus. Carrying out proofs usually needs many user interactions. On the other hand, there had been almost no applications of automated theorem provers in the area of software engineering and hardware design, because automated theorem provers had not been powerful enough and most researchers focused on mathematical theorem proving. However, recent developments in automated deduction led to systems with tremendously increased power. Furthermore, other automated reasoning systems (e.g., model checkers, special-purpose decision procedures) have found widespread use.

In this part of the book, we present a number of case studies and approaches which show what can be done with current (i.e., 1997) automated deduction technology when applied to problems in software engineering and hardware design. Equivalently important, this work reports on lessons which had to be learned and on problems encountered.

The individual chapters cover central areas of a typical software life-cycle: *program synthesis* deals with the automatic generation of running (and efficient) programs out of a given specification. Although there exist many theoretical approaches, only very few systems are being used in practice. The most prominent examples are Amphion (Lowry et al., 1994; Stickel et al., 1994) for the synthesis of astro-dynamical software, Planware (Burstein and Smith, 1996) for the synthesis of high-performance schedulers, and KIDS (Smith, 1990) for more general-purpose synthesis of algorithms. Chapter III.2.5 gives an introduction into this area and focuses on knowledge-based software synthesis.

An important question which arises during the formal development of software is whether an algorithm terminates for all legal inputs or runs into endless recursion or loops. Chapter III.2.6 presents an approach for the automatic detection of termination for a large class of functional programs and also applies the developed methods to loops in imperative programs.

When a program is implemented in a higher programming language, one expects that the compiler for that language is working correctly, i.e., that the semantics of the program is retained. As many bugs in existing compilers show, the design of a correct compiler is far from trivial. Chapter III.2.7 describes a case study from the area of compiler verification. PROLOG programs (Clocksin and Mellish, 1984) are often executed on the so-called Warren Abstract Machine (WAM), (Ait-Kaci, 1991), an efficient execution vehicle for many commercial PROLOG systems. Based on an existing formalization of the transformation steps from a PROLOG program down to instructions of the WAM (Börger and Rosenzweig, 1995), Dynamic Logic and the verification system KIV have been used to formally show the correctness of central transformation steps.

With today's boom of distributed applications (WWW, electronic commerce, etc.) the importance of communication protocols and authentication protocols has increased tremendously. Communication protocols have to make sure that data are transferred correctly from A to B, whereas authentication protocols are in charge of guaranteeing a correct identification of the communication partners. In this area, formal methods have long been used to analyze protocols (cf. (Geiger, 1995; Meadows, 1994) for authentication protocols and (Holzmann, 1991) for communication protocols). In Chapter III.2.8, we present work on verification of protocols using automated theorem provers for first order logic.

Automated theorem provers have to traverse large search spaces in order to find a proof. This search space is spanned by the logic calculus, proof procedure, and the formula itself. Therefore, formulas with a large number of axioms describing the underlying theory are extremely hard to handle, even if only few axioms are actually needed. Large sets of axioms show up in many applications (e.g., in this book in Chapter III.2.7, III.2.8, III.2.11). The application of automated theorem provers in these areas can only be successful, if there are ways to reasonably reduce the number of axioms before the search starts. Chapter III.2.9 presents a straightforward, but powerful way to reduce the number of axioms in hierarchically structured theories.

Quite often, requirements for a system are expressed as a set of rules. Then, important questions are if that set of rules is complete ("Is there a rule for each situation?") and consistent ("Are there any contradictions?"). In Chapter III.2.10 a commercial real-world example is analyzed: a set of rules to calculate the fees for banking transactions. With a combination of first-order theorem proving and constraint solving techniques, consistency and completeness can be shown automatically.

After a successful development of a piece of software, the question arises, if this piece of software (or parts thereof) can be used for further program developments. Software Reuse — for a survey see e.g. (Mili et al., 1998; Krueger, 1992) — is concerned with the preparation for reuse and finding reusable components. Although many administrative aspects are important here, one of the major technical tasks concerns the search for matching components, given a search key (e.g., a target specification). Logic-based component retrieval, as described in Chapter III.2.11 uses techniques from automated deduction to find matching software components from a library. Practical usability (e.g., short answer times: "results while-u-wait") implies strong requirements on the theorem provers explored in this case study.

Formal methods are being used in the area of hardware design and manufacturing to a high degree. Many commercial tools for transformation of representation levels ("silicon compilers"), generation of test-patterns, etc., exist. Interactive verification of complex circuits (like processors) have been made (e.g. (Boyer and Yu, 1992; Burch et al., 1990; Clarke et al., 1993)). A high degree of automatic processing has come in with the widespread use of model checking tools (Holzmann, 1991; Dill et al., 1992; Burch et al., 1992). Our final Chapter III.2.12 describes a successful application of a rewriting system to the verification of a small RISC-style processor.

The work presented in this part of this book is intended to provide a many-faceted picture of the role of automated deduction techniques in the area of software engineering and hardware design. We hope that this information will give motivation and support for exploring new possibilities of theorem proving applications.

References

Ait-Kaci, H.: 1991, *Warren's Abstract Machine: A Tutorial Reconstruction.* MIT-Press.

Börger, E. and D. Rosenzweig: 1995, 'The WAM — definition and compiler correctness'. In: *Logic Programming: Formal Methods and Practical Applications*, Vol. 11 of *Studies in Computer Science and Artificial Intelligence*. North-Holland.

Boyer, R. and Y. Yu: 1992, 'Automated Correctness Proofs of Machine Code Programs for a Commercial Microprocessor'. In: D. Kapur (ed.): *Proceedings of the 11th Conference on Automated Deduction (CADE)*, Vol. 607 of *LNAI*. Saratoga Springs, NY, pp. 416–430, Springer.

Boyer, R. S. and J. S. Moore: 1988, *A Computational Logic Handbook.* Academic Press.

Boyer, R. S. and J. S. Moore: 1990, 'A Theorem Prover for a Computational Logic'. In: M. E. Stickel (ed.): *Proceedings of the 10th Conference on Automated Deduction (CADE)*, Vol. 449 of *LNAI*. pp. 1–15, Springer.

Burch, J. R., E. M. Clarke, K. L. McMillan, D. Dill, and L. J. Hwang: 1992, 'Symbolic Model Checking: 10^{20} States and Beyond'. *Information and Computing* **98**(2), 142–170.

Burch, J. R., E. M. Clarke, K. L. McMillan, and D. L. Dill: 1990, 'Sequential Circuit Verification Using Symbolic Model Checking'. In: *Proceedings of the 27th ACM/IEEE Design Automation Conference*. IEEE Comp. Soc. Predd.

Burstein, M. B. and D. Smith: 1996, 'ITAS: A Portable Interactive Transportation Scheduling Tool Using a Search Engine Generated from Formal Specifications'. In: *Proceedings of the 3rd International Conference on AI Planning Systems (AIPS-96)*. pp. 35–44, AAAI Press.

Clarke, E. M., O. Grumberg, H. Hiraishi, S. Jha, D. E. Long, K. L. McMillan, and L. A. Ness: 1993, 'Verification of the Futurebus+ Cache Coherence Protocol'. In: *Proceedings of the 11th International Symposium on Comp. Hardware Description Lang. and their Applications*.

Clocksin, W. F. and C. S. Mellish: 1984, *Programming in Prolog*. Springer.

Crow, J., S. Owre, J. Rushby, N. Shankar, and M. Srivas: 1995, 'A Tutorial Introduction to PVS'. In: *WIFT'95 Workshop on Industrial Strength Formal Specification Techniques, Boca Raton, FL, USA*.

Dill, D. L., A. J. Drexler, A. J. Hu, and C. H. Yang: 1992, 'Protocol Verification as a Hardware Design Aid'. In: *IEEE International Conference on Computer Design: VLSI in Computers and Processors*. pp. 522–525.

Formal Methods Archive: 1998, 'The World Wide Web Virtual Library: Formal Methods'. `http://www.comlab.ox.ac.uk/archive/formal-methods.html`.

Geiger, J.: 1995, 'Formale Methoden zur Verifikation kryptographischer Protokolle'. Fortgeschrittenenpraktikum, Technische Universität München, Institut für Informatik. in German.

Gordon, M. J. C.: 1988, 'HOL: A proof generating system for higher-order logic'. In: G. Birtwistle and P. A. Subrahmanyam (eds.): *VLSI Specification, Verification and Synthesis*. Kluwer, pp. 73–128.

Holzmann, G. J.: 1991, *Design and Validation of Computer Protocols*. Prentice Hall.

Krueger, C. W.: 1992, 'Software Reuse'. *ACM Computing Surveys* **24**(2), 131–183.

Lowry, M., A. Philpot, T. Pressburger, and I. Underwood: 1994, 'AMPHION: Automatic Programming for Scientific Subroutine Libraries'. In: *Proceedings of the 8th International Symposium on Methodology for Intelligent Systems, Charlotte, NC, USA*. pp. 326–335.

Meadows, C. A.: 1994, 'Formal Verification of Cryptographic Protocols: A Survey'. In: *Proceedings AsiaCrypt*.

Mili, A., R. Mili, and R. Mittermeir: 1998, 'A Survey of Software Reuse Libraries'. *Annals of Software Engineering*. To appear.

Neumann, P. G.: 1995, *Computer Related Risks*. ACM Press.

Paulson, L. C.: 1994, *Isabelle: A Generic Theorem Prover*, Vol. 828 of *LNCS*. Springer.

Smith, D. R.: 1990, 'KIDS: A Semiautomatic Program Development System'. *IEEE Transactions on Software Engineering* **16**(9), 1024–1043.

Stickel, M., R. Waldinger, M. Lowry, T. Pressburger, and I. Underwood: 1994, 'Deductive Composition of Astronomical Software from Subroutine Libraries'. In: A. Bundy (ed.): *Proceedings of the 11th Conference on Automated Deduction (CADE)*, Vol. 814 of *LNAI*. Nancy, pp. 341–355, Springer.

Storey, N.: 1996, *Safety-critical Computer Systems*. Addison Wesley.

CHAPTER 5
PROGRAM SYNTHESIS

1. INTRODUCTION

Since almost 30 years software production has to face two major problems: the cost of non-standard software, caused by long development times and the constant need for maintenance, and a lack of confidence in the reliability of software. Recent accidents like the crash of KAL's 747 in August 1997 or the failed launch of the Ariane 5 rocket in 1996 have partially been attributed to software problems and limit the extent to which software is adopted in safety-critical areas. Both problems have become even worse over the years since they grow with the complexity of behavior that can be produced. Reliable software development techniques therefore would immediately find application in developing economically important software systems.

Attempts to elaborate such techniques have been undertaken since the upcoming of the software crisis. To a large extent programming has been identified as a reasoning process on the basis of knowledge of various kinds, an activity in which people tend to make a lot of mistakes. Therefore it is desirable to provide machine support for software development and to develop software engineering tools which are based on knowledge processing and logical deductions.

1.1. *The General Idea of Program Synthesis*

Program synthesis deals with the aspects of the software development process which can, at least in principle, be automated. Its goal is to mechanically *synthesize* correct and efficient computer code from declarative specifications. Although it does not address the problem of obtaining an accurate statement of the requirements in the first place, it makes the overall process faster and more reliable, as verification and maintenance can now be done on the specification rather than the code.

Like any computerized process, program synthesis relies on a formal representation of the objects it has to deal with. Research has been active in two areas: the development of expressive *calculi* which support formal reasoning about specifications and algorithms and the implementation of deduction *strategies* which derive programs from their specifications.

W. Bibel, P. H. Schmitt (eds.), Automated Deduction. A basis for applications. Vol. III

In principle, the synthesis of programs must be expressible in some calculus with well-known deductive rules. Thus there is a need for formal logical calculi in which mathematical specifications, algorithmic structures, and reasoning about their properties can be properly represented. This has led to a revived interest in *constructive type theories* (Martin-Löf, 1984; Constable *et al.*, 1986; Girard *et al.*, 1989; Coquand & Huet, 1988; Nordström *et al.*, 1990) and proof development systems which can perform the corresponding reasoning steps interactively or guided by proof tactics (see also chapter II.1.3). Type theories are expressive enough to represent all the fundamental concepts from mathematics and programming quite naturally. The basic inference steps in formal calculi, however, are extremely small and way below the level of reasoning used by humans during program development. To make program synthesis practically feasible the level of abstraction in formal reasoning must be raised. This research area still needs exploration.

Strategies for automated program synthesis often avoid the formal overhead of a rigorous approach. Most of them were developed independent from the above calculi. There are three basic categories. The *proofs-as-programs* paradigm understands the derivation of program as a proof process which implicitly constructs an algorithm. Strategies are based on standard theorem proving techniques, tailored towards constructive reasoning (Section 3.1), and on mechanisms for extracting algorithms from proofs. Similarly, *transformational synthesis* uses general rewriting techniques in order to transform a specification into a form that can straightforwardly be converted into a program (Section 3.2). An analysis of the structure of algorithmic classes is the basis of *knowledge based program synthesis* (Section 4). Strategies derive the parameters of program schemata and instantiate them accordingly.

1.2. *History of Program Synthesis*

The identification of algorithms with proofs has been suggested early in the history of constructive mathematics (Kolmogorov, 1932). Consequently, automatic program synthesis systems showed up shortly after the first theorem provers. Pioneer work of C. Green and R. Waldinger (Green, 1969; Waldinger, 1969; Waldinger & Lee, 1969) was followed by many approaches based on the proofs-as-programs paradigm. Transformational synthesis was introduced a few years later (Manna & Waldinger, 1975). Program synthesis was a very vivid research area until the early 80's. Some approaches were implemented and tested successfully for a number of examples but none of them scaled up very well. With the advent of logic programming the correctness problem almost appeared to be solved. The activities in the original core of the field dropped and its focus shifted towards efficiency improvements.

Large scale synthesis became possible after D. Smith began to incorporate a structural analysis of algorithmic classes into efficient synthesis strategies (Smith, 1985b; Smith, 1987b). The resulting KIDS system (Smith, 1991a) has come close to a commercial breakthrough. Its success is based on an effective use of algorithm schemata, design strategies and subsequent optimization techniques which currently have to be guided by an experienced user.

1.3. *Overview of this Chapter*

In this chapter we will describe the principal approaches to program synthesis and the related deductive techniques. We will focus mostly on general principles and derivation strategies and omit approaches which are tailored to a specific application or are only weakly related to the field of automated deduction. After introducing some notation and basic concepts in section 2 we shall discuss the proofs-as-programs paradigm and transformational synthesis in Section 3. In Section 4 we present knowledge based synthesis techniques and their relation to automated deduction. Aspects of program development which are related to program synthesis will be discussed briefly in Section 5. We conclude with a short outline of future directions.

1.4. *Other Sources of Information*

Obviously not all approaches to program synthesis can be covered equally well in this chapter and some research areas cannot be discussed at all. A reader interested in a broader overview of the field will find additional information in surveys like (Balzer, 1985; Goldberg, 1986; Steier & Anderson, 1989; Lowry & McCartney, 1991). The proceedings of the IEEE International Conference on *Automated Software Engineering* (formerly the *Knowledge Based Software Engineering Conference*), the *International Conference on Mathematics of Program Construction*, the International Workshop on *Logic Program Synthesis and Transformation*, and the Journal of *Automated Software Engineering* present recent developments on a regular basis.

2. PRELIMINARIES AND NOTATION

Program synthesis is a formal reasoning process which, in addition to first-order logic, often involves reasoning about computable functions and predicates describing their properties. Furthermore, type information is crucial in most modern programming languages since the data type of a variable determines which computations can be applied to it. In a logical formula we will provide a type for each quantified variable and allow quantification over predicates and functions. We shall use $\neg$, $\wedge$, $\vee$, $\Rightarrow$, $\Leftarrow$, $-$, $\forall$, and $\exists$ for negation, conjunction, disjunction, implication, reverse implication, equivalence, universal and existential quantifier. Quantified variables will be followed by a colon and a type expression. If available, we use standard mathematical notation for data types, functions, and predicates. The formula $\forall x:\mathbb{N}.\exists z:\mathbb{Z}.(z^2 \leq x \wedge x<(z+1)^2)$, for instance, expresses that every natural number has an integer square root.

Formal specifications and programs are the basic objects of reasoning in program synthesis. A programming problem is characterized by the *domain* D of the desired program, its *range* R, an *input condition* I on *admissible* input values x, and an *output condition* O on *feasible* output values z. A program is characterized by a such specification and a possibly partial computable function *body* from D to R. Formally, these concepts are defined as follows.

DEFINITION 1. A formal *specification spec* is a 4-tuple (D,R,I,O) where D and R are data types, I is a boolean predicate on D, and O one on $D\times R$. A formal *program p* is a tuple $((D,R,I,O),body)$ where (D,R,I,O) is a specification and *body*: $D\not\to R$ a computable function.

In the literature there is a variety of formal notations for specifications and programs. The domain, range, and input-condition may be implicitly contained in the output condition. Specific aspects of the output condition may be emphasized. The above tuple form represents the essential information that is handled during the synthesis process but is sometimes difficult to read. We will often use the following more convenient notation.

FUNCTION $f(x:D):R$ WHERE $I[x]$ RETURNS z SUCH THAT $O[x,z] \equiv body[f,x]$

where $body[f,x]$ expresses that the variable x and the function name f may occur free in *body*. The latter is a convenient way to express recursion. We will use a functional programming language mixed with mathematical notation to describe a program body.

The above notation suggests that a program computes a *single* value satisfying the output condition O. Often, however, *all* solutions of the output condition have to be computed, assuming that there are finitely many. This problem could be expressed by the specification

FUNCTION $f(x:D)$:Set(R) WHERE $I[x]$ RETURNS y SUCH THAT $y=\{z\,|\,O[x,z]\}$

but we abbreviate it by the following more elegant notation

FUNCTION $f(x:D)$:Set(R) WHERE $I[x]$ RETURNS $\{z\,|\,O[x,z]\}$

Note that here $O[x,z]$ is only a part of the output condition.

Another key concept is the notion of correctness. A program is correct if its body computes a feasible output for each admissible input.

DEFINITION 2. A computable function *body*:$D \not\to R$ *satisfies* a formal specification *spec*= (D, R, I, O) if $O(x, body(x))$ holds for all $x \in D$ satisfying $I(x)$. A program *p*=(*spec*,*body*) is *correct* if its body satisfies its specification. A formal specification *spec*= (D, R, I, O) is *satisfiable* if there is a computable function *body*:$D \not\to R$ which satisfies it.

Thus from the viewpoint of deduction program synthesis is the same as proving the satisfiability of a specification and constructing a program body during the proof process.

3. Synthesis in the Small

Originally, approaches to program synthesis were developed in the framework of automated deduction (Green, 1969; Manna & Waldinger, 1979; Manna & Waldinger, 1980; Bibel, 1980). Their synthesis strategies were built on general techniques from first-order theorem proving and rewriting which made them comparably easy to explain. Their fundamental ideas work very well for small programming examples and are, at least in principle, applicable to all kinds of programming problems. Essentially, there are two broad paradigms for program synthesis: "proofs-as-programs" (Bates & Constable, 1985), and "synthesis by transformations" (Burstall & Darlington, 1977).

1. In the *proofs-as-programs* paradigm a specification is expressed by the statement that a realization of the specification exists. To synthesize an algorithm, a constructive proof of this statement is produced. Because the proof is constructive, it embodies a method for realizing the specification which can be extracted from it This paradigm emphasizes the synthesis of algorithms (computable functions) from declarative specifications. Its main issue is correctness. Research focuses on the development of strong theorem provers and mechanisms for extracting algorithms from proofs.
2. In contrast, *synthesis by transformations* derives the algorithm from the specification by forward reasoning, usually based on rewrite rules which encode the logical laws of a particular domain. A specification is viewed as executable, though inefficient program and the emphasis lies on improving efficiency through program transformations. This paradigm is particularly well-suited for the synthesis of logic programs since a declarative formula can in fact be viewed as executable program which "only" has to be transformed into some restricted syntax like Horn logic.

In the following we shall present both approaches separately and discuss their principal advantages and limitations.

3.1. *Proofs as Programs*

The idea to use formal proofs as a method forconstructing programs from formal specifications arose from the observation that developing a program and proving it correct are just two aspects of the same problem (Floyd, 1967). While verifying a given program one has to formalize the same ideas which were originally used in the construction of the program. As a consequence, program verification essentially repeats the entire programming process except for a different degree of formality.

To avoid such double work when designing software the proofs-as-programs paradigm has focused on developing a program and its correctness proof at the same time (Green, 1969; Manna & Waldinger, 1980). A specification *spec*= (D,R,I,O) is represented by a *specification theorem* of the form $\forall x{:}D.\exists z{:}R.\ (I[x] \Rightarrow O[x,z])$. To synthesize an algorithm, a *constructive* proof of this theorem is generated. In such a proof each individual step corresponds to a step in a computation. Thus the complete proof corresponds to a program satisfying the specification described by the theorem. This program can be *extracted* from the proof in a final step. The following example illustrates the main argument behind the proofs-as-programs paradigm.

EXAMPLE 3. The theorem $\forall x{:}\mathbb{Z}.\exists z{:}\mathbb{Z}.\ (x \geq 0 \Rightarrow z^2 \leq x \wedge x < (z+1)^2)$ expresses the statement that for each natural number x there is some integer z which is the square root of x. A constructive proof for this statement shows that for each x such a z can be *constructed*. It embodies a universal method for computing z from x, i.e. an algorithm for computing integer square roots.

This algorithm strongly depends on *how* the theorem was proved. Typically one would proceed by induction on x, providing 0 as solution for the base case and performing a case analysis in the step case. This means to check if the previous x is the exact square of the previous z or not and to provide a new value for z accordingly. The resulting proof implicitly contains an algorithm which computes z via primitive recursion and a conditional in the step case. It could formally be expressed as

```
FUNCTION sqrt(x:ℤ):ℤ WHERE x≥0 RETURNS z SUCH THAT z²≤x ∧ x<(z+1)
≡ if x=0 then 0 else let z=sqrt(x-1) in if z²=x-1 then z+1 else z
```

The correspondence between programs satisfying a given specification and proofs of the specification theorem can be understood as a specific reinterpretation of the *Curry-Howard Isomorphism* (Curry *et al.*, 1958; Tait, 1967) between proofs and terms of an extended λ-calculus. For instance, *constructing* a proof for a universally quantified formula $\forall x{:}\alpha.P[x]$ corresponds to a function which takes an element x of the type α and returns (a term corresponding to) a proof for $P[x]$. *Analyzing* $\forall x{:}\alpha.P[x]$ in order to show $P[a/x]$ corresponds to applying such a function to the element a. Likewise, a proof

Table I. Proofs-as-programs correspondence for first-order logic

Connective	Construction	Analysis
Atomic Literal	Variable*	—
Conjunction	Pair	Projection
Disjunction	Injections	Conditional
Implication	Function declaration	Function application
Negation	Function declaration	Function application
Existential Quantification	Pair	Projection
Universal Quantification	Function declaration	Function application

* *The same literal will always be associated with the same variable*

for a formula $\exists x:\alpha.P[x]$ corresponds to a pair (a, p_a) where a is an element of α and p_a a proof for $P[a/x]$. A constructive proof for $P\vee Q$ corresponds either to a left injection $\texttt{inl}(p)$ or a right injection $\texttt{inr}(q)$, depending on whether P or Q is being shown. Analyzing a disjunction leads to a conditional which examines these injections.

Table I describes the correspondence between proofs and programs for first-order logic. Details depend on the individual proof calculus and the formal language in which programs are to be expressed. In sequent or natural deduction calculi the correspondence can be represented straightforwardly since programming constructs can be tied directly to proof rules. This makes it easy to define a mechanism for extracting a program from the proof of a specification theorem and to justify the proofs-as-programs principle.

THEOREM 4. *A specification spec=* (D,R,I,O) *is satisfiable iff the theorem* $\forall x{:}D.\exists z{:}R.\ (I[x] \Rightarrow O[x,z])$ *can be proved (constructively).*

Proof. According to table I the proof of a specification theorem corresponds to a computable function *pf* which on input $x\in D$ returns a pair (z, p) where $z\in R$ and p is a proof term for $I[x] \Rightarrow O[x,z]$. By projecting p away, i.e. by defining $body(x) \equiv \texttt{first}(pf(x))$ we get a correct program for the specification. Conversely, if we have a computable function $body$: $D\not\to R$ and a term *prf* corresponding to a correctness proof for the program $((D,R,I,O), body)$, then combining *body* and *prf* leads to a proof of the specification theorem. □

It should be noted that in a proof of a specification theorem only the part that shows how to build z from x needs to be constructive. The verification of $I[x] \Rightarrow O[x,z]$ does not contribute to the generated algorithm and can be performed by non-constructive means as well.

Synthesis strategies based on the proofs-as-programs principle heavily rely on the availability of theorem provers which can deal with a rich variety of application problems. In principle, any constructive proof method for which an extraction mechanism can be defined is useful for this purpose.

Skolemization and resolution, although not entirely constructive,[1] generate substitutions which can be extracted from a proof and were used in early approaches (Green, 1969; Manna & Waldinger, 1975) to program synthesis.

It has turned out that pure first-order theorem proving is too weak for program synthesis as it cannot generate more than simple substitutions which do not consider the meaning of atomic propositions. Reasoning about recursion or loops plays a vital part in the design of almost any program as well as knowledge about arithmetic and other application domains. The former has been addressed by extending theorem proving by induction techniques such as *rippling* (Bundy, 1989; Bundy *et al.*, 1992) or *constructive matching* (Franova, 1985). The latter requires specialized decision procedures since an axiomatization of the application domain would require thousands of formulas to be considered and thus be extremely inefficient.

Currently there is no integrated theorem proving system which can deal with all these proofs tasks equally well. In spite of a variety of approaches and systems which have been developed over the past 25 years the automatic syntheses of programs through proofs has only been able to create solutions for relatively small problems. Furthermore, pure proof methods do not always lead to an efficient solution. Inductive proofs of specification theorems as in example 3 can only generate primitive recursive programs with *linear* complexity although logarithmic time is possible. Faster algorithms can only be created in less rigorous frameworks which allow inductions on the output variable or by explicitly providing the algorithm structure (see Section 4.3).

3.2. *Transformational Synthesis*

While the proofs-as-programs paradigm primarily aims at the derivation of formally verified programs, efficiency considerations have been the original motivation for program synthesis by transformations. For many programming problems it is relatively easy to produce and verify a prototype implementation in a lucid, mathematical and abstract style. Correctness, comprehensibility, and easy modifications are the advantages of such programs. But a lack of efficiency makes them unacceptable for practical purposes and they can only viewed as detailed specification of the "real" program.

Initially, program development by transformations (Darlington, 1975; Burstall & Darlington, 1977) focused on systematically transforming *programs* into more intricate but efficient ones. This idea was adapted to program synthesis (Manna & Waldinger, 1975; Darlington, 1975; Manna & Waldinger,

[1] Matrix-based theorem provers have recently been extended to constructive first-order logic (Otten & Kreitz, 1995). The corresponding extraction mechanism is based on a conversion of matrix proofs into sequent proofs (Schmitt & Kreitz, 1995).

1979) as a technique for transforming *specifications*, which are not necessarily executable, into expressions which can be handled by a computer. In logic programming this distinction between declarative specifications and algorithmic programs has almost disappeared. Research in transformational synthesis today aims mostly[2] at *logic program synthesis and transformations* (Clark & Sickel, 1977; Hogger, 1981), also called "*deductive program synthesis*".

Approaches to transformational synthesis usually proceed by step-wisely rewriting the output-condition of a specification into a logically equivalent or stronger formula which algorithmically can be viewed as refinement of the specification. In a final step program formation rules are applied in order to create programming constructs from the derivation tree. In the context of logic programming this step is almost superfluous since the resulting formula only needs a few syntactical adjustments to be understood as logic program.

Individual approaches vary greatly as far as notation is concerned but essentially proceed as follows. A specification *spec*= (D, R, I, O) is represented by the formula $\forall x{:}D.\forall z{:}R.\ (I[x] \Rightarrow (P(x,z) - O[x,z]))$ which defines a new predicate P as equivalent to the given output condition. The intuition behind this is to view $O[x,z]$ as body of a program whose name P is newly introduced and to derive properties of P by *forward reasoning*. In this setting the role of x and z is not yet fixed but the transformation process will optimize $O[x,z]$ with respect to a particular designation of input and output variables. The goal of transforming $O[x,z]$ is to create a valid formula of the form

$$\forall x{:}D.\forall z{:}R.\ (I[x] \Rightarrow (P(x,z) \Leftarrow O_f[x,z,P]))$$

which is algorithmically satisfactory. This goal of transformational synthesis cannot be expressed within the logical formalism but only within the derivation *strategy*. A typical criterion in logic program synthesis, for instance, is that $O_f[x,z,P]$ must be a Horn formula and contains, apart from recursive instances of P, only predicates which are already known to be executable.

In order to transform $O[x,z]$ into $O_f[x,z,P]$ *conditional rewrite rules* are applied. These rules are based on (conditional) equivalences or implications which are used to define concepts of some application domain and to state their properties. In particular, the definition of P itself is considered as conditional equivalence which can be folded if recursion needs to be introduced.

Program formation rules determine how to convert the result of these transformations into a valid program. Logic programs, for instance, can be generated by simply dropping the universal quantifiers, the input condition, and any existential quantifier in the body O_f, by splitting disjunctions into two separate clauses with the same head, and by replacing functions by the

---

[2] Research on *program* transformations is still an active research area. Here, algebraic methods are dominating while logical deduction plays only a minor role.

corresponding predicates. Furthermore, equalities and structural decompositions in the body can be replaced by modifying the head of a clause in order to take stronger advantage of unification. The construction of functional and imperative programs is more complex since it requires a formation rule corresponding to each logical junctor and one for recursion. Some of these rules are similar to the algorithm schemata given in Sections 4.2.1 and 4.2.2.

As in the case of proofs-as-programs, the validity of transformational synthesis essentially depends on the program formation mechanism. The requirement to base the transformations on equivalences or implications only makes sure that transforming $O[x,z]$ will result in a valid formula. The correctness of the generated program is then assured by the following theorem.

THEOREM 5. *Let spec=* (D,R,I,O) *be a formal specification and P be a predicate defined by* $\forall x{:}D.\forall z{:}R.(I[x] \Rightarrow (P(x,z) - O[x,z]))$. *If the formula* $\forall x{:}D.\forall z{:}R.(I[x] \Rightarrow (P(x,z) \Leftarrow O_f[x,z,P]))$ *is valid then the program extracted from it satisfies the specification spec.*

Proof. (Sketch) In the case of logic programs, where the semantics is almost identical to the declarative meaning, the generated algorithm is logically a refinement of the original output condition whose correctness is easy to see. A full proof requires an induction about the structure of O_f and a precise definition of the program formation rules for each logical junctor. □

Research in transformational synthesis has mostly been concerned with developing efficient rewrite techniques and heuristics producing satisfactory practical results. Some approaches use folding and unfolding a fixed set of equivalences or equations (Darlington, 1975; Darlington, 1981), a modification of the Knuth-Bendix completion procedure (Dershowitz, 1985), or refinements according to given transformation and program formation (Manna & Waldinger, 1977; Manna & Waldinger, 1979) are closely related to each other. The LOPS-approach (Bibel, 1980) provides a small set of AI-oriented strategies which we shall describe in the following example.

EXAMPLE 6. In order to synthesize an algorithm computing the maximum m of a finite set S of integers we begin by defining a new predicate *max* to describe an element which is greater or equal to all other elements of S.

$$\forall S{:}Set(\mathbb{Z}).\forall m{:}\mathbb{Z}.(S{\neq}\emptyset \Rightarrow (\mathit{max}(S,m) \;-\; m{\in}S \wedge \forall x{\in}S.x{\leq}m))$$

In the first step we try to *guess* a solution, assuming that we will either find one or provide a way to reduce the problem. Formally, we introduce an existentially quantified variable g and add $g{=}m \vee g{\neq}m$ as indicator for a successful or a failing guess. To limit the possibilities for guessing we introduce a *domain condition* $g{\in}S$, which is taken from the original problem description. We add these conditions to the output condition of *max* and get

$$\forall S{:}Set(\mathbb{Z}).\forall m{:}\mathbb{Z}.(S{\neq}\emptyset \Rightarrow$$
$$(\mathit{max}(S,m) \;-\; \exists g{:}\mathbb{Z}.(g{\in}S \wedge m{\in}S \wedge \forall x{\in}S.x{\leq}m \wedge (g{=}m \vee g{\neq}m))))$$

Next, we distribute the disjunction over the initial output condition and introduce two predicates max_1 and max_2 as abbreviations. They use g as additional input variable with input condition $g \in S$. This transformation results in

$$\forall S{:}Set(\mathbb{Z}).\forall m{:}\mathbb{Z}.(S{\neq}\emptyset \Rightarrow$$
$$(\boldsymbol{max}(S,m) \;-\; \exists g{:}\mathbb{Z}.(g{\in}S \land (\boldsymbol{max_1}(S,g,m) \lor \boldsymbol{max_2}(S,g,m)))))$$

$$\forall S{:}Set(\mathbb{Z}).\forall g,m{:}\mathbb{Z}.(S{\neq}\emptyset{\land}g{\in}S \Rightarrow (\boldsymbol{max_1}(S,g,m) \;-\; m{\in}S{\land}g{=}m{\land}\forall x{\in}S.x{\leq}m))$$

$$\forall S{:}Set(\mathbb{Z}).\forall g,m{:}\mathbb{Z}.(S{\neq}\emptyset{\land}g{\in}S \Rightarrow (\boldsymbol{max_2}(S,g,m) \;-\; m{\in}S{\land}g{\neq}m{\land}\forall x{\in}S.x{\leq}m))$$

The above steps are equivalence preserving and can be executed by a strategy called *GUESS-DOMAIN*. Before continuing we analyze the three subproblems. The first one is solved once we have solutions for max_1 and max_2. The second has a trivial solution and we only have to investigate the third.

The strategy *GET-REC* tries to introduce recursion by reformulating the output condition for max_2 into an instance of the output condition of *max*. It retrieves information about well-founded recursions on finite sets and attempts a rewrite towards S-g, the set resulting from removing g from S. For this purpose it searches for lemmata involving $S\text{-}g{\neq}\emptyset$, $m{\in}S\text{-}g$, and $\forall x{\in}S\text{-}g.\,x{\leq}m$ and rewrites the problem accordingly. This leads to a disjunction in the input condition which is split into two cases.

$$\forall S{:}Set(\mathbb{Z}).\forall g,m{:}\mathbb{Z}.(S{=}\{g\} \land g{\in}S \Rightarrow$$
$$(\boldsymbol{max_2}(S,g,m) \;-\; m{\in}S \land g{\neq}m \land\forall x{\in}S.x{\leq}m))$$

$$\forall S{:}Set(\mathbb{Z}).\forall g,m{:}\mathbb{Z}.(S\text{-}g{\neq}\emptyset \land g{\in}S \Rightarrow$$
$$(\boldsymbol{max_2}(S,g,m) \;-\; m{\in}S\text{-}g \land g{\neq}m \land g{\leq}m \land \forall x{\in}S\text{-}g.x{\leq}m))$$

The output condition of the first case proves to be contradictory since the input condition states that S consists of the single element g. In the second we can fold back the definition of *max*, simplify $g{\neq}m \land g{\leq}m$, and get

$$\forall S{:}Set(\mathbb{Z}).\forall g,m{:}\mathbb{Z}.(S\text{-}g{\neq}\emptyset \land g{\in}S \Rightarrow (\boldsymbol{max_2}(S,g,m) \;-\; \boldsymbol{max}(S\text{-}g,m) \land g{<}m))$$

Now all subproblems are solved. After removing redundancies we create a logic program according to the above-mentioned formation rules and get

```
max(S,M)         :- member(G,S), max_aux(S,G,M).
max_aux(S,G,M)   :- setminus(S,G,S_G), max(S_G,M), less(G,M),!.
max_aux(S,M,M)   :- setless(S,M).
```

The strategies *GUESS-DOMAIN* and *GET-REC* are the center of all LOPS derivations. (Bibel, 1980) describes several strategies which support them if more complex examples shall be solved. Unfortunately many of them turned out to be difficult to formalize precisely and the LOPS approach, like many other good ideas, could not be turned into a system which was able to solve more than a few small examples.

3.3. *Discussion*

Despite their different origins there are no principal difference between the proofs-as-programs paradigm and transformational synthesis. Transformations can be simulated in a proof environment by using lemmata (i.e. the cut rule) and proof transformations if recursion is introduced. Proof rules can be viewed as special rewrite rules. In principle, each derivation strategy in one paradigm can be converted into one for the other. The real question is rather which environment is more convenient for a particular method. The proofs-as-programs paradigm provides a clearly defined proof goal and favors analytic methods from theorem proving. Depending on the rigorousity of the proof system correctness can be guaranteed for all inferences. The latter is not the case for transformational synthesis since the transformation rules which deal with domain knowledge are usually not verified. The advantage is a smaller formal overhead which together with the focus on forward reasoning makes the derivation of programs often somewhat easier.

Both synthesis paradigms, however, rely on general reasoning methods which operate on the level of elementary logical inferences. This makes the fundamental techniques relatively easy to explain and implement and was quite successful for solving simple programming problems like sorting or the n-queens problem. A fully automated synthesis of larger problems, however, is almost impossible due to the explosion of the search space. Interaction is also very difficult since general methods from automated deduction have little to do with the way in which a programmer would reason when developing a program. Therefore these synthesis techniques do not scale up very well.

The main problem of general approaches to program synthesis is that they force the synthesis system to derive an algorithm almost from scratch and to re-invent algorithmic principles which are well understood in computer science. Solving complex programming problems, however, heavily relies on knowledge about application domains *and* standard programming techniques. One can hardly expect a synthesis system to be successful if such expertise is not already embedded. More recent approaches therefore make programming knowledge explicit, instead of hiding it in the code of their strategies, and develop their algorithm design strategies on the basis of such knowledge.

4. KNOWLEDGE BASED PROGRAM SYNTHESIS

Knowledge based synthesis strategies arose from the observation that the development of complex algorithms requires a deep understanding of fundamental algorithmic structures. They aim at a *cooperation* between programmer and machine. A programmer shall guide the derivation process by high-level strategic decisions which, like the selection of an appropriate program structure, require a deep understanding of the problem and the intended solution. The synthesis system will fill in the formal details and ensure the correctness of the generated algorithms.

This means that much work will be invested into the development of the synthesis *system* in order to ease the burden for the synthesis *process*. The design of knowledge based synthesis systems requires an analysis of the fundamental structure of standard algorithms and of the axioms which guarantee their correctness. It also requires the development of strategies which construct algorithms on the basis of this analysis and need only a few high-level design decisions. Since the analysis usually leads to a set of theorems stating conditions on the parameters of some well-structured algorithm schema the strategies have to derive values for these parameters which satisfy these conditions. The final step consists of instantiating the algorithm schema. The result can later be refined by program transformation techniques to be made more efficient. One advantage of this technique is that it makes the synthesis strategy independent from the programming language in which the algorithm schema is formulated and allows to use *any* programming language with a well-defined semantics for this purpose.

Synthesis strategies have been developed for a number of algorithmic classes (see (Smith & Lowry, 1990) for a general exposition). In this section we will illustrate the principle by describing a strategy for the development of so-called global search algorithms and present a variety of strategies related to other classes of algorithms. We will also discuss how to integrate knowledge based algorithm design and methods from automated deduction.

4.1. *Synthesizing Global Search Algorithms*

Solving a problem by enumerating candidate solutions is a well-known concept in computer science. *Global search* is a technique that generalizes binary search, backtracking, and other methods which explore a search space by looking at whole sets of possible solutions at once. Its basic idea is to combine enumeration and elimination processes. Usually, global search investigates the complete set of output values for a given input. It systematically enumerates a search space, which must contain this set, and tests if certain elements of the search space satisfy the output-condition. The test is necessary to guarantee correctness but rather inefficient if the search space is much bigger than the set of solutions. Therefore whole regions of the search space are filtered out during the enumeration process if it can be determined that they cannot contain output values.

A careful analysis in (Smith, 1987b), later refined and formalized in (Kreitz, 1992; Kreitz, 1996), has shown that the common structure of global search algorithms can be expressed by the pair of abstract programs presented in Figure 1. These programs contain placeholders D, R, I, and O for a specification and seven additional components $S, J, s_0, sat, split, ext, \Phi$ which are specific

```
FUNCTION f(x:D):Set(R) WHERE I[x]  RETURNS {z | O[x,z]}
≡ if Φ[x,s0[x]]  then fgs(x,s0[x])  else []
FUNCTION fgs(x,s:D×S):Set(R)  WHERE I[x]∧J[x,s]∧Φ[x,s]
   RETURNS {z | O[x,z]∧ sat[z,s]}
≡ let immediate_solutions = filter (\z.O[x,z]) (ext[s])
   and recursive_solutions =
       let filtered_subspaces = filter (\t.Φ[x,t]) (split[x,s]) in
         flatten (map (\t.fgs(x,t)) filtered_subspaces)
   in  append immediate_solutions recursive_solutions
```

Figure 1. Structure of global search algorithms

for a global search algorithm. On input x this algorithm starts investigating an *initial search space* $s_0[x]$ and passes it through the *filter* Φ which checks globally whether a search region s contains solutions. Using an auxiliary function f_{gs} the algorithm then repeatedly *extracts* ($ext[s]$) candidate solutions for testing and *splits* a search space s into a set $split[x,s]$ of subspaces which again are passed through the filter Φ. Subspaces which survive filtering contain solutions and are investigated recursively.

For the sake of efficiency search spaces are represented by *space descriptors* $s\in S$ instead of sets of values. The fact that a value $z\in R$ belongs to the search space described by s is denoted by $sat[z,s]$ while $J[x,s]$ expresses that s is a *meaningful* space descriptor for the input x. Formally, S must be a data type. J and Φ must be predicates on $D\times S$ and sat one on $R\times S$. $s_0{:}D\not\to S$, $split{:}D\times S\not\to S$, and $ext{:}S\not\to \texttt{Set}(R)$ must be computable functions.

$\forall x{:}D.\ \forall z{:}R.\ \forall s{:}S.$

1. $I(x) \Rightarrow J(x,s_0(x))$
2. $I(x)\wedge J(x,s) \Rightarrow \forall t\in split(x,s).\,J(x,t)$
3. $I(x)\wedge O(x,z) \Rightarrow sat(z,s_0(x))$
4. $I(x)\wedge J(x,s)\wedge O(x,z) \Rightarrow sat(z,s)\ -\ \exists\, k{:}\mathbb{N}.\,\exists t\in split^k(x,s).\,z\in ext(t)$
5. $I(x)\wedge J(x,s) \Rightarrow \Phi(x,s)\ \Leftarrow\ \exists z{:}R.\,sat(z,s)\wedge O(x,z)$
6. $I(x)\wedge J(x,s) \Rightarrow \exists k{:}\mathbb{N}.\ split_\Phi^k(x,s)=\emptyset$

where $split_\Phi(x,s)=\{t\in split(x,s)\mid \Phi(x,t)\}$
and $split_\Phi^0(x,s)=s,\ split_\Phi^{k+1}(x,s)=\bigcup_{t\in split_\Phi^k(x,s)} split_\Phi(x,t)$.

Figure 2. Axioms of global search

Six requirements, formalized in Figure 2, ensure the correctness of the global search algorithms. The initial descriptor must be meaningful (1). Splitting must preserve meaningfulness (2). All solutions must be contained in the initial search space (3) and be extractable after splitting finitely many times (4). Subspaces containing solutions must pass the filter (5). Filtered splitting must eventually terminate (6).

THEOREM 7. *If $D,R,I,O,S,J,sat,s_0,split,ext,$ and Φ fulfill the axioms of global search then the pair of programs in Figure 1 is correct.*

Proof. (see (Smith, 1987b; Kreitz, 1996)*for details)* The correctness of f follows from the correctness of f_{gs} and axioms 1 and 3. The *validity* of f_{gs} follows from axioms 2, 4, and 5. To prove this fact one has to show by induction that the result of computing $f_{gs}(x,s)$ is $\bigcup\{\ \{z{\in}ext(t)\,|\,O(x,z)\}\ |\ t\in\bigcup_{j<i}split^j_\Phi(x,s)\}\ \}$ where i is the smallest number with $split^i_\Phi(x,s)=\emptyset$, i.e. the number of steps until $f_{gs}(x,s)$ terminates. *Termination* and thus total correctness of f_{gs} follows from axiom 6. □

Thus a global search algorithm for a given specification can be synthesized by deriving seven components S, J, sat, s_0, $split$, ext, and Φ which satisfy the six axioms and instantiating the abstract programs accordingly. A direct derivation of global search algorithms, however, is still a difficult task since Theorem 7 does not show how to *find* the additional components. Also, a verification of the axioms, particularly of axioms 4 and 6, would put a heavy load on the derivation process. It is much more meaningful to base the construction of global search algorithms on additional knowledge about algorithmic structures. For each range type R, for instance, there are usually only a few general methods to enumerate search spaces. Each global search algorithm will use an instance of such a method. Therefore it makes sense to store information about generic enumeration processes in a knowledge base and to develop techniques for adapting them to a particular programming problem.

The investigations in (Smith, 1987b) have shown that standard enumeration structures for some range type R can be represented by objects of the form $((D_G,R,I_G,O_G),\ S,J,s_0,sat,split,ext)$ which satisfy the axioms 1 to 4. Such objects are called *GS-theories*. A *problem reduction* mechanism will make sure that these axioms are preserved when the enumeration structure is *specialized* to a specification which has the same range type.

Specializing a standard GS-theory G works as follows. Its specification $spec_G=(D_G,R_G,I_G,O_G)$ characterizes a general enumeration method f_G which explores the space R_G as far as possible. Specialization simply means to truncate the search space such that superfluous elements will not be enumerated. It is possible if $spec_G$ is more general than the given specification.

DEFINITION 8. A specification $spec_G=(D_G,R_G,I_G,O_G)$ *generalizes* the specification $spec=(D,R,I,O)$ if the following condition holds.

$$R{=}R_G\wedge\forall x{:}D.\,\exists x_G{:}D_G.(I(x)\Rightarrow(I_G(x_G)\wedge\forall z{:}R.(O(x,z)\Rightarrow O_G(x_G,z))))$$

We also say that $spec_G$ can be *specialized* to *spec*.

Thus specialization restricts the *output* of f_G to values which satisfy the stronger condition O. Furthermore, it allows to adapt the *input* x of a specified problem since the x is mapped to a value x_G which serves as input for the search performed by f_G. According to the proofs-as-programs paradigm (theorem 4) a proof of the generalization implicitly contains a substitution $\theta{:}D{\rightarrow}D_G$ which maps x to x_G. θ can be extracted from the proof and then be used for refining f_G into a search method with inputs from D instead of D_G.

COROLLARY 9. *If $spec_G = (D_G, R_G, I_G, O_G)$ generalizes the specification $spec = (D,R,I,O)$ then there is a function $\theta{:}D{\rightarrow}D_G$ which satisfies the condition $\forall x{:}D.(I(x) \Rightarrow (I_G(\theta(x)) \wedge \forall z{:}R.(O(x,z) \Rightarrow O_G(\theta(x),z))))$.*

Altogether problem reduction allows us to create a global search algorithm f for *spec* by defining $f(x) = \{z \in f_G(\theta(x)) \mid O(x,z)\}$. For the sake of efficiency, the modifications caused by θ and O are moved directly into the components of the global search algorithm. By an index θ as in J_θ or $split_\theta$ we indicate that θ is applied to all arguments expecting a domain value from D, e.g. $split_\theta(x,s) = split(\theta(x),s)$.

Specializing predefined GS-theories allows us to derive six components of a global search algorithm which satisfy axioms 1 to 4 with little effort. Similarly, we can avoid having to prove the sixth axiom explicitly. For each enumeration structure there are only a few standard methods which ensure termination through an elimination process. In (Kreitz, 1992; Kreitz, 1996) it has been shown that these can be represented by filters for a GS-theory G which satisfy axiom 6. Such filters are called *well-founded wrt.* G and this property is preserved during specialization as well. Thus specialization reduces the proof burden to checking that, after specialization, the selected filter is *necessary* wrt. the GS-theory, i.e. that it satisfies axiom 5. The process of adapting the search space to the specific problem can be completely formalized and expressed in a single theorem.

THEOREM 10. *Let $G = ((D_G,R,I_G,O_G), S, J, s_0, sat, split, ext)$ be a GS-theory such that (D_G,R,I_G,O_G) generalizes $spec = (D,R,I,O)$. Let θ be the substitution extracted from the generalization proof.*
Then $G_\theta = ((D,R,I,O), S, J_\theta, s_{0\theta}, sat, split_\theta, ext)$ is a GS-theory. Furthermore if Φ is a well-founded filter wrt. G then Φ_θ is well-founded wrt. G_θ

Proof. Axioms 1 to 3 for G_θ can be shown directly by combining the properties of θ (Corollary 9) with the global search axioms. Axiom 4 and the well-foundedness of Φ_θ can be proved by a straightforward induction over k and the number of iterations of *split*, again using the properties of θ and the corresponding axioms for G and Φ. □

Adapting standard algorithmic knowledge to a given problem moves most of the proof burden into the creation of the knowledge base and keeps the synthesis process itself comparably easy. Information retrieved from the knowledge base will provide all the basic components and guarantee that axioms 1 to 4 and 6 are satisfied. Only the specialization property and the necessity of the specialized filter – conditions which are much easier to prove than axioms 4 and 6 – need to be checked. These insights led to the following strategy.

STRATEGY 11. *Given the specification*

`FUNCTION` $F(x{:}D)$`:Set(`R`) WHERE` $I[x]$ `RETURNS` $\{z \mid O[x,z]\}$

1. *Select a GS-theory* $G=((D_G,R,I_G,O_G),\ S,J,s_0,sat,split,ext)$ *for* R.
2. *Prove that* $spec_G=(D_G,R,I_G,O_G)$ *generalizes the specification. Derive a substitution* $\theta : D \rightarrow D_G$ *from the proof and specialize* G *with* θ.
3. *Select a well-founded filter* Φ *for* G *and specialize it with* θ.
4. *Prove that the specialized filter is necessary for the specialized GS-theory.*
5. *Instantiate the program schema given in Figure 1.*

In steps 1 and 3 of the above strategy, the selection of the GS-theory G and the well-founded filter Φ is mostly a matter of design. If there are choices, the decision should be made by a human programmer rather than by a predefined heuristic. In step 4 the specialized filter could be further refined heuristically by adding conditions which preserve its necessity. In some cases this improves the efficiency of the generated algorithm drastically. The global search strategy has been successfully applied to a number of programming problems (Smith, 1987b; Smith, 1991a). In particular it has been used for the synthesis of commercial transportation scheduling algorithms (Smith & Parra, 1993; Gomes *et al.*, 1996) which turned out to be much more efficient than any hand-coded implementation.

4.2. *Design Strategies for Other Algorithmic Classes*

Schema-based synthesis strategies have been developed for a variety of algorithmic classes. The can create generate & test algorithms (Smith, 1987a), static algorithms such as the formation of conditionals (Smith, 1985a), divide & conquer algorithms (Smith, 1985b), local search algorithms (Lowry, 1991), and general problem reduction generators (Smith, 1991b). In the following we shall briefly describe the central ideas, their justification, and the corresponding synthesis strategies.

Static Algorithms

Static algorithms, i.e. algorithms without loops or recursion, are a means for adapting already existing programs to similar tasks. Such algorithms are usually designed during the synthesis of complex programming problems but can also created separately. The corresponding design strategies occur both as independent method and as part of more sophisticated synthesis techniques.

Operator Match is a reduction technique that solves a programming problem $spec = (D,R,I,O)$ by transforming its inputs into inputs of a known problem $spec' = (D',R',I',O')$ and transforming outputs of that problem back into its own range type. This technique works if, after transformation, the input condition I is stronger that I' and the output condition O is weaker than O':

$$\forall x{:}D.\, \exists x'{:}D'.\, (I(x) \Rightarrow (I'(x') \wedge \forall z'{:}R'.\exists z{:}R.(O'(x',z') \Rightarrow O(x,z))))$$

We say that *spec reduces to spec'* if the above condition can be proved. In this case, according to the proofs-as-programs paradigm, the proof contains a substitution $\theta{:}D \rightarrow D'$ which maps x to x' and a substitution $\sigma{:}D' \times R' \rightarrow D \times R$ which maps (x',z') to (x,z). Thus a program satisfying the original specification can be generated by composing θ, a solution g for *spec'*, and σ.

THEOREM 12. *Let* $spec = (D,R,I,O)$ *and* $spec' = (D',R',I',O')$ *be specifications,* $\theta{:}D \rightarrow D'$, $g : D' \not\rightarrow R'$, *and* $\sigma{:}D' \times R' \rightarrow D \times R$. *If (1) spec reduces to spec', (2)* θ *and* σ *are substitutions extracted from the proof of (1), and (3)* g *satisfies spec' then the following program is correct*

FUNCTION $f(x{:}D){:}R$ WHERE $I[x]$ RETURNS z SUCH THAT $O[x,z]$
$\equiv \sigma(\theta(x), g(\theta(x)))$

Specialization, already discussed in the context of global search (see definition 8), can be seen as a variant of operator matching. The difference is that specialization deals with set-valued problems and *restricts* the output.

THEOREM 13. *Let* $spec{=}(D,R,I,O)$ *and* $spec'{=}(D',R',I',O')$ *be specifications,* $\theta{:}D \rightarrow D'$, *and* $g{:}D' \not\rightarrow Set(R')$. *If (1) spec' generalizes spec, (2)* θ *is the substitution extracted from the proof of (1), and (3)* g *satisfies the specification* FUNCTION $G(x'{:}D'){:}Set(R')$ WHERE $I'[x']$ RETURNS $\{z' \mid O'[x',z']\}$ *then the following program is correct*

FUNCTION $f(x{:}D){:}Set(R)$ WHERE $I[x]$ RETURNS $\{z \mid O[x,z]\}$
$\equiv \{z \in g(\theta(x)) \mid O(x,z)\}$

Reasoning by Cases is a technique which generates conditional programs if a more direct approach to solving a programming problem fails. The essential idea (Smith, 1985a) is to split the input into two or more cases which are analyzed separately and then composed again into a correct program.

THEOREM 14. *Let* $spec = (D,R,I,O)$ *be a specification,* $g,h{:}D \not\rightarrow R$, *and* p *be a boolean function on* D. *If (1)* g *satisfies the specification* $(D,R,I \wedge p,O)$[3] *and (2)* h *satisfies the specification* $(D,R,I \wedge \neg p,O)$ *then the following program is correct*

FUNCTION $f(x{:}D){:}R$ WHERE $I[x]$ RETURNS z SUCH THAT $O[x,z]$
$\equiv$ if $p(x)$ then $g(x)$ else $h(x)$

[3] We use $I \wedge p$ as abbreviation for a predicate defined by $(I \wedge p)(x) = I(x) \wedge p(x)$.

There are data types D' and R', predicates O_D on $D \times D' \times D$, O' on $D' \times R'$, and O_C on $R \times R' \times R$, and a well-founded ordering relation $\succ$ on D such that

1. *Directly-solve* satisfies the specification
   FUNCTION $F_p(x{:}D){:}R$ WHERE $I[x] \wedge prim[x]$ RETURNS z SUCH THAT $O[x,z]$
2. $\forall x, x_1{:}D. \forall x_2{:}D'. \forall z_2{:}R'. \forall z_1, z{:}R.$
   $O_D(x,x_2,x_1) \wedge O'(x_2,z_2) \wedge O(x_1,z_1) \wedge O_C(z_2,z_1,z) \Rightarrow O(x,z)$
3. *Decompose* satisfies the specification
   FUNCTION $F_d(x{:}D){:}D' \times D$ WHERE $I[x] \wedge \neg prim[x]$
   RETURNS x_2,x_1 SUCH THAT $I'[x_2] \wedge I[x_1] \wedge x_1 \succ x \wedge O_D[x,x_2,x_1]$
4. *Compose* satisfies the specification
   FUNCTION $F_c(z_2,z_1{:}R' \times R){:}R$ RETURNS z SUCH THAT $O_C(z_2,z_1,z)$
5. g satisfies the specification
   FUNCTION $F_G(x_2{:}D'){:}R'$ WHERE $I'[x_2]$ RETURNS z_2 SUCH THAT $O'(x_2,z_2)$

Figure 3. Axioms of divide & conquer

While the justification of this method is trivial, the difficulty lies in obtaining a proper predicate which makes a solution of the individual cases feasible. One way to do this is creating *derived antecedents* (Smith, 1982; Smith, 1985a) by trying to find the requirements for the validity of a given formula. This mechanism rewrites a formula (e.g. the satisfiability condition for a specification) by applying domain lemmata until it is proved or some preconditions for its validity remain. These preconditions are used as additional input-condition for one subproblem, which now is solved, while their negation yields a second subproblem, which has to be investigated further. Eventually, this results in a cascade of conditionals which solve the original problem.

Divide & Conquer

Divide & conquer is one of the most common techniques for processing recursively defined data structures. It solves a problem by dividing it into subproblems whose solutions will be computed ('conquered') independently and composed into a *single* solution for the main problem. Divide & conquer algorithms proceed by *decomposing* an input value into "smaller" inputs, which are processed recursively, and possibly other values, which are processed by some auxiliary algorithm. The resulting output values are *composed* into an output for the original problem. Input values which cannot be decomposed anymore are considered *primitive* and will be solved *directly*. Formally, the common structure of divide & conquer algorithms can be expressed as

FUNCTION $f(x{:}D){:}R$ WHERE $I[x]$ RETURNS z SUCH THAT $O[x,z]$
$\equiv$ if $prim[x]$ then $Directly\text{-}solve[x]$ else $(Compose \circ g \times f \circ Decompose)[x]$

This abstract program contains placeholders *prim*, *Directly-solve*, *Compose*, *g*, and *Decompose* which have to be instantiated to create a concrete divide

& conquer algorithm. The analysis in (Smith, 1985b) has shown that five requirements, formalized in Figure 3, ensure the correctness of this algorithm. *Directly-solve* must compute a correct solution for primitive inputs (1). A *Strong Problem Reduction Principle* must hold (2) saying that the output condition O can be recursively decomposed into subproblems O_D, O_C, O', and O. O_D is to be solved by *Decompose* which also "reduces" the input values[4] w.r.t. some well-founded ordering $\succ$ on D (3) to ensure the termination of f. O_C is to be solved by *Compose* (4) and O' by the auxiliary function g (5).

THEOREM 15. *If D, R, I, O, prim, Directly-solve, Compose, g, and Decompose fulfill the five axioms of divide & conquer then the program*

FUNCTION $f(x{:}D){:}R$ WHERE $I[x]$ RETURNS z SUCH THAT $O[x,z]$
≡ if $prim[x]$ then $Directly\text{-}solve[x]$ else $(Compose \circ g{\times}f \circ Decompose)[x]$

is correct.

Proof. For primitive input values f computes a correct output because of axiom 1. Otherwise f is $Compose \circ g{\times}f \circ Decompose$. Its partial correctness is shown by composing axioms 3, 4, and 5, the induction hypotheses for smaller input values, and the strong problem reduction principle. f terminates since $\succ$ is well-founded. □

Again, the formal theorem is the foundation of a schema-based synthesis strategy. This strategy has to derive the five additional components, make sure that the axioms are satisfied, and instantiate the program schema accordingly. Obviously, it cannot fulfill its task without referring to knowledge about application domains. A knowledge base has to provide *standard decomposition techniques* on input-domains (e.g. splitting sequences in two halves or into a first element and the rest), *standard well-founded orderings* (e.g. ordering lists according to their length), and *standard composition techniques* on output-domains (e.g. appending lists or prepending an element before a list). These informations support a synthesis of divide & conquer algorithms according to the following strategy.

STRATEGY 16. *Given the specification*

FUNCTION $f(x{:}D){:}R$ WHERE $I[x]$ RETURNS z SUCH THAT $O[x,z]$

1. *Select $\succ$ and Decompose.*
2. *Construct the auxiliary function g.*
3. *Verify Decompose and derive the input condition for its correctness. Choose prim as negation of the additional input condition.*
4. *Verify the strong problem reduction principle and construct Compose.*
5. *Construct Directly-solve.*
6. *Instantiate the divide & conquer schema.*

[4] If the domain D' of g is identical to D then *both* values generated by *Decompose* must be smaller than the original input value. This allows binary recursion.

In (Smith, 1985b) this strategy is accompanied by a series of heuristics which help solving the individual steps. Step 1, which gives us D' and O_D, will look up the knowledge base. In step 2, which yields O',I',R' and axiom 5, g is usually chosen as f if D'=D and as identity function otherwise. This simplifies the synthesis process since the effects of g can always be captured in the function *Compose*. Step 3, which leads to axiom 3, uses the mechanism for deriving antecedents (page 123) to determine the requirements for the correctness of *Decompose*. An admissible input value that does not fulfill these requirements must be *primitive*. In step 4 the same mechanism is used to generate O_C as precondition for Axiom 2. Constructing *Compose* is usually the most difficult step. It requires a new synthesis on the basis of axiom 4. Often *Compose* will be another divide & conquer algorithm but in simpler cases we can also use *operator match* on the basis of known library programs. Step 5 will usually try *operator match* to satisfy axiom 1.

There are several variants of this strategy. One may proceed in reverse order by selecting *Compose* first and successively constructing g, $\succ$, *Decompose*, *prim*, and *Directly-solve*. In either of the two strategies one might construct *Compose* or *Decompose* before constructing g.

Each of these strategies will lead to different algorithms since the selected operator is usually less complicated than the constructed one. When applied to the problem of sorting lists, for instance, strategy 16 eventually yields Mergesort, if *Decompose* splits lists in halves, and Insertion Sort, if it splits into first element and rest. If the reverse strategy is used, Quicksort will be generated, if *Compose* appends two lists, and Selection Sort, if *Compose* prepends elements to lists. Thus selecting the synthesis strategy and the operations in its first step is not just a strategic issue. It is a design decision which should be made by a programmer interacting with the system.

Local search

Local search is a technique that is used for dealing with optimization problems like scheduling or travelling salesman problems. It investigates a set of possible output values for a given specification. But the focus is not on *finding* them but on selecting the *best*, e.g. a schedule that can be executed in the shortest amount of time. Formally, the specification is associated with a cost function on the range and the goal is to find a solution with minimal costs.

Since optimization is often NP-complete, *Hillclimbing* algorithms like the simplex algorithm or linear programming approximate optimal solutions by exploring the local *neighborhood* of some initial solution which was easy to compute. By incremental variations the elements of this neighborhood are searched. The algorithms proceeds until a *local minimum* has been found. In (Lowry, 1988; Lowry, 1991) the general structure of these *local search*

$$
\begin{array}{l}
\texttt{FUNCTION } f_{opt}\,(x{:}D){:}R \;\; \texttt{WHERE } I[x] \;\; \texttt{RETURNS } z\\
\quad \texttt{SUCH THAT } O[x,z] \;\wedge\; \forall t{:}R.\, O[x,t] \;\Rightarrow\; cost[x,z]\leq cost[x,t]\\
\equiv\; f_{ls}(x, f(x))\\[1ex]
\texttt{FUNCTION } f_{ls}\,(x{:}D,y{:}R){:}R \;\; \texttt{WHERE } I[x]\wedge O[x,y] \;\; \texttt{RETURNS } z\\
\quad \texttt{SUCH THAT } O[x,z] \;\wedge\; \forall t\in N[x,z].\, O[x,t] \;\Rightarrow\; cost[x,z]\leq cost[x,t]\\
\equiv\; \texttt{if } \forall t\in N[x,y].\, O[x,t] \;\Rightarrow\; cost[x,y]\leq cost[x,t]\\
\qquad \texttt{then } y \;\; \texttt{else } f_{ls}\,(x, \texttt{arb}(\{t\in N[x,y] \mid O[x,t]\wedge cost[x,y]>cost[x,t]\}))
\end{array}
$$

Figure 4. Structure of local search algorithms

algorithms has been described as a pair of programs which is presented in Figure 4. Besides placeholders D, R, I, and O for the basic specification these programs contain three additional components. $cost{:}D\times R\to\mathcal{R}$ is a function which assigns to each pair of inputs and outputs a *cost-value* in some ordered domain $(\mathcal{R},\leq)$. It occurs both in the (extended) specification and the program. $N{:}D\times R\to\texttt{Set}(R)$ describes the neighborhood structure by computing a set of output values which, depending on the input, are accessible from a given output. $f{:}D\not\to R$ computes the initial solution.

$$
\begin{array}{lll}
\multicolumn{3}{l}{\forall x{:}D.\;\forall z,t{:}R.}\\
1.\; I[x]\wedge O[x,z] & \Rightarrow & z\in N(x,z)\\
2.\; I[x]\wedge O[x,z]\wedge O[x,t] & \Rightarrow & \exists k{:}\mathbb{N}.\, t\in N_O^k(x,z)\\
3.\; I[x]\wedge O[x,z] & \Rightarrow & \forall t\in N(x,z).\, O[x,t] \Rightarrow cost[x,z]\leq cost[x,t]\\
 & & \Rightarrow\; \forall t{:}R.\, O[x,t] \Rightarrow cost[x,z]\leq cost[x,t]\\
\multicolumn{3}{l}{4.\; f \text{ satisfies the base specification } spec=(D,R,I,O)}
\end{array}
$$

where $N_O^0(x,z)=\{z\}$ and $N_O^{k+1}(x,z)=\bigcup_{t\in\{t\in N(x,z)\mid O[x,t]\}} N^k(x,t)$

Figure 5. Axioms of local search

Four axioms, listed in Figure 5, ensure the correctness of local search algorithms. The neighborhood structure must be reflexive (1) and *connected* w.r.t. the solutions of the base specification (2): by exploring local neighborhoods all solution must be reachable from each other. If the algorithm shall be exact, local minima must be global minima (3). Finally f has to satisfy the base specification without the optimality conditions. The following theorem has been proved in (Lowry, 1988).

THEOREM 17. *If D, R, I, O, cost, N, and f fulfill the four axioms of local search then the pair of programs in Figure 4 is correct.*

Theorem 17 describes the requirements on local search algorithms which compute *optimal* solutions. They are met by some algorithms but usually the third axiom is dropped to allow suboptimal solutions. In this case local search algorithms compute local minima whose quality depends solely on the neighborhood structure. Local effects will dominate if the neighborhood structure is too fine-grained and the algorithm is likely to compute a poor approximation of the global optimum. A coarse neighborhood structure, however, may lead to inefficiency since large search spaces will be explored in each step.

Finding good neighborhood structures is the only crucial aspect in a derivation of local search algorithms. The other components are either already given (*cost*) or easy to determine (f). As usual, the construction of local search algorithms depends on general knowledge. For a given range type R, for instance, there are only a few fundamental neighborhood structures. These can be represented by formal objects, called *LS-theories*, and will be refined by specialization and filters.

The strategy for designing local search algorithms is similar to the one for global search. We first select a LS-theory and specialize it to the given specification. We then modify the neighborhood structure by filters which eliminate output values which are either no solutions for the base specification (*feasibility constraints*) or more expensive than the current one (*optimality constraints*). Both constraints roughly correspond to the notion of necessary filters in global search. If exactness is required, then a filter for *global optimality constraints* can be added as well. All filters can be derived by forward inference. Finally a solution f for the base specification is synthesized and the schematic algorithm is instantiated. A detailed description of this strategy and application examples can be found in (Lowry, 1988; Lowry, 1991).

Problem Reduction Generators

Problem reduction generators deal with problems that require a set of individual solutions to be computed independently by composing several partial solutions. Typical examples are dynamic programming and branch & bound algorithms. From a theoretical point of view they can be seen as generalization of global search and divide & conquer. Structurally they are similar to divide & conquer algorithms except for the fact that now a series of composition and decomposition functions are used and that the primitive case is contained in some of them.

$$\texttt{FUNCTION } f(x{:}D)\texttt{:Set}(R) \texttt{ WHERE } I[x] \texttt{ RETURNS } \{z \mid O[x,z]\}$$
$$\equiv \bigcup_{i,k\in\mathbb{N}} (Compose_i \circ (F_{i_1}\times..\times F_{i_k}) \circ Decompose_i)[x]$$

The requirements for correctness of this algorithm generalize the axioms 2 to 5 of divide & conquer (Figure 3) and include a complex version of the strong problem reduction principle. The derivation strategy is similar to the one for synthesizing divide & conquer algorithms but puts a heavier burden on the inference mechanism. A detailed exposition can be found in (Smith, 1991b).

4.3. *Integration into Automated Deduction*

In an implementation of design strategies for knowledge based program synthesis, the semi-formal descriptions leave much room for interpretation. The KIDS system (Smith, 1991a), which aims at tools for semi-automatic software development, encodes these strategies directly in a high-level programming language which includes a reasoning mechanism based on formula transformations. Many steps in these strategies, however, involve solving deductive problems such as extracting substitutions from proofs of logical properties, verifying program components, or deriving the preconditions for the correctness of some formula. Standard techniques from automated deduction could be used if knowledge based program synthesis would take place in the framework of a proof system. This would also make the derivations safer since the possibility of invalid inferences, which is always present in a hand-coded synthesis system, is ruled out by the proof environment.

Integrating design strategies into a proof based system requires a very rigorous approach. Each step in a derivation must be completely formal, such that it can be controlled by the prover, but remain on the high level of abstraction which we have used so far.

Formally verified theorems stating the requirements for the correctness of an abstract program schema (Kreitz, 1996) are the key for an integration of schema-based synthesis strategies into a deductive framework. These theorems can be applied as high-level inference rules which reduce the synthesis task to the task of proving instances of the axioms. The latter can be solved by further theorem applications, knowledge base queries, first-order theorem proving, or simple inductions. The conceptually difficult problem – generating the algorithm and proving it correct – has been solved once and for all while proving the formal theorem and requires only a single derivation step.

Technically, such an integration of knowledge based program synthesis into proof systems involves a complete formalization of all the aspects involved in the derivation of programs. Standard application domains and their laws have to be represented by formal definitions and theorems in some logical calculus. Formula transformations can then be expressed as application of verified lemmata. Next, in order to reason about programs as such, we must formalize concepts like *program*, *specification*, *correctness* and related notions. Concepts like specialization, operator match, GS-theories, filters etc. need to be formalized. Concrete GS-theories, filters, etc. have to be formalized and verified. Finally, the theorems underlying the derivation strategies have to be stated and proved. This last step may require a lot of interaction but remove the proof burden from the synthesis process.

On this basis, strategies for automated program derivation can be programmed as proof tactics which apply verified theorems and lemmata and call standard proof procedures (see e.g. chapter I.1.5) to check properties like specialization, operator match, or necessity. It has been shown in (Bibel *et al.*, 1997, Section 4) that such an integrated approach, despite its formal overhead, can make program synthesis practically feasible even in the rigorous setting of proof systems.

4.4. *Discussion*

Program synthesis based on abstract algorithm schemata requires a greater amount of preparation than the techniques described in Section 3. But these efforts pay off during the synthesis: both the synthesis process and the generated programs are much more efficient. The successful synthesis of transportation scheduling algorithms reported in (Smith & Parra, 1993; Gomes *et al.*, 1996) shows that knowledge based program synthesis, if properly guided and supported by optimization techniques (Section 5.2), can produce commercially competetive software.

Furthermore, the schematic approach to program synthesis can also be used for teaching systematic programming. Students can investigate the effects of choosing different algorithmic structures as solution for the same problem and gain insights into the consequences of their own decisions. The four sorting algorithms resulting from different design decisions in a divide & conquer synthesis (cf. Section 4.2.2) are a good example for this (a fifth algorithm, Bubble Sort, can be generated via local search). Studying the formal foundations of the corresponding strategies, on the other hand, will lead to a deeper understanding of the algorithmic structures and their specific strengths and weaknesses.

Thus knowledge based synthesis has several advantages over approaches based entirely on general proof or transformation techniques. Since it can be represented in both frameworks it should not be viewed as competitor but as the next step in the evolution of automated programming which indicates a major direction for future research.

5. Related Topics

In the previous sections we have described methods for deriving well-structured algorithms from *formal* specifications. These methods represent the core of program synthesis. But systematic program development also involves methods for obtaining proper specifications and for transforming the derived algorithm into efficient program code. We shall briefly discuss both subjects.

5.1. *Acquiring Formal Specifications*

Formally specifying a given programming problem is by no means a trivial task. It requires a deep analysis of the informal requirements, the problem domain, the environment in which the program shall be executed, implicit assumptions, etc. before a system architecture can be designed and individual modules and algorithms can be specified.

Over the past there has been a substantial amount of work on developing computer support for acquiring specifications (see (Hayes, 1987; Jones & Shaw, 1990; Kelly & Nonnenmann, 1991) and (Lowry & Duran, 1989, Sections B & D)). In particular, computers are used to visualize a design or as support for a certain design methodology. *Formal methods* such as VDM (Jones, 1990) or OMT (Rumbaugh et. al., 1991) provide guidelines and visualization techniques for specifying large software sytems and are also used for communicating design decisions.

Nevertheless, acquiring formal specifications is mostly a design process that involves creativity and design decisions in order to balance between conflicting requirements. Although informal reasoning is necessary to justify the overall design of the system the deductive aspects are comparably small. Computers can only serve as support tool when making decisions but it will be difficult to go beyond this level.

5.2. *Algorithmic Optimizations*

Automatically generated algorithms are hardly the most efficient solution for a given problem. Often a human programmer immediately detects components that can be optimized. Program expressions may be *simplified* or even completely eliminated if their context, particularly the input condition, is taken into account. Expressions containing constants may be *partially evaluated.* Expensive computations may be replaced by iterative updates of a new variable. Finally, it may pay off to select a non-standard implementation of certain data structures. All these steps can be represented by *program transformations* or *rewrite techniques* and integrated into a synthesis system. The real issue is, however, to automate these transformations such that a user only has to select an expression and a particular technique for optimizing it.

Simplification transforms a program expression into an equivalent one which is easier to compute. The transformations are based on knowledge about equivalences within some application domain. These equivalences are given a *direction* which allows to apply them as long as possible. Simplification strongly depends on an efficient organization of the knowledge base which may contain thousands of equivalences.

Partial evaluation (Bjørner *et al.*, 1988) symbolically evaluates subexpressions which contain constants or other fixed structures. Logically this corresponds to unfolding definitions and simplifying the result.

Finite differencing (Paige & Koenig, 1982) replaces a subexpression by a new variable which will be updated in each recursive call of the program. It can result in a significant speedup and supports parallel computations since depth is being replaced by breadth.

Data Type Refinement (Blaine & Goldberg, 1991) selects from several implementations of an abstract data type the one that is most efficient for the program. It depends on an analysis of the operations which refer to these data types and requires automatic conversions between different representations.

The SPECWARE approach (Srinivas & Jüllig, 1995), which implements concepts from category theory for establishing relations between algebraic theories, provides an environment in which the modular construction of formal specifications, knowledge based algorithm design stategies, algorithmic optimizations, and the construction of executable code from mathematical algorithms can be integrated. A first implementation of such an integrated program synthesis system, called PLANWARE, is currently being developed.

6. CONCLUSION

Despite its long history program synthesis is still a very active research area. Over the years its focus has shifted from the fundamental questions in the field of automated deduction to the practical aspects of systematic program development. It appears that knowledge based program synthesis, coupled with automated deduction techniques, is the most promising approach since it can be made theoretically sound and practically useful at the same time. The integration of optimization strategies, techniques for adapting generic programs (like transportation scheduling algorithms or communication systems) to specific applications, and methods for acquiring and structuring domain knowledge is becoming more and more important. Furthermore, general methodologies for creating efficient and reliable implementations of synthesis systems are an important research issue.

There is a commercial interest in computer systems which can rapidly develop efficient and reliable software, whose key properties are guaranteed but not necessarily formally verified. Program synthesis systems have come close to the point where they can be used for that purpose, provided they are guided by programmers with a strong background in formal reasoning.

References

R. Balzer. A 15 year perspective on automatic programming. *IEEE Transactions on Software Engineering*, SE-11(11):1257–1268, 1985.

J. Bates & R. Constable. Proofs as programs. *ACM Transactions on Programming Languages and Systems*, 7(1):113–136, 1985.

W. Bibel, D. Korn, C. Kreitz, F. Kurucz, J. Otten, S. Schmitt, G. Stolpmann. A multi-level approach to program synthesis. *7th International Workshop on Logic Program Synthesis and Transformation*, LNAI, Springer, 1998.

W. Bibel. Syntax–directed, semantics–supported program synthesis. *Artificial Intelligence*, 14(3):243–261, 1980.

D. Bjørner, A. Ershov, N. Jones, eds. *Partial evaluation and mixed computation.* North–Holland, 1988.

L. Blaine & A. Goldberg. DTRE—a semi-automatic transformation system. *IFIP TC2 Conference on Constructing Programs from Specifications.* Elsevier, 1991.

A. Bundy, F. van Harmelen, A. Ireland, A. Smaill. Rippling: a heuristic for guiding inductive proofs. *Artificial Intelligence*, 1992.

A. Bundy. Automatic guidance of program synthesis proofs. In *Workshop on Automating Software Design, 11th International Joint Conference on Artificial Intelligence*, pp. 57–59. Morgan Kaufman, 1989.

R. Burstall & J. Darlington. A transformation system for developing recursive programs. *Journal of the ACM*, 24(1):44–67, 1977.

K. Clark & S. Sickel. Predicate logic: A calculus for the formal derivation of programs. *5th International Joint Conference on Artificial Intelligence*, pp. 419–420. Morgan Kaufman, 1977.

R. Constable, et. al. *Implementing Mathematics with the NuPRL proof development system.* Prentice Hall, 1986.

T. Coquand & G. Huet. The calculus of constructions. *Information and Computation*, 76:95–120, 1988.

H. Curry, R. Feys, W. Craig. *Combinatory Logic*, vol. 1. North–Holland, 1958.

J. Darlington. Application of program transformation to program synthesis. *IRIA Symposium on Proving and Improving programs*, pp. 133–144, 1975.

J. Darlington. An experimental program transformation and synthesis system. *Artificial Intelligence*, 16(1):1–46, 1981.

N. Dershowitz. Synthesis by completion. *9th International Joint Conference on Artificial Intelligence,*, pp. 208–214. Morgan Kaufman, 1985.

R. Floyd. Assigning meaning to programs. *Symposia in Applied Mathematics*, 19:19–32, 1967.

M. Franova. A methodology for automatic programming based on the constructive matching strategy. *EUROCAL 85*, pp. 568–570. Springer, 1985.

J.-Y. Girard, Y. Lafont, P. Taylor. *Proofs and Types.* Cambridge University Press, 1989.

A. Goldberg. Knowledge-based programming: A survey of program design and construction techniques. *IEEE Transactions on Software Engineering*, SE-12(7):752–768, 1986.

C. Gomes, D. Smith, S. Westfold. Synthesis of schedulers for planned shutdowns of power plants. *11th Conference on Knowledge-Based Software Engineering*, 1996.

C. Green. An application of theorem proving to problem solving. *1st International Joint Conference on Artificial Intelligence*, pp. 219–239. Morgan Kaufman, 1969.

I. Hayes. *Specification Case Studies*. Prentice-Hall, 1987.

C. Hogger. Derivation of logic programs. *Journal of the ACM*, 28(2):372–392, 1981.

C. Jones & R. Shaw. *Case Studies in Systematic Software Development*. Prentice-Hall, 1990.

C. Jones. *Systematic software development using VDM*. Prentice-Hall, 1990.

V. Kelly & U. Nonnenmann. Reducing the complexity of formal specification acquisition. *Automating Software Design*, pp. 41–64, AAAI/MIT Press, 1991.

A. Kolmogorov. Zur Deutung der intuitionistischen Logik. *Mathematische Zeitschrift*, 35:58–65, 1932.

C. Kreitz. *METASYNTHESIS: Deriving Programs that Develop Programs*. Thesis for Habilitation, TU Darmstadt, 1992.

C. Kreitz. Formal mathematics for verifiably correct program synthesis. *Journal of the IGPL*, 4(1):75–94, 1996.

M. Lowry & R. Duran. Knowledge-based software engineering. *Handbook of Artificial Intelligence, Vol. IV*, chapter XX. Addison Wesley, 1989.

M. Lowry & R. McCartney. *Automating Software Design*, AAAI/MIT Press, 1991.

M. Lowry. The structure and design of local search algorithms. *Workshop on Automating Software Design. AAAI-88*, pp. 88–94, 1988.

M. Lowry. Automating the design of local search algorithms. *Automating Software Design*, pp. 515–546, AAAI/MIT Press, 1991.

Z. Manna & R. Waldinger. Knowledge and reasoning in program synthesis. *Artificial Intelligence*, 6(2):175–208, 1975.

Z. Manna & R. Waldinger. The automatic synthesis of systems of recursive programs. *5th International Joint Conference on Artificial Intelligence*, pp. 405–411. Morgan Kaufman, 1977.

Z. Manna & R. Waldinger. Synthesis: Dreams ⇒ programs. *IEEE Transactions on Software Engineering*, SE-5(4):294–328, 1979.

Z. Manna & R. Waldinger. A deductive approach to program synthesis. *ACM Transactions on Programming Languages and Systems*, 2(1):90–121, 1980.

P. Martin-Löf. *Intuitionistic Type Theory*. Bibliopolis, 1984.

B. Nordström, K. Petersson, J. Smith. *Programming in Martin-Löfs Type Theory. An introduction*. Clarendon Press, 1990.

J. Otten & C. Kreitz. A connection based proof method for intuitionistic logic. *4th Workshop on Theorem Proving with Analytic Tableaux and Related Methods*, LNAI 918, pp. 122–137. Springer, 1995.

R. Paige & S. Koenig. Finite differencing of computable expressions. *ACM Transactions on Programming Languages and Systems*, 4(3):402–454, 1982.

J. Rumbaugh et. al. *Object-oriented Modeling and Design*. Prentice Hall, 1991.

S. Schmitt & C. Kreitz. On transforming intuitionistic matrix proofs into standard-sequent proofs. *4th Workshop on Theorem Proving with Analytic Tableaux and Related Methods*, LNAI 918, pp. 106–121. Springer, 1995.

D. Smith & M. Lowry. Algorithm theories and design tactics. *Science of Computer Programming*, 14(2-3):305–321, 1990.

D. Smith & E. Parra. Transformational approach to transportation scheduling. *8th Knowledge-Based Software Engineering Conference*, pp. 60–68, 1993.

D. Smith. Derived preconditions and their use in program synthesis. *6th Conference on Automated Deduction*, LNCS 138, pp. 172–193. Springer, 1982.

D. Smith. Reasoning by cases and the formation of conditional programs. *9th International Joint Conference on Artificial Intelligence*, pp. 215–218. Morgan Kaufman, 1985.

D. Smith. Top-down synthesis of divide-and-conquer algorithms. *Artificial Intelligence*, 27(1):43–96, 1985.

D. Smith. On the design of generate-and-test algorithms: Subspace generators. *IFIP TC2 Conference on Program Specification and Transformation*, pp. 207–220. North–Holland, 1987.

D. Smith. Structure and design of global search algorithms. Technical Report KES.U.87.12, Kestrel Institute, November 1987.

D. Smith. KIDS: A semi-automatic program development system. *IEEE Transactions on Software Engineering*, 16(9):1024–1043, 1990.

D. Smith. KIDS — a knowledge-based software development system. *Automating Software Design*, pp. 483–514, AAAI/MIT Press, 1991.

D. Smith. Structure and design of problem reduction generators. *IFIP TC2 Conference on Constructing Programs from Specifications*, pp. 91–124. Elsevier, 1991.

Y. Srinivas & R. Jüllig. SPECWARE: Formal Support for composing software. *Conference on the Mathematics of Program Construction*, 1995.

D. Steier & A. Anderson. *Algorithm Synthesis: A comparative study*. Springer, 1989.

W. Tait. Intensional interpretation of functionals of finite type. *Journal of Symbolic Logic*, 32(2):187–199, 1967.

R. Waldinger & R. Lee. PROW: A step toward automatic program writing. *1st International Joint Conference on Artificial Intelligence*, pp. 241–252. Morgan Kaufman, 1969.

R. Waldinger. Constructing programs automatically using theorem proving. PhD Thesis, Carnegie-Mellon University, 1969.

JÜRGEN GIESL, CHRISTOPH WALTHER, JÜRGEN BRAUBURGER

CHAPTER 6

TERMINATION ANALYSIS FOR FUNCTIONAL PROGRAMS

1. INTRODUCTION

Proving termination is a central problem in software development and formal methods for termination analysis are essential for program verification. However, since the halting problem is undecidable and totality of functions is not even semi-decidable, there is no procedure to prove or disprove the termination of *all* algorithms.

While most work on the automation of termination proofs has been done in the areas of *term rewriting systems* (for surveys see e.g. (Dershowitz, 1987; Steinbach, 1995b)) and of *logic programs* (e.g. (Ullman and van Gelder, 1988; Plümer, 1990; De Schreye and Decorte, 1994)), in this chapter we focus on *functional programs* and we also investigate the application of our methods for termination analysis of *loops* in *imperative programs*.

To prove termination of a functional algorithm, one has to find a well-founded relation such that the arguments in each recursive call are smaller than the corresponding inputs. (A relation $\prec$ is *well founded* iff there is no infinite descending chain $\ldots \prec t_2 \prec t_1$.) A semi-automatic method for termination proofs of LISP functions has been implemented in the NQTHM system of *R. S. Boyer* and *J S. Moore* (1979). For an algorithm $f(x)$ their prover shows that in each recursive call $f(r)$ a given *measure* is decreased. The system uses a *measure function* $|.|$ which maps data objects to natural numbers and it verifies that the number $|r|$ is smaller than the number $|x|$.

For this proof the *system user* has to supply so-called *induction lemmata* of the form $\Delta \to |r| < |x|$, which assert that certain operations are measure decreasing if some hypotheses Δ are satisfied. For the termination proof of an algorithm $f(x)$ with a recursive call $f(r)$, the system searches among the known induction lemmata for some lemma with the conclusion $|r| < |x|$. Then the prover verifies $b \to \Delta$, where b is the condition under which the recursive call $f(r)$ is performed.

Of course instead of using induction lemmata, one could also try to prove inequalities like $|r| < |x|$ *directly*. But the advantage of induction lemmata is

Authors' address: Fachbereich Informatik, Technische Universität Darmstadt, Alexanderstr. 10, 64283 Darmstadt, Germany, E-mail: {giesl,walther,brauburger}@ informatik.tu-darmstadt.de

W. Bibel, P. H. Schmitt (eds.), Automated Deduction. A basis for applications. Vol. III

that difficult facts are established once and for all (by an *induction* proof). Then by applying induction lemmata, subsequent termination proofs are usually quite simple, i.e. they can often be accomplished without induction.

Boyer and Moore's technique for proving termination is very powerful, as the methods works for *arbitrary measure functions* $|.|$. But its disadvantage is a low degree of automation. Induction lemmata have to be *formulated by the system user* and therefore a human has to find the idea why an algorithm terminates. Moreover, to ensure their soundness, these lemmata have to be verified by the system before use. This verification may be hard, as in general an induction proof is needed.

In this chapter, we present an alternative method for automated termination proofs of algorithms (Section 2). With this method a certain class of induction lemmata can be synthesized *automatically* and the soundness of these induction lemmata is *guaranteed by construction*. So compared to the technique of Boyer and Moore the advantage of our method is a much higher degree of automation.

But a limitation of this method is that it is restricted to *one single fixed* measure function, viz. the so-called *size measure function*, while the system of Boyer and Moore can use arbitrary measure functions for termination proofs. Using the size measure function, data objects t are compared by their *size* $|t|_{\#}$, i.e. lists are compared by their length, trees are compared by the number of their nodes etc. Although this approach is successful for many examples, there are numerous relevant algorithms whose termination proofs require a measure different from *size* and for these algorithms the method of Section 2 must fail.

Therefore, in Section 3 we extend our method in order to handle *arbitrary measure functions*. To determine suitable measures automatically we use approaches from the area of *term rewriting systems* for the generation of well-founded *term orderings*. But unfortunately term orderings cannot be directly used for termination proofs of *functional algorithms* which call other algorithms in the arguments of their recursive calls. The reason is that for termination of term rewriting systems orderings between *terms* are needed, whereas for functional programs orderings between the *evaluated terms* are required. Our method solves this problem and enables *term orderings* to be used for *functional programs*. In this way, we obtain an *automated* method for termination proofs where suitable measure functions and induction lemmata are synthesized by machine. This combines a high degree of automation with the powerful generality of Boyer and Moore's method.

The approaches of Section 2 and 3 aim to prove that an algorithm terminates for *each* input ("*total* termination"). Thus, if the termination proof fails then they cannot find a (sub-)domain where termination is provable. However, this is necessary for termination proofs of algorithms which call *partial* auxiliary algorithms, i.e. auxiliary algorithms which do not terminate for all inputs. In particular, this problem arises when examining the termination behaviour of loops in imperative programs. Therefore, in Section 4 we extend our techniques to termination analysis of partial functions.

2. Termination Proofs with Argument-Bounded Algorithms

In this section we illustrate a first approach for automated termination proofs using a *fixed* ordering. After an introduction to termination proofs in Section 2.1, Section 2.2 shows how inequalities are proved by the technique of *estimation*. For this technique we need certain knowledge about the algorithms under consideration and in Section 2.3 we discuss how this knowledge can be acquired automatically.

2.1. *Termination Proofs of Functional Programs*

We regard an eager first-order functional language with (free) algebraic data types and pattern matching (where the patterns have to be exhaustive and exclusive)[1]. For example, consider the data type bool with the nullary *constructors* true and false and the data type nat (for naturals) whose objects are built with the constructors 0 and s : nat → nat.

The following algorithms compute predecessor, subtraction, and division (the boolean function lt is the usual "less than" relation on nat).

function $\mathsf{p} : \mathsf{nat} \to \mathsf{nat}$

$$\mathsf{p}(0) = 0$$
$$\mathsf{p}(\mathsf{s}(x)) = x$$

function $\mathsf{minus} : \mathsf{nat} \times \mathsf{nat} \to \mathsf{nat}$

$$\mathsf{minus}(x,0) = x$$
$$\mathsf{minus}(0,\mathsf{s}(y)) = 0$$
$$\mathsf{minus}(\mathsf{s}(x),\mathsf{s}(y)) = \mathsf{minus}(x,y)$$

function $\mathsf{quot} : \mathsf{nat} \times \mathsf{nat} \to \mathsf{nat}$

$$\mathsf{quot}(x,0) = 0$$
$$\mathsf{quot}(x,\mathsf{s}(y)) = \mathsf{if}(\mathsf{lt}(x,\mathsf{s}(y)),\ 0,\ \mathsf{s}(\mathsf{quot}(\mathsf{minus}(\mathsf{p}(x),y),\mathsf{s}(y))))$$

For each data type s there is a pre-defined conditional $\mathsf{if} : \mathsf{bool} \times s \times s \to s$. These conditionals are the only algorithms with non-eager semantics, i.e. when evaluating $\mathsf{if}(b,t_1,t_2)$, the (boolean) term[2] b is evaluated first and de-

[1] An eager language evaluates the arguments of a function call *before* application (this corresponds to the *call-by-value* parameter passing discipline). The use of pattern matching instead of *selectors* (or *destructors*) has no real impact on the difficulty of the termination proof, but it eases the readability of our presentation.

[2] In the following we often refer to boolean terms as "formulas", where ¬, ∧, ∨, and → are pre-defined boolean functions with obvious semantics.

pending on the result of its evaluation either t_1 or t_2 is evaluated afterwards yielding the result of the whole conditional.

To prove *termination* of algorithms one has to show that in each recursive call a certain *measure* is decreased w.r.t. a well-founded relation. For that purpose a *measure function* $|.|$ is used which maps each data object t to a natural number $|t|$. For minus' termination proof we can use the *size* measure $|.|_\#$, where the size of an object of type nat is the number it represents (i.e. the number of s-occurrences it contains). So we have $|0|_\# = 0$, $|\mathsf{s}(0)|_\# = 1$ etc. Termination of the (binary) function minus can now be verified by regarding its first argument only. Hence, we have to show that $|x|_\#$ is smaller than $|\mathsf{s}(x)|_\#$ for all instantiations of x, i.e. we have to verify the following *termination formula*[3] for minus.

$$|x|_\# < |\mathsf{s}(x)|_\#$$

So for a function f with several arguments we try to prove termination by comparing the i-th argument of the recursive calls with the i-th input argument (for a fixed i). For every recursive call in a defining equation $f(\ldots t_i \ldots) = \ldots f(\ldots r_i \ldots) \ldots$ we obtain a *termination formula*

$$(6.1) \quad b \rightarrow |r_i|_\# < |t_i|_\#$$

where b is the condition under which the recursive call $f(\ldots r_i \ldots)$ is evaluated. For instance, the recursive call of quot is evaluated under the condition $\neg \mathsf{lt}(x, \mathsf{s}(y))$. So, by regarding the first argument of quot, we obtain the termination formula

$$\neg \mathsf{lt}(x, \mathsf{s}(y)) \rightarrow |\mathsf{minus}(\mathsf{p}(x), y)|_\# < |x|_\#.$$

The approach of comparing data objects t by their size $|t|_\#$ is frequently used in termination proofs. For any data object of type s, we can define an abstract notion of *size* by counting the number of *reflexive*[4] constructors of type s, where substructures are ignored.

For example, let list be a data type with the constructors empty : list and add : nat $\times$ list $\rightarrow$ list, where $\mathsf{add}(n, l)$ represents the insertion of the number n in front of the list l. Then we have $|\mathsf{add}(\mathsf{s}(0), \mathsf{empty})|_\# = 1$, because this term (of type list) contains only one occurrence of a reflexive list-constructor,

[3] We only regard universally closed formulas of the form $\forall \ldots \varphi$ for verification, where φ is quantifier free and we omit the quantifiers to ease readability.

[4] A function symbol $f : s_1 \times \ldots \times s_n \rightarrow s$ is *reflexive* if its range type s is among its domain types s_i, and otherwise it is *irreflexive*. For instance, 0 is an irreflexive function symbol, whereas s, p, minus, and quot are reflexive.

whereas $|\mathsf{add}(0,\mathsf{add}(0,\mathsf{empty}))|_\# = 2$. Now we can define the *size ordering* $\prec_\#$ which compares data objects by their size, i.e. $t_1 \prec_\# t_2$ iff $|t_1|_\# < |t_2|_\#$. The *non-strict* size ordering $\preceq_\#$ is defined as $t_1 \preceq_\# t_2$ iff $|t_1|_\# \leq |t_2|_\#$. Hence, the termination formulas of minus and quot can be restated as

$$(6.2) \qquad x \prec_\# \mathsf{s}(x),$$

$$(6.3) \qquad \neg\mathsf{lt}(x,\mathsf{s}(y)) \rightarrow \mathsf{minus}(\mathsf{p}(x),y) \prec_\# x.$$

2.2. *Proving Inequalities by Estimation*

Termination formula (6.2) of minus holds by definition of the size ordering since $\mathsf{s}(x)$ contains one more occurrence of s than x for all ground instantiations of x. Note that for the termination proof of minus a pair of terms built only with *variables* and *constructors* has to be compared. But for an algorithm like quot which calls other *algorithms* (viz. minus and p), a termination formula like (6.3) is more difficult to verify. The reason is that *size* is defined in terms of the constructors 0 and s only. Hence, we cannot directly compute the size of the data object resulting from $\mathsf{minus}(\mathsf{p}(x),y)$ by evaluation of the algorithms minus and p.

A common verification technique is to use *estimations* for proving such inequalities. Assume that we know the estimations

$$\mathsf{minus}(x,y) \preceq_\# x$$
$$\mathsf{p}(x) \preceq_\# x$$

which state that minus and p are *1-bounded* operations, i.e. the results of $\mathsf{minus}(x,y)$ and $\mathsf{p}(x)$ are always size-smaller or have the same size as their first argument x.

In general, a function g is *p-bounded* iff its result is always size-smaller than or size-equal to its input x_p on argument position p, i.e. $g(x_1,\ldots,x_n) \preceq_\# x_p$ for all ground instantiations of $x_1,\ldots,x_n$, and it is *argument-bounded* iff it is p-bounded for some argument position p. From the 1-boundedness of minus and p we can conclude

$$(6.4) \qquad \mathsf{minus}(\mathsf{p}(x),y) \preceq_\# \mathsf{p}(x) \preceq_\# x.$$

Hence, by transitivity of $\preceq_\#$ the *non-strict* version of the termination formula (6.3) is verified. But for the termination proof of quot, $\mathsf{minus}(\mathsf{p}(x),y)$ has to be *strictly* smaller than x under the condition of quot's recursive call.

Assume also that we know *difference predicates* $\Delta^1_{\mathsf{minus}}$ and Δ^1_{p} which indicate whether $\mathsf{minus}(x,y)$ resp. $\mathsf{p}(x)$ are *strictly* smaller than x:

function $\Delta^1_{\mathsf{minus}} : \mathsf{nat} \times \mathsf{nat} \to \mathsf{bool}$

$$\begin{aligned} \Delta^1_{\mathsf{minus}}(x,0) &= \mathsf{false} \\ \Delta^1_{\mathsf{minus}}(0,\mathsf{s}(y)) &= \mathsf{false} \\ \Delta^1_{\mathsf{minus}}(\mathsf{s}(x),\mathsf{s}(y)) &= \mathsf{true} \end{aligned}$$

function $\Delta^1_{\mathsf{p}} : \mathsf{nat} \to \mathsf{bool}$

$$\begin{aligned} \Delta^1_{\mathsf{p}}(0) &= \mathsf{false} \\ \Delta^1_{\mathsf{p}}(\mathsf{s}(x)) &= \mathsf{true} \end{aligned}$$

Then the following *induction lemmata* hold for minus and p:

$$\begin{aligned} \Delta^1_{\mathsf{minus}}(x,y) &\leftrightarrow \mathsf{minus}(x,y) \prec_{\#} x, \\ \Delta^1_{\mathsf{p}}(x) &\leftrightarrow \mathsf{p}(x) \prec_{\#} x. \end{aligned}$$

To finish the termination proof of quot we have to establish that under the condition of quot's recursive call at least one of the two inequalities in (6.4) is strict, i.e. that $\mathsf{minus}(\mathsf{p}(x),y) \prec_{\#} x$ holds. Since $\Delta^1_{\mathsf{minus}}(\mathsf{p}(x),y)$ is equivalent to the strictness of the first inequality and $\Delta^1_{\mathsf{p}}(x)$ is equivalent to the strictness of the second inequality,

$$\Delta^1_{\mathsf{minus}}(\mathsf{p}(x),y) \vee \Delta^1_{\mathsf{p}}(x)$$

is an equivalent requirement for $\mathsf{minus}(\mathsf{p}(x),y) \prec_{\#} x$. Hence, termination of quot is proved by verifying the *termination hypothesis*

$$(6.5) \quad \neg \mathsf{lt}(x,\mathsf{s}(y)) \to \Delta^1_{\mathsf{minus}}(\mathsf{p}(x),y) \vee \Delta^1_{\mathsf{p}}(x),$$

i.e. by verifying $x \geq y+1 \;\to\; x-1 \neq 0 \wedge y \neq 0 \vee x \neq 0$.

In general, each p-bounded function $g : s_1 \times \ldots \times s_n \to s$ is associated with a *difference predicate* $\Delta^p_g : s_1 \times \ldots \times s_n \to \mathsf{bool}$ such that

$$(6.6) \quad \Delta^p_g(x_1,\ldots,x_n) \leftrightarrow g(x_1,\ldots,x_n) \prec_{\#} x_p$$

holds for all ground instantiations of $x_1,\ldots,x_n$. So if a p-bounded function g is applied to some input $t_1,\ldots,t_n$ and it returns something size-smaller than t_p, then the difference predicate Δ^p_g applied to $t_1,\ldots,t_n$ yields true and it returns false iff g returns something size-equal to t_p.

The notion of argument-bounded functions is the key concept for our formalization of estimation proofs. Suppose that for each $p \in \mathbb{N}$ we know a set Γ_p of p-bounded function symbols and let Γ be the family of all these sets. Then a *decidable* so-called *estimation relation* $\preceq_\Gamma$ on terms can be defined, such that for all terms t_1 and t_n we have

$$(6.7) \quad t_1 \preceq_\Gamma t_n \text{ implies } t_1 \preceq_{\#} t_n.$$

The relation $\preceq_\Gamma$ mirrors the proof technique of estimation and it is based on the knowledge about data types and the argument-bounded functions in Γ.

Suppose we know that minus and p are 1-bounded, i.e. $\{\mathsf{minus},\mathsf{p}\} \subseteq \Gamma_1$. Then the non-strict inequality $\mathsf{minus}(\mathsf{p}(x),y) \preceq_{\#} x$ can be verified directly by estimating minus and p, cf. (6.4). So the general idea to establish $t_1 \preceq_\Gamma t_n$ is to test whether t_n is a subterm of t_1, as e.g. x is a subterm of $\mathsf{minus}(\mathsf{p}(x),y)$, where only subterms in argument-bounded positions p are inspected.

The estimation relation $\preceq_\Gamma$ provides a deductive requirement for the *non-strict* semantic relation $\preceq_{\#}$, cf. (6.7). But as termination proofs are based on the relation $\prec_{\#}$, we need a deductive means for the *strict* size ordering $\prec_{\#}$.

Here, the general idea is to scan an estimation $t_1 \preceq_\Gamma t_2 \preceq_\Gamma \ldots \preceq_\Gamma t_{n-1} \preceq_\Gamma t_n$ step by step. For each estimation step $t_i \preceq_\Gamma t_{i+1}$, where $t_i = g_i(\ldots t_{i+1} \ldots)$ and t_{i+1} is the p_i-th argument of the p_i-bounded function g_i, the corresponding "call" of the difference predicate $\Delta_{g_i}^{p_i}(\ldots t_{i+1} \ldots)$ is collected. From all these "calls", one computes the disjunction

$$\Delta_{g_1}^{p_1}(\ldots t_2 \ldots) \vee \Delta_{g_2}^{p_2}(\ldots t_3 \ldots) \vee \ldots \vee \Delta_{g_{n-1}}^{p_{n-1}}(\ldots t_n \ldots).$$

This formula is called the *difference equivalent* $\Delta_\Gamma(t_1,t_n)$ of t_1 and t_n. Since each literal of $\Delta_\Gamma(t_1,t_n)$ is built with a difference predicate, we have generated an equivalent requirement for $t_1 \prec_{\#} t_n$, i.e.

$$t_1 \preceq_\Gamma t_n \ \text{ implies } \ \Delta_\Gamma(t_1,t_n) \leftrightarrow t_1 \prec_{\#} t_n.$$

Apart from the estimation rule for argument-bounded functions, we may also define additional estimation rules based on data types. For instance, the estimation step

$$t_p \preceq_\Gamma cons(\ldots t_p \ldots)$$

can be performed for each constructor *cons* with a reflexive argument position p, and the corresponding difference equivalent $\Delta_\Gamma(t_p, cons(\ldots))$ is true. So by this rule we can conclude inequalities like $t_2 \preceq_\Gamma \mathsf{add}(t_1,t_2)$ and $t \preceq_\Gamma \mathsf{s}(t)$ (cf. the termination formula (6.2) of minus). Moreover, for each data type, "$t \preceq_\Gamma t$" can be used as an estimation step where $\Delta_\Gamma(t,t)$ is defined as false.

Using the estimation relation and the difference equivalent, our procedure for automated termination proofs is straightforward. For each recursive call in a defining equation $f(\ldots t \ldots) = \ldots f(\ldots r \ldots) \ldots$ we have to prove a termination formula of the form

$$b \rightarrow r \prec_{\#} t,$$

where b is the condition under which $f(\ldots r \ldots)$ is evaluated, cf. (6.1). Now we first test whether $r \preceq_\Gamma t$ holds (otherwise the termination proof fails). If $r \preceq_\Gamma t$ can be established, then the *termination hypothesis*

$$b \rightarrow \Delta_\Gamma(r,t)$$

is generated. Subsequently, an *induction theorem proving system* (e.g. one of those described in (Boyer and Moore, 1979; Bundy et al., 1990; Walther, 1994a; Bouhoula and Rusinowitch, 1995; Kapur and Zhang, 1995; Hutter and Sengler, 1996)) is used to verify the generated termination hypotheses. This verification is usually quite simple, i.e. the proof often succeeds using case analysis and propositional reasoning only. If all termination hypotheses can be proved, then the termination formulas hold for each recursion and consequently, the algorithm under consideration terminates.

For example, to verify termination of quot, $\mathsf{minus}(\mathsf{p}(x),y) \preceq_\Gamma x$ is established first. Then the difference equivalent $\Delta_\Gamma(\mathsf{minus}(\mathsf{p}(x),y),x)$ is computed as $\Delta^1_{\mathsf{minus}}(\mathsf{p}(x),y) \vee \Delta^1_{\mathsf{p}}(x)$ and the termination hypothesis (6.5) is generated. Finally, an induction theorem prover is called with this termination hypothesis, which is easily verified by case analysis and symbolic evaluation.

2.3. *Argument-Bounded Functions*

The generation of termination hypotheses is based on the argument-bounded functions in Γ and on their difference predicates. Hence, for an automation of our approach, we have to determine argument-boundedness and to synthesize difference predicates[5] automatically.

To recognize argument-bounded functions we define a decidable *algorithm schema* such that every instance of this schema computes an argument-bounded function. The main idea embodied in this schema is to construct a meta-induction proof to verify that a function is argument-bounded. The defining equations of the difference predicate are generated in parallel to the proof steps of the meta-induction.

For example, to verify that minus is 1-bounded, the right-hand side of minus' first defining equation, viz. x, is compared with the corresponding first input argument x. Since $x \preceq_\Gamma x$ holds, minus is 1-bounded at least for the first case. The corresponding difference equivalent $\Delta_\Gamma(x,x)$ is false, which means that in this case the result is never size-smaller than the input argument x. Therefore, we obtain $\Delta^1_{\mathsf{minus}}(x,0) = \mathsf{false}$ as the first defining equation for the difference predicate $\Delta^1_{\mathsf{minus}}$.

For the second equation of minus the result 0 has to be compared with the corresponding input 0. We have $0 \preceq_\Gamma 0$ and $\Delta_\Gamma(0,0) = \mathsf{false}$. Hence, minus is also 1-bounded in this case and we obtain $\Delta^1_{\mathsf{minus}}(0,\mathsf{s}(y)) = \mathsf{false}$ as the second defining equation for the difference predicate. This completes the base case of the meta-induction.

[5] Strictly speaking, we synthesize *algorithms* which compute difference predicates. For the sake of brevity, we also refer to these algorithms as "difference predicates".

Finally the result $\mathsf{minus}(x,y)$ of the third (recursive) equation is compared with $\mathsf{s}(x)$. Because we assume $\mathsf{minus}(x,y) \precsim_\Gamma x$ as the induction hypothesis, it remains to verify $x \precsim_\Gamma \mathsf{s}(x)$. As this obviously holds (where the corresponding difference equivalent $\Delta_\Gamma(x,\mathsf{s}(x))$ is true), minus is 1-bounded for the recursive case, too. Now the difference equivalent $\Delta_\Gamma(\mathsf{minus}(x,y),\mathsf{s}(x))$ is computed as $\Delta^1_{\mathsf{minus}}(x,y) \vee \mathsf{true}$ which is simplified to true (here we may already use $\Delta^1_{\mathsf{minus}}(x,y)$ because we are in an inductive construction). Thus, $\mathsf{minus}(x,y)$ is size-smaller than $\mathsf{s}(x)$, and therefore, the last defining equation of $\Delta^1_{\mathsf{minus}}$ is $\Delta^1_{\mathsf{minus}}(\mathsf{s}(x),\mathsf{s}(y)) = \mathsf{true}$. This completes the step case of the meta-induction. In this way, minus is recognized as a 1-bounded function and the algorithm for $\Delta^1_{\mathsf{minus}}$ as given in Section 2.2 is synthesized automatically.

Note that the above meta-induction is based on the recursions of minus. Hence, the proof for the argument-boundedness of minus is only sound if minus terminates, because otherwise the induction relation used is not well founded. So before proving the termination of quot we must have proved termination of minus. Therefore, in this chapter we always demand that (apart from recursive calls) algorithms only call other algorithms whose termination has been verified *before*, which excludes mutually recursive algorithms (see (Giesl, 1997) for an extension of termination analysis to mutual recursion).

2.4. *Refinements and Summary*

We have presented a method to generate termination hypotheses, such that the truth of these hypotheses implies termination of the algorithm under consideration. The termination hypotheses represent the *idea* why an algorithm terminates. Hence, with our method the *creativity* for finding the right argument for termination has been mechanized for a certain class of algorithms.

Our approach is based on induction lemmata stating that certain operations are argument-bounded w.r.t. the size-ordering. Unlike the method of Boyer and Moore, our method is able to synthesize such induction lemmata automatically. The obtained lemmata are sound by construction, i.e. they do not have to be verified by a theorem prover.

The general procedure to compute the relevant knowledge for termination proofs works as follows. For each new algorithm f we first prove termination. Afterwards, for each argument position p it is tested whether f is p-bounded and if so, a difference predicate Δ^p_f is synthesized. In this way, the knowledge about argument-bounded algorithms is increased gradually by machine.

For functions with several arguments we proved termination by examining just one argument position. However, our approach is generalized in a straightforward way by taking *several* argument positions into account, cf. (Walther, 1994b). Numerous refinements of our method have been developed, e.g. techniques for the simplification of difference predicates which ease the proof of termination hypotheses considerably. Further details and refinements can be found in (Walther, 1988; Walther, 1991; Walther, 1994b).

Our method works for polymorphic types as well and an adaptation to non-free data types can be found in (Sengler, 1996; Sengler, 1997). Based on our schema for argument-bounded functions, *McAllester* and *Arkoudas* (1996) suggest a programming discipline such that only terminating programs can be defined. Termination analysis can also be extended to languages with lazy evaluation strategy (Panitz and Schmidt-Schauß, 1997) and to higher-order functions by inspecting the decrease of their first-order arguments, cf. (Nielson and Nielson, 1996).

3. Termination Proofs with Term Orderings

The termination proof method just presented is restricted to *one single fixed* measure function. However, there are many useful algorithms which require a measure function $|.|$ different from the *size* measure function $|.|_\#$. Therefore, we now extend our approach to *arbitrary* well-founded *term orderings*.

3.1. *Term Orderings and Functional Programs*

Consider the data type tree for binary trees with the constructors nil and cons. The nullary function nil represents leafs and $\mathsf{cons}(t_1, t_2)$ is the tree whose root has the direct subtrees t_1 and t_2. The algorithm flatten linearizes trees such that all left subtrees are leaves, cf. Figure 1.

function flatten : tree → tree

$$\begin{aligned}
\mathsf{flatten}(\mathsf{nil}) &= \mathsf{nil}\\
\mathsf{flatten}(\mathsf{cons}(\mathsf{nil}, y)) &= \mathsf{cons}(\mathsf{nil}, \mathsf{flatten}(y))\\
\mathsf{flatten}(\mathsf{cons}(\mathsf{cons}(u, v), w)) &= \mathsf{flatten}(\mathsf{cons}(u, \mathsf{cons}(v, w)))
\end{aligned}$$

In the third defining equation, the size of the argument is not decreasing in the recursive call, but it remains the same (where the size of a tree is the number of its inner nodes, i.e. the number of occurrences of cons). So the size ordering cannot be used for the termination proof of flatten and therefore the method of Section 2 fails. To prove the termination of flatten we have to find a well-founded ordering on terms (a so-called *term ordering*) which satisfies the termination formulas

$$(6.8)\qquad y \prec \mathsf{cons}(\mathsf{nil}, y),$$

$$(6.9)\qquad \mathsf{cons}(u, \mathsf{cons}(v, w)) \prec \mathsf{cons}(\mathsf{cons}(u, v), w).$$

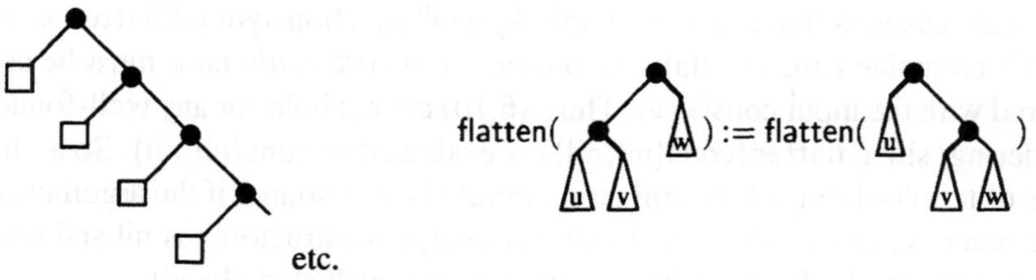

Figure 1. A tree in linear form and rotation of trees as performed by flatten.

Techniques for the automated generation of well-founded term orderings have been developed in the area of *term rewriting systems*. For instance, (6.8) and (6.9) are satisfied by a *polynomial ordering* (Lankford, 1979), where every n-ary function symbol is associated with an n-ary polynomial over the natural numbers. Then a ground term r is smaller than a ground term t with respect to the polynomial ordering iff the number $|r|$ corresponding to r is smaller than the number $|t|$ corresponding to t. As all ground terms are associated with natural numbers, polynomial orderings are well founded.

If nil is associated with the nullary polynomial 0 and $\mathsf{cons}(x,y)$ is associated with the polynomial $1+2x+y$ (i.e. $|\mathsf{cons}(x,y)| = 1+2|x|+|y|$), then $|\mathsf{cons}(\mathsf{nil},y)| = 1+|y|$. As $|y| < 1+|y|$ holds for all natural numbers $|y|$, this polynomial ordering satisfies the first requirement (6.8). Analogously, $|\mathsf{cons}(u,\mathsf{cons}(v,w))| = 2+2|u|+2|v|+|w|$ and $|\mathsf{cons}(\mathsf{cons}(u,v),w)| = 3+4|u|+2|v|+|w|$. As $2+2|u|+2|v|+|w| < 3+4|u|+2|v|+|w|$ is true for all naturals $|u|,|v|,|w|$, the polynomial ordering also satisfies (6.9) and the termination of flatten is proved.

However, in general this naive approach of using term orderings for termination proofs is not sound. For example, the algorithm

function f : tree → tree
f(nil) = nil
f(cons(u,v)) = f(flatten(cons(u,v)))

is obviously not terminating. Nevertheless there exists a well-founded term ordering $\prec$ satisfying the termination formula

(6.10) $\mathsf{flatten}(\mathsf{cons}(u,v)) \prec \mathsf{cons}(u,v).$

For instance, let $r \prec t$ hold iff the leading function symbol of r is flatten and the leading symbol of t is cons. So the existence of a well-founded term ordering such that terms in the recursive calls are smaller than the corresponding input terms is *not sufficient* for the termination of a functional program.

The reason is that a term with the *defined*[6] function symbol flatten is used in f's recursive call, viz. flatten(cons(u, v)), whose *evaluation* must be compared with the input cons(u, v). Thus, (6.10) cannot hold for any well-founded ordering, since flatten(cons(nil, nil)) is evaluated to cons(nil, nil). So a direct use of term orderings for termination proofs is only sound, if the arguments of the recursive calls contain no functions except constructors (as nil and cons), because terms built only with constructors are evaluated already.

A straightforward attempt to enable the use of term orderings for functional algorithms which call other algorithms is the restriction to term orderings which *respect the semantics* of the called algorithms. Under such orderings, *different terms* which denote the *same data object* (like flatten(cons(nil, nil)) and cons(nil, nil)) are equivalent.

But in general this restriction is too strong. Consider the following purge-and-sort algorithm sort (i.e. sort also eliminates duplicates).

function sort : list $\rightarrow$ list
sort(x) $=$ if(emptyp(x), empty, add(min(x),sort(rm(min(x),x))))

This algorithm calls three other algorithms emptyp, min, and rm, where emptyp(x) checks if x is empty, min(x) computes the minimum of a non-empty list x, and rm(n,x) removes *all* occurrences of n from x.

However, none of the orderings typically used in the area of term rewriting systems both respects the semantics of min and rm and makes inputs greater than the corresponding recursive calls of sort. (In fact, most term orderings that are amenable to automation[7] are *simplification orderings* (Dershowitz, 1987; Steinbach, 1995b), i.e. orderings which possess the *subterm property* ($t \prec f(\ldots t \ldots)$). Such orderings do not respect the semantics of the algorithm min, because min(add(0, empty)) evaluates to 0, but min(add(0, empty)) can never be equivalent to its subterm 0 w.r.t. a simplification ordering.)

Another straightforward attempt for the termination proof of the algorithm sort is to transform sort and the auxiliary algorithms min and rm into a *term rewriting system* and to prove its termination instead. Functional programs in our language can be regarded as a special kind of (conditional) term rewriting systems with *innermost evaluation* strategy. But due to their special form and this evaluation strategy we may use a different approach for termination proofs of functional programs than it is necessary for term rewriting systems.

[6] Functions whose semantics are determined by algorithms are called *defined*.

[7] There also exist techniques which can orient *every* terminating term rewriting system (e.g. semantical path orderings (Kamin and Levy, 1980), transformation orderings (Bellegarde and Lescanne, 1990), or semantic labelling (Zantema, 1995)). But the disadvantage of these powerful approaches is that up to now there are only few suggestions for their automated generation (Steinbach, 1995a).

For instance, for functional programs it is sufficient to compare the input ***arguments*** with the ***arguments*** in the recursive calls, while for term rewriting systems left- and right-hand sides of ***all rules*** have to be compared (and moreover, the ordering has to be ***monotonic***), cf. (Dershowitz, 1987).

Therefore by transforming functional programs into term rewriting systems we impose unnecessarily strong requirements for the termination proof. For instance, no simplification ordering can prove termination of the term rewriting system corresponding to sort, min, and rm, since it contains a rule

$$\mathsf{sort}(\mathsf{add}(n,y)) \to \mathsf{add}(\ldots,\mathsf{sort}(\mathsf{rm}(\ldots,\mathsf{add}(n,y))))$$

whose left-hand side is embedded in its right-hand side (and the same holds for the system resulting from quot, minus, and p from Section 2).

3.2. *Elimination of Defined Function Symbols*

In the preceding section we showed that term orderings can be used for termination proofs of algorithms like flatten, where the size ordering (and therefore the approach of Section 2) fails. But on the other hand, the estimation technique of Section 2 proves termination of algorithms like sort which call other algorithms in the arguments of their recursive calls, whereas for such algorithms the direct use of term orderings is unsound. To benefit from the advantages of both approaches, we need a method which combines the handling of such algorithms with the flexibility of arbitrary term orderings (i.e. we need a proper extension of our approach in Section 2). Since the straightforward solutions impose too strong requirements such that termination proofs *often fail*, we now develop a method which overcomes these problems. To prove the termination of sort we have to show the existence of a well-founded ordering $\prec$ (on the *evaluated* terms) satisfying the termination formula

$$(6.11)\quad \neg\mathsf{emptyp}(x) \;\to\; \mathsf{rm}(\mathsf{min}(x),x) \prec x,$$

where $r \prec t$ abbreviates "evaluation of r" $\prec$ "evaluation of t".

As demonstrated, the application of methods for the synthesis of well-founded term orderings is only sound if the termination formulas do not contain defined function symbols (like rm and min). Therefore we develop a calculus to transform termination formulas like (6.11) into formulas *without defined function symbols.*

This calculus transforms a set of termination formulas TF_0 of an algorithm into sets TF_1, TF_2, etc. until we obtain a set of formulas TF_n containing no defined function symbols any more, cf. Figure 2. This transformation is an *abduction* process, i.e. $\mathrm{TF}_{i+1} \models \mathrm{TF}_i$ holds for all i. Hence, if a relation $\prec$ satisfies TF_{i+1}, then it also satisfies TF_i.

$$\begin{array}{ccccccccc} TF_0 & \mapsto & TF_1 & \mapsto & TF_2 & \mapsto & \ldots & \mapsto & TF_n \\ \Perp & & \Perp & & \Perp & & & & \Perp \\ \prec & \curvearrowleft & \prec & \curvearrowleft & \prec & \curvearrowleft & \ldots & \curvearrowleft & \prec \end{array}$$

Figure 2. Elimination of defined function symbols from termination formulas.

The formulas in TF_n resulting from the transformation process contain no defined function symbols. Therefore the naive approach of using a term ordering $\prec$ satisfying TF_n is sound. As TF_n implies TF_0, this ordering $\prec$ also satisfies the original termination formulas in TF_0 and therefore the existence of a term ordering satisfying the requirements in TF_n is sufficient for the termination of the algorithm.

The derivation tree in Figure 3 illustrates the transformation of sort's termination formula (6.11). Every node in the tree is transformed into its successors by application of one transformation rule. Leaves of the tree are formulas that do not contain defined function symbols (and therefore no transformation rule is applicable to them). As each transformation rule is an abduction step, every node is implied by the formulas of all its successors. Hence, the formulas at the leaves imply the termination formula at the root of the tree. Thus, the existence of a well-founded term ordering satisfying the requirements at the leaves of the tree in Figure 3 is sufficient for termination of sort. In the following our termination proof method is presented in three steps.

3.2.1. *Estimation and Generalization*

The termination formula (6.11) of sort contains the defined function symbols rm and min. The central idea of our procedure is an *estimation* of defined function symbols by new *free* function symbols (i.e. function symbols like constructors which are not defined by algorithms). Therefore rm is replaced

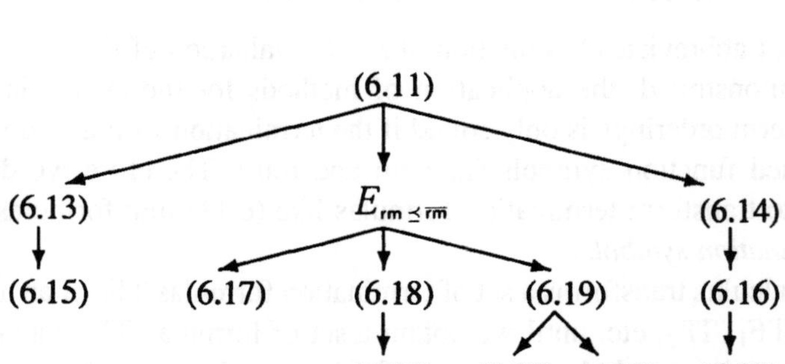

Figure 3. Termination proof of sort.

by a new free symbol $\overline{\mathsf{rm}}$ and we demand that the result of $\overline{\mathsf{rm}}$ is always greater or equal than the result of rm, i.e.

(6.12) $\mathsf{rm}(n,x) \preceq \overline{\mathsf{rm}}(n,x)$.

Unlike rm, the free function symbol $\overline{\mathsf{rm}}$ has *no fixed semantics* (i.e. there is no algorithm for $\overline{\mathsf{rm}}$) and our aim is to transform the termination formula (6.11) into inequalities which contain free symbols like $\overline{\mathsf{rm}}$, but no defined symbols like rm. If these resulting inequalities are satisfied by a well-founded term ordering, then the termination of sort is proved.

This generalizes the approach of Section 2 where argument-bounded functions were *estimated* by their *arguments* according to the *size ordering*. For example, (6.12) holds for $\overline{\mathsf{rm}}(n,x) := x$ and $\preceq := \preceq_{\#}$, since rm is 2-bounded. However, now we perform estimations without fixing the *upper bounds* and the *ordering used*. Instead we generate constraints and *postpone* the search for suitable upper bounds and orderings satisfying these constraints to the end of the termination proof.

Assume that we know a set of so-called *estimation inequalities* $E_{\mathsf{rm} \preceq \overline{\mathsf{rm}}}$ (without defined function symbols) which imply (6.12). Then demanding

(6.13) $\overline{\mathsf{rm}}(\mathsf{min}(x),x) \preceq x$

and $E_{\mathsf{rm} \preceq \overline{\mathsf{rm}}}$ implies the *non-strict* version of termination formula (6.11):

$$\mathsf{rm}(\mathsf{min}(x),x) \preceq \overline{\mathsf{rm}}(\mathsf{min}(x),x) \preceq x.$$

But to ensure the *strict* inequality (6.11), one of the inequalities (6.13) and $\mathsf{rm}(\mathsf{min}(x),x) \prec \overline{\mathsf{rm}}(\mathsf{min}(x),x)$ must be strict, whenever $\neg\mathsf{emptyp}(x)$ holds.

Assume also that we know a *strictness predicate* $\delta_{\mathsf{rm} \prec \overline{\mathsf{rm}}}$ which is true iff the result of rm is strictly smaller than the result of $\overline{\mathsf{rm}}$. Then $\delta_{\mathsf{rm} \prec \overline{\mathsf{rm}}}(\mathsf{min}(x),x)$ indicates whether the estimation of $\mathsf{rm}(\mathsf{min}(x),x)$ by $\overline{\mathsf{rm}}(\mathsf{min}(x),x)$ is strict. Therefore we can replace the termination formula (6.11) of sort by (6.13), $E_{\mathsf{rm} \preceq \overline{\mathsf{rm}}}$, and the formula

(6.14) $\neg\mathsf{emptyp}(x) \rightarrow \delta_{\mathsf{rm} \prec \overline{\mathsf{rm}}}(\mathsf{min}(x),x) \vee \overline{\mathsf{rm}}(\mathsf{min}(x),x) \prec x$.

The transformation of (6.11) into (6.13), $E_{\mathsf{rm} \preceq \overline{\mathsf{rm}}}$, and (6.14) corresponds to the step from the root of the derivation tree in Figure 3 to its direct successors. This transformation is an abduction step, because (6.13), $E_{\mathsf{rm} \preceq \overline{\mathsf{rm}}}$, and (6.14) are sufficient for the termination formula (6.11), and it is obtained by the *estimation rule* in Figure 4.

This rule[8] embodies a main principle of our transformation. Similar to Section 2, for a termination formula $b \rightarrow r \prec t$ we ensure that the *non-strict*

[8] We also use an estimation rule for *non-strict* inequalities $g(\ldots) \preceq t$ which only contains the first two consequences of the rule in Figure 4.

$$\frac{b \rightarrow g(r_1,\ldots,r_n) \prec t}{\begin{array}{c}\bar{g}(r_1,\ldots,r_n) \preceq t \\ E_{g \preceq \bar{g}} \\ b \rightarrow \delta_{g \prec \bar{g}}(r_1,\ldots,r_n) \vee \bar{g}(r_1,\ldots,r_n) \prec t\end{array}}$$

Figure 4. Estimation rule to estimate a defined function symbol g.

unconditional inequality $r \preceq t$ holds. The advantage of omitting the condition b (which contains semantical information) is that defined symbols in $r \preceq t$ can be eliminated by repeated application of *purely syntactical* inference rules (like estimation). A *semantical* rule which considers the condition b is only necessary to guarantee that one of the estimation steps in the derivation is strict, cf. Section 3.2.3.

The resulting formulas (6.13) and (6.14) still contain the defined symbol min. To eliminate min we can again use estimation and replace it by a new free symbol $\overline{\text{min}}$. But apart from estimation there is another (obvious) method to eliminate defined symbols like min, viz. replacing terms like $\text{min}(x)$ in inequalities[9] by fresh variables. In this way, (6.13) and (6.14) are generalized to

(6.15) $\qquad \overline{\text{rm}}(z,x) \preceq x,$

(6.16) $\qquad \neg\text{emptyp}(x) \rightarrow \delta_{\text{rm} \prec \overline{\text{rm}}}(\text{min}(x),x) \vee \overline{\text{rm}}(z,x) \prec x.$

Each *generalization* is also an abduction, because if (6.15) resp. (6.16) hold for all z, then (6.13) resp. (6.14) hold as well. As (6.15) contains no defined symbols, this inequality corresponds to a leaf in the derivation tree.

3.2.2. *Estimation Inequalities and Strictness Predicates*

Our estimation technique is based on *estimation inequalities* like $E_{\text{rm} \preceq \overline{\text{rm}}}$ and on *strictness predicates* like $\delta_{\text{rm} \prec \overline{\text{rm}}}$ and in this section we show how they are computed. The estimation inequalities $E_{\text{rm} \preceq \overline{\text{rm}}}$ guarantee that $\overline{\text{rm}}$ is an upper bound for rm, where rm is defined as

function rm : nat × list → list
$\quad \text{rm}(n,\text{empty}) \quad = \text{empty}$
$\quad \text{rm}(n,\text{add}(m,y)) = \text{if}(\text{eq}(n,m),\ \text{rm}(n,y),\ \text{add}(m,\text{rm}(n,y)))$.

Here, eq computes the equality predicate on naturals.

---

[9] To perform estimation and generalization repeatedly, these techniques may also be applied to *inequalities* $g(\ldots) \prec t$ occurring in disjunctions of the form $b \rightarrow \delta_1(\ldots) \vee \ldots \vee \delta_n(\ldots) \vee g(\ldots) \prec t$. A method to eliminate the defined symbols from the terms $\delta(\ldots)$ is presented in Section 3.2.3.

To construct $E_{\mathsf{rm} \preceq \overline{\mathsf{rm}}}$ we consider each *result term* of rm separately. Instead of $\mathsf{rm}(n,x) \preceq \overline{\mathsf{rm}}(n,x)$ we therefore demand (again omitting *conditions* like $\mathsf{eq}(n,m)$ in *non-strict* inequalities)

$$\begin{array}{lrcl} (6.17) & \mathsf{empty} & \preceq & \overline{\mathsf{rm}}(n,\mathsf{empty}), \\ (6.18) & \mathsf{rm}(n,y) & \preceq & \overline{\mathsf{rm}}(n,\mathsf{add}(m,y)), \\ (6.19) & \mathsf{add}(m,\mathsf{rm}(n,y)) & \preceq & \overline{\mathsf{rm}}(n,\mathsf{add}(m,y)). \end{array}$$

We cannot define $E_{\mathsf{rm} \preceq \overline{\mathsf{rm}}} = \{(6.17),(6.18),(6.19)\}$ as (6.18) and (6.19) still contain the defined symbol rm. Defined function symbols occurring in such inequalities have to be eliminated by *estimation* or *generalization* again.

But the problem here is that rm *itself* appears in the inequalities (6.18) and (6.19). Therefore we cannot use the (non-strict version of the) estimation rule in Figure 4 for the estimation of rm, because we do not know the estimation inequalities $E_{\mathsf{rm} \preceq \overline{\mathsf{rm}}}$ yet.

The construction of *estimation inequalities* corresponds to the problem of proving *argument-boundedness* in Section 2. Therefore, we solve it in a similar way by computing $E_{\mathsf{rm} \preceq \overline{\mathsf{rm}}}$ *inductively*. The base case of the inductive construction corresponds to rm's non-recursive equation. Inequality (6.17) ensures that in this case $\overline{\mathsf{rm}}$ is an upper bound for rm.

In the step case we have to guarantee that (6.18) and (6.19) hold, i.e. for inputs of the form $(n,\mathsf{add}(m,y))$ the result of rm must be smaller or equal than the result of $\overline{\mathsf{rm}}$. As *induction hypothesis* we assume that this estimation is already correct for the arguments (n,y), i.e. $\mathsf{rm}(n,y) \preceq \overline{\mathsf{rm}}(n,y)$. Therefore

$$(6.20) \quad \overline{\mathsf{rm}}(n,y) \preceq \overline{\mathsf{rm}}(n,\mathsf{add}(m,y))$$

is sufficient for (6.18) and we can replace (6.18) by inequality (6.20) which does not contain defined function symbols.

So to eliminate the defined symbol rm from (6.18) we use the (non-strict) estimation rule (Figure 4) and due to an inductive argument we can omit the second consequence, viz. $E_{\mathsf{rm} \preceq \overline{\mathsf{rm}}}$, of this inference rule.

For the transformation of (6.19) we again use the induction hypothesis $\mathsf{rm}(n,y) \preceq \overline{\mathsf{rm}}(n,y)$. However, to eliminate the defined symbol rm from (6.19) by estimation, we would need

$$(6.21) \quad \mathsf{add}(m,\mathsf{rm}(n,y)) \preceq \mathsf{add}(m,\overline{\mathsf{rm}}(n,y)).$$

To imply (6.21), in addition to the induction hypothesis we have to demand that the result of add should also be decreasing if the second argument of add is decreasing. In other words, add should be *monotonic* in its second argument, i.e. we demand the requirement

$$(6.22) \quad u \prec v \rightarrow \mathsf{add}(m,u) \prec \mathsf{add}(m,v).$$

Now (by the induction hypothesis), (6.19) is implied by (6.22) and

(6.23) $\mathsf{add}(m, \overline{\mathsf{rm}}(n,y)) \preceq \overline{\mathsf{rm}}(n, \mathsf{add}(m,y))$.

The monotonicity problem always appears except when estimating the leading function symbol. To estimate a function within a term at a position π, the rule in Figure 4 is extended by a consequence which demands that the term is monotonic in the position π. Subsequently any defined symbols in monotonicity formulas like (6.22) are eliminated by generalization.

Note that the monotonicity problem results from the consideration of *arbitrary* term orderings. If we restrict ourselves to the size ordering of Section 2, i.e. $\preceq := \preceq_{\#}$, then monotonicity requirements like (6.22) always hold for reflexive arguments of constructors.

Now we have finished our inductive construction of $E_{\mathsf{rm} \preceq \overline{\mathsf{rm}}}$ (as illustrated in Figure 3) and obtain

(6.17)	$E_{\mathsf{rm} \preceq \overline{\mathsf{rm}}}$	$=$	$\{\mathsf{empty} \preceq \overline{\mathsf{rm}}(n, \mathsf{empty}),$
(6.20)			$\overline{\mathsf{rm}}(n,y) \preceq \overline{\mathsf{rm}}(n, \mathsf{add}(m,y)),$
(6.22)			$u \prec v \rightarrow \mathsf{add}(m,u) \prec \mathsf{add}(m,v),$
(6.23)			$\mathsf{add}(m, \overline{\mathsf{rm}}(n,y)) \preceq \overline{\mathsf{rm}}(n, \mathsf{add}(m,y))\}.$

So in general *estimation inequalities* $E_{g \preceq \bar{g}}$ are computed as follows.

1. For each result r of the algorithm g we build the formula $r \preceq \bar{g}(\ldots)$.
2. Then the defined symbols in $r \preceq \bar{g}(\ldots)$ are eliminated by estimation or generalization. When g *itself* is estimated, the second consequence ($E_{g \preceq \bar{g}}$) of the (non-strict) estimation rule is omitted.

The algorithm for the *strictness predicate* $\delta_{\mathsf{rm} \prec \overline{\mathsf{rm}}}$ is also constructed by induction (corresponding to the construction of *difference predicates* in Section 2). The predicate $\delta_{\mathsf{rm} \prec \overline{\mathsf{rm}}}(n,x)$ has to return true iff $\mathsf{rm}(n,x)$ is strictly smaller than $\overline{\mathsf{rm}}(n,x)$. By analysis according to the result terms of rm we obtain the following defining equations for $\delta_{\mathsf{rm} \prec \overline{\mathsf{rm}}}$:

$$\delta_{\mathsf{rm} \prec \overline{\mathsf{rm}}}(n, \mathsf{empty}) = \mathsf{empty} \prec \overline{\mathsf{rm}}(n, \mathsf{empty})$$
$$\delta_{\mathsf{rm} \prec \overline{\mathsf{rm}}}(n, \mathsf{add}(m,y)) = \mathsf{if}(\mathsf{eq}(n,m),$$
$$\text{(6.24)} \qquad \mathsf{rm}(n,y) \prec \overline{\mathsf{rm}}(n, \mathsf{add}(m,y)),$$
$$\text{(6.25)} \qquad \mathsf{add}(m, \mathsf{rm}(n,y)) \prec \overline{\mathsf{rm}}(n, \mathsf{add}(m,y)))$$

However, the inequalities (6.24) and (6.25) still contain the defined function symbol rm. Therefore we eliminate this symbol by estimation again. So inequality (6.24) is transformed into (6.20), $E_{\mathsf{rm} \preceq \overline{\mathsf{rm}}}$, and

(6.26) $\delta_{\mathsf{rm} \prec \overline{\mathsf{rm}}}(n,y) \vee \overline{\mathsf{rm}}(n,y) \prec \overline{\mathsf{rm}}(n, \mathsf{add}(m,y))$.

We construct $\delta_{\mathsf{rm} \prec \overline{\mathsf{rm}}}$ *inductively*, i.e. when defining $\delta_{\mathsf{rm} \prec \overline{\mathsf{rm}}}(n, \mathsf{add}(m,y))$ we use "$\delta_{\mathsf{rm} \prec \overline{\mathsf{rm}}}(n,y) = \mathsf{rm}(n,y) \prec \overline{\mathsf{rm}}(n,y)$" as an induction hypothesis. This results in the recursive call $\delta_{\mathsf{rm} \prec \overline{\mathsf{rm}}}(n,y)$ of (6.26).

Note that (6.20) is already included in $E_{\mathsf{rm} \preceq \overline{\mathsf{rm}}}$. Hence, as long as $E_{\mathsf{rm} \preceq \overline{\mathsf{rm}}}$ holds we only have to consider the third consequence of the estimation rule in Figure 4 and replace (6.24) by (6.26). We proceed in the same way for inequality (6.25) and obtain the following algorithm for $\delta_{\mathsf{rm} \prec \overline{\mathsf{rm}}}$.

function $\delta_{\mathsf{rm} \prec \overline{\mathsf{rm}}} : \mathsf{nat} \times \mathsf{list} \to \mathsf{bool}$

$$
\begin{aligned}
\delta_{\mathsf{rm} \prec \overline{\mathsf{rm}}}(n, \mathsf{empty}) &= \mathsf{empty} \prec \overline{\mathsf{rm}}(n, \mathsf{empty}) \\
\delta_{\mathsf{rm} \prec \overline{\mathsf{rm}}}(n, \mathsf{add}(m,y)) &= \mathsf{if}(\mathsf{eq}(n,m), \\
&\qquad \delta_{\mathsf{rm} \prec \overline{\mathsf{rm}}}(n,y) \vee \overline{\mathsf{rm}}(n,y) \prec \overline{\mathsf{rm}}(n, \mathsf{add}(m,y)), \\
&\qquad \delta_{\mathsf{rm} \prec \overline{\mathsf{rm}}}(n,y) \vee \mathsf{add}(m, \overline{\mathsf{rm}}(n,y)) \prec \overline{\mathsf{rm}}(n, \mathsf{add}(m,y)))
\end{aligned}
$$

So *algorithms* for *strictness predicates* $\delta_{g \prec \bar{g}}$ are constructed as follows.

1. Each result r of the algorithm g is replaced by $r \prec \bar{g}(\ldots)$.
2. Then the defined function symbols in $r \prec \bar{g}(\ldots)$ are eliminated by estimation or generalization. When using estimation all consequences of the estimation rule except the third one are omitted.

3.2.3. *Elimination of Strictness Predicates*

By estimation and generalization of defined function symbols each termination formula $b \to r \prec t$ is transformed into a (possibly empty) set of non-strict inequalities and a formula of the form

$$(6.27) \quad b \to \delta_1(\ldots) \vee \ldots \vee \delta_n(\ldots) \vee r' \prec t.$$

For example, the termination formula (6.11) of sort has been transformed into inequality (6.15), the estimation inequalities $E_{\mathsf{rm} \preceq \overline{\mathsf{rm}}}$, and the formula

$$(6.16) \quad \neg \mathsf{emptyp}(x) \to \delta_{\mathsf{rm} \prec \overline{\mathsf{rm}}}(\mathsf{min}(x), x) \vee \overline{\mathsf{rm}}(z,x) \prec x.$$

The formula (6.27) corresponds to the notion of *termination hypotheses* from Section 2 and the unconditional part of (6.27), viz. $\delta_1(\ldots) \vee \ldots \vee \delta_n(\ldots) \vee r' \prec t$, is the *difference equivalent* $\Delta_{\prec}(r,t)$ of r and t. Hence, this extends the notion of difference equivalents to arbitrary term orderings $\prec$. Similar to Section 2 we have

$$r \preceq t \text{ implies } \Delta_{\prec}(r,t) \leftrightarrow r \prec t.$$

While (6.15) and $E_{\mathsf{rm} \preceq \overline{\mathsf{rm}}}$ contain no defined symbols, formula (6.16) contains the strictness predicate $\delta_{\mathsf{rm} \prec \overline{\mathsf{rm}}}$ which is *defined* by an algorithm. To complete the elimination of defined symbols we now have to eliminate the strictness predicate $\delta_{\mathsf{rm} \prec \overline{\mathsf{rm}}}$ from requirement (6.16).

For that purpose we choose some of the *inequalities occurring in* (6.16) *and in the algorithm* $\delta_{\mathsf{rm} \prec \overline{\mathsf{rm}}}$, i.e. inequalities from

$$\begin{array}{lrcl} (6.28) & \overline{\mathsf{rm}}(z,x) & \prec & x, \\ (6.29) & \mathsf{empty} & \prec & \overline{\mathsf{rm}}(n,\mathsf{empty}), \\ (6.30) & \overline{\mathsf{rm}}(n,y) & \prec & \overline{\mathsf{rm}}(n,\mathsf{add}(m,y)), \\ (6.31) & \mathsf{add}(m,\overline{\mathsf{rm}}(n,y)) & \prec & \overline{\mathsf{rm}}(n,\mathsf{add}(m,y)). \end{array}$$

Our method selects a subset of these inequalities which is sufficient for (6.16), where *small* subsets are preferable in order to minimize the number of resulting constraints. Of course, (6.28) is sufficient for (6.16). However, if (6.16) is transformed into (6.28) then the termination proof of sort fails. The reason is that there exists no well-founded ordering satisfying both $E_{\mathsf{rm} \preceq \overline{\mathsf{rm}}}$ and (6.28).

But (6.16) is also implied by inequality (6.30) from the second defining equation of $\delta_{\mathsf{rm} \prec \overline{\mathsf{rm}}}$. Essentially, the reason is that every non-empty list x contains its minimum. Hence when evaluating $\delta_{\mathsf{rm} \prec \overline{\mathsf{rm}}}(\mathsf{min}(x),x)$, after a finite number of recursive calls $\delta_{\mathsf{rm} \prec \overline{\mathsf{rm}}}$ is called with a list which begins with its minimum. Then the if-condition of the second defining equation is satisfied and therefore $\delta_{\mathsf{rm} \prec \overline{\mathsf{rm}}}(\mathsf{min}(x),x)$ returns true if (6.30) is true.

Different to the syntactical inference rules in Section 3.2.1 and 3.2.2 we now had to consider the *semantics* of the strictness predicate $\delta_{\mathsf{rm} \prec \overline{\mathsf{rm}}}$. To eliminate strictness predicates we have to *prove* that certain inequalities (like (6.30)) are sufficient for formulas like (6.16). To perform such proofs automatically an induction theorem proving system is used (similar to the verification of termination hypotheses in Section 2).

So to *eliminate defined strictness predicates* from a formula of the form $b \to \delta_1(\ldots) \vee \ldots \vee \delta_n(\ldots) \vee r' \prec t$ our method proceeds as follows.

1. Let Φ be the set containing $r' \prec t$ and all inequalities from the algorithms $\delta_1, \ldots, \delta_n$ and their auxiliary algorithms.
2. Then the formula is replaced by a minimal subset of Φ that is sufficient for $b \to \delta_1(\ldots) \vee \ldots \vee \delta_n(\ldots) \vee r' \prec t$.

By replacing (6.16) with (6.30) we have finished the transformation of sort's termination formula into inequalities without defined symbols, i.e. we have constructed the derivation tree in Figure 3. To prove sort's termination one now has to find a well-founded term ordering satisfying the constraints (6.15), (6.17), (6.20), (6.22), (6.23), (6.30) at the leaves of the tree:

(6.15) $\overline{\mathsf{rm}}(z,x) \preceq x$,

(6.17) $\mathsf{empty} \preceq \overline{\mathsf{rm}}(n,\mathsf{empty})$,

(6.20) $\overline{\mathsf{rm}}(n,y) \preceq \overline{\mathsf{rm}}(n,\mathsf{add}(m,y))$,

(6.22) $u \prec v \rightarrow \mathsf{add}(m,u) \prec \mathsf{add}(m,v)$,

(6.23) $\mathsf{add}(m,\overline{\mathsf{rm}}(n,y)) \preceq \overline{\mathsf{rm}}(n,\mathsf{add}(m,y))$,

(6.30) $\overline{\mathsf{rm}}(n,y) \prec \overline{\mathsf{rm}}(n,\mathsf{add}(m,y))$.

For instance, these constraints are satisfied by the polynomial ordering associating empty with 0, $\mathsf{add}(n,x)$ with $x+1$, and $\overline{\mathsf{rm}}(n,x)$ with x. Therefore the termination of sort is proved.

Note that this polynomial ordering is just the *size* ordering used in Section 2. In fact, sort's termination can also be proved with the approach of Section 2. However, the technique just presented is a proper extension of Section 2, i.e. every termination proof with the method of Section 2 can also be performed with the method from this section, but not vice versa. The reason is that the extended method also proves termination for algorithms like flatten where orderings different from *size* are needed. While flatten did not use auxiliary algorithms, our method also handles examples like the following reachability algorithm on directed graphs, which calls another algorithm union in its recursive call and which also requires an ordering different from size.

function rch : nat × nat × graph × graph → bool
$\mathsf{rch}(x,y,\varepsilon,h) = \mathsf{false}$
$\mathsf{rch}(x,y,\mathsf{edge}(u,v,i),h) = \mathsf{if}(\mathsf{eq}(x,u), \mathsf{if}(\mathsf{eq}(y,v), \mathsf{true}, \mathsf{rch}(x,y,i,h) \vee \mathsf{rch}(v,y,\mathsf{union}(i,h),\varepsilon)), \mathsf{rch}(x,y,i,\mathsf{edge}(u,v,h)))$

Here, ε (the empty graph) and $\mathsf{edge}(x,y,g)$ (the graph g extended by an edge from x to y) are the constructors of type graph and $\mathsf{rch}(x,y,g,\varepsilon)$ returns true iff there is a path from node x to node y in the directed graph g. Termination of rch can be proved with our method where the resulting constraints are satisfied by a polynomial ordering of degree 2, whereas there is no termination proof using the size ordering. A collection of numerous such algorithms can be found in (Giesl, 1995d).

For the computation of well-founded term orderings which satisfy the derived constraints we use procedures developed in the area of term rewriting systems and computer algebra. For instance, the algorithm of *G. E. Collins* (1975) decides whether there exists a polynomial ordering over the reals of a given degree satisfying a set of constraints. A procedure to generate polynomial orderings using an efficient, incomplete modification of Collins' algorithm is given in (Giesl, 1995a).

3.3. *Refinements and Summary*

We have presented a method for automated termination proofs of functional programs which extends the approach of Section 2 by allowing *arbitrary* term orderings. When using term orderings for termination proofs of *functional programs*, *defined* function symbols in the recursive calls have to be eliminated. This elimination proceeds in three steps. First, defined function symbols g in the termination formulas are *generalized* or *estimated* by new free function symbols $\bar{g}$ (Section 3.2.1). To guarantee that $\bar{g}$ is an upper bound for g we demand *estimation inequalities* $E_{g \preceq \bar{g}}$. By the estimation of a function symbol g we also obtain a formula containing the *strictness predicate* $\delta_{g \prec \bar{g}}$. This predicate indicates whether the result of $\bar{g}$ is *strictly* greater than the result of g (Section 3.2.2). Finally the (defined) strictness predicate has to be eliminated (using an induction theorem prover) (Section 3.2.3).

Our method is easily extended to algorithms with *several* arguments. For that purpose we introduce a new free *tuple* function symbol v and instead of two *tuples* $(r_1 \ldots r_n)$ and $(t_1 \ldots t_n)$ we compare the *terms* $\mathsf{v}(r_1 \ldots r_n)$ and $\mathsf{v}(t_1 \ldots t_n)$. So for instance, we demand $\mathsf{v}(x,y) \prec \mathsf{v}(\mathsf{s}(x),\mathsf{s}(y))$ to prove termination of the algorithm minus.

The automatic transformation of termination formulas into formulas without defined function symbols is performed in finitely many steps as each transformation rule decreases the number of defined symbols. Therefore derivation trees only contain paths of *finite* length. But our procedure contains two *choice points*. First, defined function symbols can be eliminated by estimation *or* by generalization. Second, for disjunctions like (6.16) there may be more than one subset of inequalities sufficient for the elimination of strictness predicates. So there can be several different derivation trees for one termination formula. But as the number of derivation trees for one termination formula is also *finite* (and *small*), one can backtrack if no well-founded term ordering satisfying the constraints at the leaves can be found. To improve the efficiency of the method, we have also developed heuristics for choosing the "right" derivation tree which have proved successful in practice. (We tested our method on a large data base of examples and for the vast majority, termination could be proved within a few seconds.) For a more detailed description of our approach and further refinements see (Giesl, 1995b; Giesl, 1995c; Giesl, 1995d; Giesl, 1997).

Our method has also been adapted for termination proofs in other areas. In particular, a method for termination proofs of *term rewriting systems* based on the comparison of *arguments* (instead of rules) and on our estimation technique can be found in (Arts and Giesl, 1996; Arts and Giesl, 1997a; Arts and Giesl, 1997b; Arts and Giesl, 1998).

4. Termination Analysis for Partial Functions

A functional program P (without mutual recursion) can be written as a sequence $\langle f_1,\ldots,f_k\rangle$ of algorithms, such that no algorithm f_i is called by an algorithm f_j preceding f_i in the sequence. A *functional program P terminates* iff its last algorithm, viz. the "main" algorithm f_k, terminates for *all* inputs (*total termination*). If all "auxiliary" algorithms f_i with $i<k$ also terminate totally, then termination of P is verified by proving total termination for all functional algorithms $f_1,\ldots,f_k$ step by step, e.g. by using one of the methods developed so far. However, demanding total termination for all "auxiliary" algorithms is much too strong for termination of P, cf. e.g. (Manna, 1974; Boyer and Moore, 1988). One only has to verify total termination of f_k, whereas the "auxiliary" algorithms f_i with $i<k$ only have to terminate for *some* inputs (*partial termination*), viz. the inputs which result from the (direct or indirect) calls of f_i by f_k. In particular, this approach is necessary when extending termination analysis to *imperative* programs, cf. Section 4.1.

To prove partial termination of an algorithm $f: s_1\times\ldots\times s_n\to s$, a totally terminating algorithm $\theta_f: s_1\times\ldots\times s_n\to \mathsf{bool}$ (a *termination predicate* for f) is synthesized, such that the evaluation of $\theta_f(x_1,\ldots,x_n)$ to true implies that evaluation of $f(x_1,\ldots,x_n)$ terminates. Then one only has to verify the truth of $\theta_f(t_1,\ldots,t_n)$ for each call $f(t_1,\ldots,t_n)$ which can be executed in the functional program P. So for

function $\mathsf{f}_1:\mathsf{nat}\to\mathsf{nat}$	*function* $\mathsf{f}_2:\mathsf{nat}\to\mathsf{nat}$
$\mathsf{f}_1(0) \quad = \mathsf{f}_1(0)$	$\mathsf{f}_2(0) \quad = 0$
$\mathsf{f}_1(\mathsf{s}(x)) = 0$	$\mathsf{f}_2(\mathsf{s}(x)) = \mathsf{f}_1(\mathsf{s}(x))$

and $P=\langle \mathsf{f}_1,\mathsf{f}_2\rangle$, we only have to prove that $\theta_{\mathsf{f}_1}(\mathsf{s}(t))$ evaluates to true for all data objects t, i.e. we only have to prove partial termination of f_1. This is sufficient for the total termination of the main function f_2 since $\mathsf{f}_1(0)$ is never called when executing the functional program P.

Considered in this framework, our methods for proving total termination from Section 2 and 3 aim to synthesize termination predicates such that $\theta_f(t_1,\ldots,t_n)$ holds for *all* $t_1,\ldots,t_n$. Therefore, in order to obtain termination predicates for algorithms which do not terminate totally, further developments of our termination methods are required.

4.1. *Termination Analysis for Imperative Programs*

A straightforward approach to prove termination of imperative programs is to transform them into functional ones. See e.g. (Henderson, 1980) for an automated translation. If termination of the resulting functions can be proved, then termination of the original imperative program is verified. As an example consider the following imperative program which adds the multiplication of x's and y's initial values to r, i.e. $r := x * y + r$. For that purpose, the program uses the total auxiliary functions even, plus, and double.

```
while  ¬eq(x,0)
  do   if even(x)  then  z := 0;
                         while ¬eq(x,z)                 }
                           do  x := p(x);               }  sets x to x/2
                               z := s(z)   od;          }
                         y := double(y)
                   else  x := p(x);
                         r := plus(r,y)   fi   od
```

To translate this imperative program into a functional one, every *while*-loop is transformed into a separate "loop-function". For instance, for the *inner while*-loop we obtain the following function half which takes the input values of the variables x and z as arguments and returns the output value of x. (Of course, a similar function returning the output value of z could also be constructed if required.) If the loop-condition is satisfied, i.e. if $\mathsf{eq}(x,z)$ is false, then half is called recursively with the new values of x and z. Otherwise, half returns the value of x.

function half : nat × nat → nat

$$\mathsf{half}(x,z) \;=\; \mathsf{if}(\,\mathsf{eq}(x,z),\; x,\; \mathsf{half}(\mathsf{p}(x),\mathsf{s}(z))\,)$$

The algorithm $\mathsf{half}(x,z)$ computes the arithmetical mean $\frac{x+z}{2}$ of two numbers if it terminates. Using the auxiliary function half, the whole imperative program is translated into the function times.

function times : nat × nat × nat → nat

$$\mathsf{times}(x,y,r) = \mathsf{if}(\,\mathsf{eq}(x,0),\, r,\, \mathsf{if}(\,\mathsf{even}(x), \mathsf{times}(\mathsf{half}(x,0), \mathsf{double}(y), r),\; \mathsf{times}(\mathsf{p}(x), y, \mathsf{plus}(r,y))\,)\,)$$

Obviously, the function times is equivalent to the original imperative program and in particular, termination of times is equivalent to the termination of the imperative program. Note that in general the auxiliary functions resulting from such a translation are *partial* even if the original imperative program terminates totally. The reason is that in imperative programs, termination of *while*-loops often depends on their contexts, i.e. on the preconditions that hold before entering the *while*-loop.

For instance, in our example the inner *while*-loop is only entered with an *even* input x. However, this restriction on x is not present in the function half. Therefore times terminates totally, but its auxiliary function half is partial, as

$\mathsf{half}(x,z)$ only terminates iff x is not smaller than z and if $x-z$ is even. Hence, to prove (total) termination of times one needs a method for termination analysis of *partial* functions.

4.2. *Generation of Termination Predicates*

In this section we present a procedure for the synthesis of termination predicates. Our aim is to generate termination predicates "as weak as possible", i.e. predicates which return true as often as possible, but of course in general this goal cannot be reached as the domains of functions are undecidable. Recall that there are two requirements for termination predicates θ_f.

(a) θ_f is *partially correct*, i.e. if $\theta_f(t_1,\ldots,t_n)$ evaluates to true, then the evaluation of $f(t_1,\ldots,t_n)$ terminates.
(b) θ_f *terminates totally*.

For instance, to satisfy requirement (a) a termination predicate θ_{half} for the algorithm half from Section 4.1 can be generated by replacing the result of the non-recursive case by true and by replacing the recursive call of half by a corresponding recursive call of θ_{half}.

function $\theta_{\mathsf{half}} : \mathsf{nat} \times \mathsf{nat} \to \mathsf{bool}$
$\theta_{\mathsf{half}}(x,z) \;=\; \mathsf{if}(\mathsf{eq}(x,z),\ \mathsf{true},\ \theta_{\mathsf{half}}(\mathsf{p}(x),\mathsf{s}(z)))$

This algorithm returns true iff it terminates. As the recursions of θ_{half} correspond to the recursions of half, the algorithm half terminates whenever θ_{half} yields true. Hence, the above algorithm for θ_{half} is *partially correct*.

However, the above algorithm does not satisfy requirement (b), i.e. it is not totally terminating. In other words, there is no well-founded ordering $\prec$ satisfying the termination formula

$$\neg\mathsf{eq}(x,z) \;\to\; \mathsf{p}(x) \prec x$$

of this algorithm. The central idea for the transformation of the above algorithm into a totally terminating one is to choose some well-founded ordering $\prec$ and to enter the recursive call $\theta_{\mathsf{half}}(\mathsf{p}(x),\mathsf{s}(z))$ only for those inputs x and z where x is greater than the corresponding argument $\mathsf{p}(x)$ in the recursive call and to return false for all other inputs.

A successful heuristic for the choice of $\prec$ is to use a well-founded ordering which satisfies at least the *non-strict* version of the termination formula $\mathsf{p}(x) \preceq x$ (and which satisfies the strict inequality $\mathsf{p}(x) \prec x$ "as often as possible"). To determine such an ordering we use the methods of Section 2 or 3. For example, by estimation of the defined symbol p, the inequality $\mathsf{p}(x) \preceq x$ is transformed into the constraints $\overline{\mathsf{p}}(x) \preceq x$ and $E_{\mathsf{p}\preceq\overline{\mathsf{p}}}$ which are satisfied by the size ordering (where $\overline{\mathsf{p}}(x)$ is associated with the polynomial x). Hence, we demand that $\theta_{\mathsf{half}}(x,z)$ may only enter its recursive call if $\mathsf{p}(x) \prec_{\#} x$ holds.

Now the methods of Section 2 or 3 are used to compute an equivalent requirement for $\mathsf{p}(x) \prec_{\#} x$, viz. the *difference equivalent* $\Delta_{\prec_{\#}}(\mathsf{p}(x),x)$ which corresponds to $\neg\mathsf{eq}(x,0)$. In this way we obtain the following algorithm for the termination predicate of half.

function $\theta_{\mathsf{half}} : \mathsf{nat} \times \mathsf{nat} \to \mathsf{bool}$
$\theta_{\mathsf{half}}(x,z) \;=\; \mathsf{if}(\mathsf{eq}(x,z), \mathsf{true}, \mathsf{if}(\mathsf{eq}(x,0), \mathsf{false}, \theta_{\mathsf{half}}(\mathsf{p}(x),\mathsf{s}(z))))$

The incorporation of the difference equivalent guarantees that the evaluation of $\theta_{\mathsf{half}}(x,z)$ can only lead to a recursive call $\theta_{\mathsf{half}}(\mathsf{p}(x),\mathsf{s}(z))$ if $\mathsf{p}(x)$ is smaller than the input x.

The above algorithm defines a termination predicate for half, i.e. θ_{half} computes a *total* function and the truth of $\theta_{\mathsf{half}}(\ldots)$ is sufficient for the termination of $\mathsf{half}(\ldots)$. In fact, $\theta_{\mathsf{half}}(x,z)$ is true iff x is greater than or equal to z and if $x-z$ is even. Hence, in this example we have even generated the weakest possible termination predicate, i.e. θ_{half} returns true not only for a subset but for *all* elements of the domain of half.

To automate the above synthesis of termination predicates we associate a boolean term $\Theta(t)$ with each term t such that evaluation of $\Theta(t)$ always terminates and evaluation of t terminates whenever $\Theta(t)$ evaluates to true[10]. $\Theta(t)$ is called a *termination condition* for t and its definition is based on the termination predicates:

(i) $\Theta(x) \quad :\equiv \mathsf{true},$ for variables x,
(ii) $\Theta(\mathsf{if}(b,r_1,r_2)) :\equiv \Theta(b) \wedge \mathsf{if}(b, \Theta(r_1), \Theta(r_2)),$
(iii) $\Theta(g(r_1...r_n)) \quad :\equiv \Theta(r_1)\wedge...\wedge\Theta(r_n) \wedge \theta_g(r_1...r_n),$ for functions $g \neq \mathsf{if}$,

where "$t_1 \wedge t_2$" abbreviates "$\mathsf{if}(t_1,t_2,\mathsf{false})$" and "$t_1 \wedge t_2 \wedge \ldots$" abbreviates "$t_1 \wedge (t_2 \wedge \ldots)$" to ease readability.

For example, the termination condition $\Theta(f(g(h(t))))$ is computed as $\Theta(t) \wedge \theta_h(t) \wedge \theta_g(h(t)) \wedge \theta_f(g(h(t)))$, because due to our eager language in this term h and g are evaluated before evaluating f. Now if $f(t_1,\ldots,t_n) = r$ is a defining equation of f, then our aim is to use $\theta_f(t_1,\ldots,t_n) = \Theta(r)$ as a defining equation for the corresponding termination predicate.

However, for total termination of θ_f the arguments of recursive calls have to be smaller than the corresponding inputs. Therefore, for recursive calls in a defining equation $f(\ldots t_i \ldots) = \ldots f(r_1 \ldots r_i \ldots r_n) \ldots$ we add a requirement to the termination condition of $f(r_1 \ldots r_i \ldots r_n)$ which ensures $r_i \prec t_i$ (for some

[10] More precisely, this implication holds for each substitution σ of t's variables by data objects. For all such σ, evaluation of $\sigma(\Theta(t))$ is terminating and $\sigma(\Theta(t)) = \mathsf{true}$ implies that the evaluation of $\sigma(t)$ is also terminating.

fixed i). To this end, instead of (iii) we use

(iv) $\Theta(f(r_1...r_n)) :\equiv \Theta(r_1) \wedge ... \wedge \Theta(r_n) \wedge \Delta_{\prec}(r_i, t_i) \wedge \theta_f(r_1...r_n)$

when computing the termination condition for a result term of f.

So for constructors and functions f whose total termination has been verified we define $\theta_f(x_1, \ldots, x_n) :\equiv \mathsf{true}$. For all other functions f we use the following construction principle for termination predicates.

From each defining equation $f(t_1, \ldots, t_n) = r$ of f construct a defining equation $\theta_f(t_1, \ldots, t_n) = \Theta(r)$ for θ_f.

This construction of termination predicate algorithms can be automated directly. For instance, θ_{half} was built according to this procedure (where we simplified expressions like $\mathsf{if}(\mathsf{true}, t, \ldots)$ to t).

Now termination of the imperative program from Section 4.1 can be verified automatically. For the function times, our method synthesizes

function $\theta_{\mathsf{times}} : \mathsf{nat} \times \mathsf{nat} \times \mathsf{nat} \rightarrow \mathsf{bool}$

$$\begin{aligned}\theta_{\mathsf{times}}(x, y, r) = \mathsf{if}(&\mathsf{eq}(x, 0),\\ &\mathsf{true},\\ &\mathsf{if}(\mathsf{even}(x), \mathsf{if}(\theta_{\mathsf{half}}(x, 0), \theta_{\mathsf{times}}(\mathsf{half}(x, 0), \mathsf{double}(y), r), \mathsf{false}),\\ &\qquad\qquad \theta_{\mathsf{times}}(\mathsf{p}(x), y, \mathsf{plus}(r, y)))).\end{aligned}$$

In the above algorithm, the difference equivalents $\Delta_{\prec_\#}(\mathsf{half}(x, 0), x)$ and $\Delta_{\prec_\#}(\mathsf{p}(x), x)$ (both of which correspond to $\neg\mathsf{eq}(x, 0)$) have been simplified to true, because they are called under the condition $\neg\mathsf{eq}(x, 0)$.

Subsequently, the conjecture $\theta_{\mathsf{times}}(x, y, r) = \mathsf{true}$ can be shown by an induction proof. In this way, total termination of times and thereby termination of the original imperative program is verified.

4.3. *Refinements and Summary*

To analyze the termination of an algorithm f, a *termination predicate* θ_f is synthesized, which represents a sufficient condition for f's termination. For this synthesis, the methods of Section 2 or 3 are used to generate a well-founded ordering satisfying all non-strict termination formulas and as many strict termination formulas as possible. (If the ordering satisfies *all* termination formulas, then total termination of f is proved. So the approaches of Section 2 and 3 are a special case of this termination analysis procedure.)

Our approach is successfully applied for termination analysis of *imperative programs*. Such a program can be translated into an equivalent functional program, where each *while*-loop corresponds to a (partially terminating) functional algorithm. Now the termination predicates for these partial "loop functions" can be used to prove termination of the whole imperative program.

A refinement of our method such that *several* arguments of an algorithm are considered is obtained by using *tuple* symbols, cf. Section 3.3. Moreover, to ease subsequent reasoning we developed several techniques to *simplify* the generated termination predicates. Further details can be found in (Brauburger and Giesl, 1996; Brauburger, 1997).

5. CONCLUSION

We have presented methods for a fully automated termination analysis of functional programs. We started with a technique to prove termination by estimating argument-bounded functions w.r.t. the size ordering. This approach has been implemented in the induction theorem prover INKA (Biundo et al., 1986; Hutter and Sengler, 1996) and it proved successful on many examples. A collection of 60 such algorithms (including classical sorting algorithms and algorithms for standard arithmetical operations) can be found in (Walther, 1991). Moreover, this method also proves termination of almost all examples from the data base of (Boyer and Moore, 1979) automatically (where one algorithm (greatest.factor) must be slightly modified). However, for three algorithms of this data base (viz. normalize, gopher, samefringe) this technique fails, as they terminate with a well-founded ordering different from size.

Hence, we generalized our first approach to allow an automatic synthesis of other well-founded relations. The resulting method is the most powerful technique for automated termination proofs developed so far, because in previous approaches the orderings used were either fixed or had to be provided by the user. We also implemented this method within the INKA prover (using a procedure for the automated generation of polynomial orderings (Giesl, 1995a)) and it performed successfully on a large collection of benchmarks (including all 82 algorithms from (Boyer and Moore, 1979) and all 60 examples from (Giesl, 1995d)). This collection contains many practically relevant examples from different areas of computer science (e.g. algorithms on numbers, lists, graphs, trees, strings, terms etc.) and several well-known challenge problems from the literature (such as *McCarthy*'s f_91 function).

Finally, we showed that our approach can also be used for partial functions by synthesizing termination predicates which approximate the domains of such functions. See (Brauburger and Giesl, 1996) for a collection of 32 examples which demonstrate that for many algorithms this approach is able to generate sophisticated termination predicates (which are not only sufficient for termination, but which even describe the *exact* domains of the functions).

ACKNOWLEDGEMENTS

This work was supported by the *Schwerpunktprogramm Deduktion* of the *Deutsche Forschungsgemeinschaft* under grants Wa 652/7-1,2.

REFERENCES

Arts, T. and J. Giesl: 1996, 'Termination of Constructor Systems'. In: *Proc. RTA '96*. New Brunswick, NJ, pp. 63–77. LNCS 1103.

Arts, T. and J. Giesl: 1997a, 'Automatically Proving Termination Where Simplification Orderings Fail'. In: *TAPSOFT '97*. Lille, France, pp. 261–272. LNCS 1214.

Arts, T. and J. Giesl: 1997b, 'Proving Innermost Normalisation Automatically'. In: *Proc. RTA '97*. Sitges, Spain, pp. 157–171. LNCS 1232.

Arts, T. and J. Giesl: 1998, 'Modularity of Termination Using Dependency Pairs'. In: *Proc. RTA '98*. Tsukuba, Japan. LNCS.

Bellegarde, F. and P. Lescanne: 1990, 'Termination by Completion'. *Applicable Algebra in Engineering, Communication and Computing* **1**, 79–96.

Biundo, S., B. Hummel, D. Hutter, and C. Walther: 1986, 'The Karlsruhe Induction Theorem Proving System'. In: *Proc. CADE-8*. Oxford, pp. 672–674. LNCS 230.

Bouhoula, A. and M. Rusinowitch: 1995, 'Implicit Induction in Conditional Theories'. *Journal of Automated Reasoning* **14**, 189–235.

Boyer, R. S. and J. S. Moore: 1979, *A Computational Logic*. Academic Press.

Boyer, R. S. and J. S. Moore: 1988, 'The Addition of Bounded Quantification and Partial Functions to A Computational Logic and Its Theorem Prover'. *Journal of Automated Reasoning* **4**, 117–172.

Brauburger, J. and J. Giesl: 1996, 'Termination Analysis for Partial Functions'. In: *Proc. SAS '96*. Aachen, Germany, pp. 113–127. LNCS 1302. Extended version available as Report IBN 96/33, TU Darmstadt. http://www.inferenzsysteme.informatik.tu-darmstadt.de/~reports/notes/ibn-96-33.ps.

Brauburger, J.: 1997, 'Automatic Termination Analysis for Partial Functions Using Polynomial Orderings'. In: *SAS '97*. Paris, France, pp. 330–344. LNCS 1145.

Bundy, A., F. van Harmelen, C. Horn, and A. Smaill: 1990, 'The OYSTER-CLAM System'. In: *Proc. CADE-10*. Kaiserslautern, Germany, pp. 647–648. LNAI 449.

Collins, G. E.: 1975, 'Quantifier Elimination for Real Closed Fields by Cylindrical Algebraic Decomposition'. In: *Proc. 2nd GI Conf. on Automata Theory and Formal Languages*. Kaiserslautern, Germany, pp. 134–183. LNCS 33.

De Schreye, D. and S. Decorte: 1994, 'Termination of Logic Programs: The Never-Ending Story'. *Journal of Logic Programming* **19, 20**, 199–260.

Dershowitz, N.: 1987, 'Termination of Rewriting'. *J. Symb. Comp* **3**, 69–115.

Giesl, J.: 1995a, 'Generating Polynomial Orderings for Termination Proofs'. In: *Proc. RTA '95*. Kaiserslautern, Germany, pp. 426–431. LNCS 914.

Giesl, J.: 1995b, 'Automated Termination Proofs with Measure Functions'. In: *Proc. 19th Ann. German Conf. on AI*. Bielefeld, Germany, pp. 149–160. LNAI 981.

Giesl, J.: 1995c, 'Termination Analysis for Functional Programs using Term Orderings'. In: *Proc. SAS '95*. Glasgow, Scotland, pp. 154–171. LNCS 983.
Giesl, J.: 1995d, *Automatisierung von Terminierungsbeweisen für rekursiv definierte Algorithmen*. Doctoral Dissertation, Infix-Verlag, Sankt Augustin, Germany.
Giesl, J.: 1997, 'Termination of Nested and Mutually Recursive Algorithms'. *Journal of Automated Reasoning* **19**, 1–29.
Henderson, P.: 1980, *Functional Programming*. Prentice-Hall, London.
Hutter, D. and C. Sengler: 1996, 'INKA: The Next Generation'. In: *Proc. CADE-13*. New Brunswick, NJ, pp. 288–292. LNAI 1104.
Kamin, S. and J.-J. Levy: 1980, 'Two Generalizations of the Recursive Path Ordering'. Unpublished Note, Dept. of Computer Sc., Univ. Illinois, Urbana, IL.
Kapur, D. and H. Zhang: 1995, 'An Overview of Rewrite Rule Laboratory (RRL)'. *J. Computer Math. Appl.* **29**, 91–114.
Lankford, D. S.: 1979, 'On Proving Term Rewriting Systems are Noetherian'. Technical Report Memo MTP-3, Math. Dept., Louisiana Tech. Univ., Ruston, LA.
Manna, Z.: 1974, *Mathematical Theory of Computation*. McGraw-Hill.
McAllester, D. and K. Arkoudas: 1996, 'Walther Recursion'. In: *Proc. CADE-13*. New Brunswick, NJ, pp. 643–657. LNAI 1104.
Nielson, F. and H. R. Nielson: 1996, 'Operational Semantics of Termination Types'. *Nordic Journal of Computing* **3**, 144–187.
Panitz, S. E. and M. Schmidt-Schauß: 1997, 'TEA: Automatically Proving Termination of Programs in a Non-Strict Higher-Order Functional Language'. In: *Proc. SAS '97*. Paris, France, pp. 345–360. LNCS 1145.
Plümer, L.: 1990, *Termination Proofs for Logic Programs*. Springer. LNAI 446.
Sengler, C.: 1996, 'Termination of Algorithms over Non-Freely Generated Data Types'. In: *Proc. CADE-13*. New Brunswick, NJ, pp. 121–135. LNAI 1104.
Sengler, C.: 1997, *Induction on Non-Freely Generated Data Types*. Doctoral Dissertation, Infix-Verlag, Sankt Augustin, Germany.
Steinbach, J.: 1995a, 'Automatic Termination Proofs with Transformation Orderings'. In: *Proc. RTA '95*. Kaiserslautern, Germany, pp. 11–25. LNCS 914.
Steinbach, J.: 1995b, 'Simplification Orderings: History of Results'. *Fundamenta Informaticae* **24**, 47–87.
Ullman, J. D. and A. van Gelder: 1988, 'Efficient Tests for Top-Down Termination of Logical Rules'. *Journal of the ACM* **35**, 345–373.
Walther, C.: 1988, 'Argument–Bounded Algorithms as a Basis for Automated Termination Proofs'. In: *Proc. CADE-9*. Argonne, IL, pp. 602–621. LNCS 310.
Walther, C.: 1991, *Automatisierung von Terminierungsbeweisen*. Vieweg, Germany.
Walther, C.: 1994a, 'Mathematical Induction'. In: D. M. Gabbay, C. J. Hogger, and J. A. Robinson (eds.): *Handbook of Logic in Artificial Intelligence and Logic Programming*, Vol. 2. Oxford University Press.
Walther, C.: 1994b, 'On Proving the Termination of Algorithms by Machine'. *Artificial Intelligence* **71**, 101–157.
Zantema, H.: 1995, 'Termination of Term Rewriting by Semantic Labelling'. *Fundamenta Informaticae* **24**, 89–105.

CHAPTER 7
THE WAM CASE STUDY: VERIFYING COMPILER CORRECTNESS FOR PROLOG WITH KIV

1. INTRODUCTION

This chapter describes the first half of the formal, machine-supported verification of a Prolog compiler with the KIV system.

Our work is based on the mathematical analysis given in (Börger and Rosenzweig, 1995), where an operational semantics (an "interpreter") for Prolog is defined as an Abstract State Machine (ASM). This interpreter is then transformed in 12 systematic refinements to an ASM which executes machine code of the Warren Abstract Machine (WAM).

The goal of our case study was to formalize ASMs and the proof techniques given in (Börger and Rosenzweig, 1995), and to give machine-checked correctness proofs for the correctness of the refinements. So far we have verified the first 6 refinements, and we will give a detailed account on the problems we found in verification.

Our motivations for beginning such a large case study — based on our current experience we estimate the necessary effort to develop a verified compiler to be around a person year — are the following

- Mathematical analysis is an indispensable prerequisite for formal verification to be applicable. Nevertheless mathematical analysis will always omit details and have minor errors. These errors are due to the large complexity of correctness proofs, which is easily underestimated at first glance. The errors usually do not invalidate the analysis, but would still result in erroneous compilers. We want to demonstrate that the absence of errors can be guaranteed by formal correctness proofs, making them a suitable counterpart to mathematical analysis.
- We want to show that Dynamic Logic (DL) as it is used in the KIV system can serve as a suitable starting point for the verification of Abstract State Machine refinements. In particular, the proof technique of commuting diagrams of Proof Maps, used informally in (Börger and Rosenzweig, 1995), can be formalized in DL.

W. Bibel, P. H. Schmitt (eds.), Automated Deduction. A basis for applications. Vol. III

– Finally, the requirements a system for the development of correct software must cope with are only discovered in ambitious case studies. Solving these requirements always leads to significant system improvements.

This chapter is organized as follows: Sect. 2 introduces the formalism of Abstract State Machines (ASMs,(Gurevich, 1995)).

Sect. 3 shortly described the KIV system, which was used to do the verification. Dynamic Logic (DL), the logic used in KIV to express properties of imperative programs, is given.

Sect. 4 defines a general translation of sequential ASMs to algebraic specifications and imperative programs. The proof task of verifying the refinement between two ASMs is identified as a problem of program equivalence in DL. The deduction problem is reduced to the development of correct *coupling invariants*, which are formulas corresponding to *proof maps* in ASMs.

Sect. 5 first introduces the Abstract State Machine (ASM) which is used to define an operational semantics of Prolog. We assume the reader to be familiar with the core of Prolog: clauses with *! (cut)* and *fail* (see e.g. (Sterling and Shapiro, 1986)). Then the 6 refinements towards the Warren Abstract Machine (WAM) we verified so far are given. Each refinement introduces orthogonal concepts of the WAM. Parallel to the refinements, the Prolog program and the query are compiled to machine instructions. On intermediate levels the input of the ASM are machine instructions interspersed with uncompiled Prolog syntax. The compilation steps are not given as a concrete program, but specified by *compiler assumptions*. This still leaves some freedom for the implementation of a compiler, in particular several variants of the final WAM are still possible. Sect. 5 closely follows (Börger and Rosenzweig, 1995). It deviates only in some minor notational issues, and will give a precise formalization of the *chain*-function to be used in one of the compiler assumptions.

In Sect. 6 we give an overview over the verification. Problems specific to the verification of each refinement will be discussed. The coupling invariant for the first compilation step (the 4th refinement) will be shown. This formula is very complex and we will describe how it was developed by iterated proof attempts with the KIV system.

Sect. 7 gives some related work and Sect. 8 concludes with an outlook on the continuing work on this case study.

2. Abstract State Machines

A (Gurevich) Abstract State Machine (ASM, also known as 'Evolving Algebra') basically consists of a number of rules which are applied to transform an initial state to a final one. The possible states of an ASM are defined to be the class of algebras over a fixed first-order signature SIG.

A rule is given by its applicability test ε (a ground boolean expression over SIG) and a set of function updates. A function update is of the form

$$f(t_1,\ldots,t_n) := t,$$

where f is a function symbol from SIG, $t_1,\ldots,t_n$ and t are ground terms. A rule is applicable in an algebra A, if the applicability test holds ($A \models \varepsilon$). Firing a rule executes all function updates in parallel resulting in a new algebra (a "successor state"). Execution of a function update like the one above changes the semantics of f at $(t_1,\ldots,t_n)$ to be t. The case $n = 0$ is admitted, and we will talk of updating a "0-ary function" f then, since to call f a "constant" would be rather misleading. The updates of a rule are always assumed to be consistent, i.e. no two updates of a rule change the same function at the same argument.

Repeated execution of rules results in traces of 'evolving' algebras (A_0, A_1, ...), where each A_{i+1} is the result of firing an applicable rule in A_i. *Computations* are defined to be traces starting in an algebra taken from some predefined set of initial algebras. A computation is either infinite (non-terminating computation) or terminates in a state in which no further rule is applicable. The semantics of an ASM is defined to be the set of all computations.

There are several extensions of the basic formalism and the semantics e.g. for parallel execution of rules (see (Gurevich, 1995)). For our case study we will only need *sort updates* for many-sorted signatures,

extend s **by** c

This sort update extends the domain of sort *s* with a new element, which is assigned to the 0-ary function *c*. Sort updates can be used just like function updates in rules (in an unsorted setting, where sorts are boolean valued functions, a sort update can be viewed as a function update, see again (Gurevich, 1995)).

In our application, the definition of a Prolog interpreter, an initial algebra will contain a Prolog program and a query as the value of two predefined constants. Execution of the rules of the first ASM will build up Prolog search trees, as they can be found in many Prolog textbooks (e.g. (Sterling and Shapiro, 1986)). If computation terminates, the answer substitution (which may be *failure*) can be read off as the value of a 0-ary function.

Since Prolog interpreters are deterministic, at most one rule of the ASM should be applicable in every state, so there will be exactly one computation of the ASM for every initial state.

3. KIV

The KIV system ((Reif, 1995), (Reif et al., 1995), (Reif and Stenzel, 1995), (Reif et al., 1997)) is an advanced tool for engineering high assurance systems. It supports the entire design process from formal, algebraic specifications to executable verified code. KIV relies on first-order algebraic specifications to describe hierarchically structured systems. Details on syntax and semantics are given in Chapter II.1.1.

Specification components can be implemented using modules which contain imperative programs. The programs contain the usual constructs found in imperative languages: assignment $x := t$ (also parallel assignments $\underline{x} := \underline{t}$), conditional, compound, local variables, while-loops and recursive procedures with both value- and reference parameters. Although the programs are written in a PASCAL-like notation, they are *abstract* programs. The operations used in them are those given in an algebraic specification, not the concrete ones available in PASCAL.

To reason about abstract programs, KIV uses Dynamic Logic (DL). DL is an extension of first-order logic by formulas $\langle\alpha\rangle\,\varphi$ (read "diamond α φ"), where α is an imperative program, and φ is again a DL-formula. The informal meaning of this formula: "α terminates and φ holds afterwards" should be sufficient for the purpose of this chapter. A formal definition of DL can be found in (Harel, 1984), (Goldblatt, 1982).

DL can be used to express total correctness of a program α with precondition φ and postcondition ψ as $\varphi \rightarrow \langle\alpha\rangle\,\psi$. Partial correctness is expressed by $\varphi \rightarrow \neg\,\langle\alpha\rangle \neg\,\psi$. Program inclusion (and equivalence) with respect to some program variables $\underline{x}$ can be formalized as $\langle\alpha\rangle\,\underline{x} = \underline{x}_0 \rightarrow \langle\beta\rangle\,\underline{x} = \underline{x}_0$. This will be important for our case study.

To deduce properties of specifications and to verify program modules, KIV offers an advanced interactive theorem prover. Details on this prover can again be found in Chapter II.1.1. Since frequently the problems found in the development of correct software are not to verify proof obligations affirmatively but rather to interpret failed proof attempts, KIV offers a number of *proof engineering* facilities to support the iterative process of (failed) proof attempts, error detection, error correction and re-proof, see (Reif and Stenzel, 1995).

4. From Abstract State Machines to Dynamic Logic

In this section we will give a translation of sequential Abstract State Machines to Algebraic Specifications and Dynamic Logic. The translation is essentially one on one, because both ASM and DL feature imperative programs. Therefore no encoding of programs (as functions or relations over a state) is required. This makes DL a good starting point for verifying properties of sequential ASMs. The translation is done in three steps. First, we will give a translation of the abstract data used into an algebraic specification. The second step translates the set of rules of an ASM into an imperative program. In the third step we will identify equivalence of two ASMs as program equivalence in DL, and give a proof technique corresponding to the use of "Proof Maps". The three steps are described in the following three subsections.

4.1. *Translation of Specifications*

To translate the abstract data types of an Abstract State Machine into an algebraic specification, we first have to separate the *static* and the *dynamic* part of the signature. The dynamic part contains those functions and sorts, for which the set of rules contains updates. The other, static part typically contains data types like lists, numbers and suitable operations on them. These can be specified algebraically.

The central idea for translating the dynamic part is to encode the domains of *dynamic* sorts and the semantics of *dynamic* functions as the values of (ordinary first-order) variables. Updates then are translated to assignments in DL.

Since the domains of dynamic sorts in an ASM usually are *finite* sets of elements (in our case: finite sets of nodes) a (standard) specification of finite sets is used. A variable *s* stores the current domain, and a sort update

extend s **with** c

corresponds to the two assignments

$$c := new(s);\ s := s \cup \{c\}$$

in DL, where for function *new : set* $\rightarrow$ *node* the axiom: $\neg$ *new(s)* $\in$ *s* is given.

0-ary functions are simply translated to ordinary first-order variables. For other dynamic functions, the case with $n > 1$ arguments can be reduced to the case with one argument by adding an appropriate tuple sort. For unary functions we essentially have to encode the (second-order) datatype of a *function* into a first-order datatype, so that a function can become the value of a variable. This can be achieved with the data type shown in Fig. 1, which defines functions from domain *dom* to codomain *codom*.

generic specification
parameter sorts dom, codom;
target sorts dynfun;
functions cf : codom → dynfun;
. ^ . : dynfun × dom → codom;
f . + (. / .) : dynfun × dom × codom → dynfun;
variables f : dynfun; x, y : dom; z : codom;
axioms cf(z) ^ x = z,
(f + (x / z)) ^ x = z,
x ≠ y → (f + (x / z)) ^ y = f ^ y
end generic specification

Fig. 1: Algebraic specification of dynamic functions

The data type contains a constant function *cf(z)* for every codomain element *z*. Application of this function to any domain element *x* (with an apply-operation, for convenience written as an infix-circumflex) just gives *z*, as stated by the first axiom. With a suitable "dummy" element *z*, constant functions are used as initial values.

A function update *f(x) := t* in the ASM-formalism becomes an assignment *f := f + (x / t)* to variable *f* in DL. It sets the new value of *f* to the result of mixfix-operation *f + (x / t)*, which is defined by the last two axioms to be the appropriately modified function.

Finally note that we did not add an extensionality axiom

$$f = g \leftrightarrow \forall x.\, f \,\hat{}\, x = g \,\hat{}\, x$$

to the specification, in contrast to the usual methodology used in KIV to specify non-free data types. Such an axiom would have allowed us to deduce (higher order) equalities between functions like *f = f + (x / f ^ x)*, but such equations did not show up in verification.

The specification can be viewed as an abstract version of a store structure. It could be implemented e.g. by association lists. In our case study the domain will be pointers (addresses, locations), and the final implementation of the dynamic functions in the WAM will be a part of computer memory.

4.2. *Translation of Programs*

Given the translation of the static and dynamic part of the ASM, it remains to translate the set of rules to an imperative program. For simplicity of the following presentation, we will assume that the test, whether any ASM rule is applicable, is *stop = run* (this can always be achieved by requiring *stop* to be initialized with *run*, and by adding a suitable final rule, which sets *stop* to *halt*, if the conjunction of all other rule tests is false). Then the result of the translation is the procedure *ASM* shown in Fig. 2.

```
ASM(in; var out)              BODY(var x)
begin                         begin
var x := t in                 if {test of rule_1} then {updates of rule_1} else
    while stop = run          ...
    do BODY(x)                if {test of rule_n} then {updates of rule_n}
end                           end
```

Fig. 2: ASM as imperative program

In our case study, the inputs $\underline{in}$ of the first ASM will be the Prolog program and the query, the output value will be the answer substitution. The variable declarations $\underline{x} = \underline{t}$ initialize the variables $\underline{x}$ the interpreter computes on with suitable values $\underline{t}$. They correspond to the definition of the initial algebra in the ASM. Application of rules is done in a while loop. The body has been put in a separate procedure *BODY* to have a suitable abbreviation for the formulas below. *BODY* is called with reference parameters $\underline{x}$, which are used as input and output. It consists of a case analysis, which selects an applicable rule and executes its updates.

4.3. *From Proof Maps to Coupling Invariants*

Correctness and completeness of the refinement of one ASM to another (ASM') is formalized in DL as the assertion of the following program equivalence between the two corresponding procedures *ASM* and *ASM'*.

$$\begin{aligned}&\text{CompAssum}(\underline{\text{in}},\underline{\text{in'}})\\&\rightarrow (\langle \text{ASM}(\underline{\text{in}};\text{out})\rangle\, \text{out=val} \leftrightarrow \langle \text{ASM'}(\underline{\text{in'}};\text{out'})\rangle\, \text{out'=val})\end{aligned} \tag{1}$$

This formula states that if the two interpreters are given inputs related by the compiler assumption *CompAssum*, then the first interpreter *ASM* terminates, if and only if the second *ASM'* terminates with the same answer substitution. Variable *val* is used to store the common result of both interpreters (this variable is not modified by *ASM* and *ASM'*).

The compiler assumption relates the two inputs of the interpreter. For optimization steps, the relation is just identity ($\underline{in} = \underline{in'}$), for compilation steps, the assumption gives sufficient criteria for the correct compilation of the Prolog program. The notions of *Correctness* and *Completeness* from (Börger and Rosenzweig, 1995) directly correspond to the implication from right to left and from left to right.

Proof maps F are defined in (Börger and Rosenzweig, 1995) to map static algebras of a 'concrete' ASM to algebras of an 'abstract' ASM. They are used to sketch correctness proofs for the equivalence of two ASMs. The basic argument is as follows: the three proof obligations

PO1: The initial states (algebras) of both ASMs are related via F

PO2: Fig. 3 commutes for every corresponding pair of rules R and R' of the two interpreters (with A_0 and B_0 being the results of rule application to A and B)

PO3: Two final states which are related via F store the same answer substitution

imply (by induction on the number of rule applications) that both ASMs are equivalent.

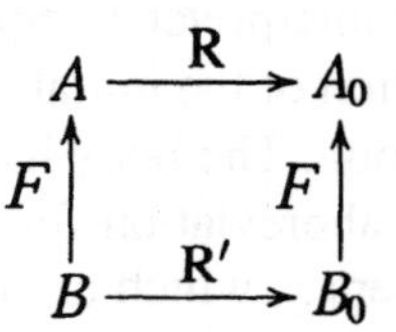

Fig. 3.

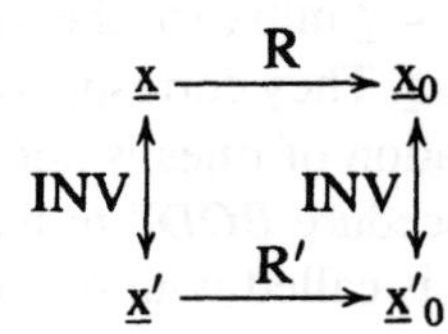

Fig. 4.

This informal argument can be formalized in DL to prove (1)as follows: the (dynamic parts of the) algebras involved in a computation have been replaced by the values of the vector of program variables. If we name the program variables, *ASM* and *ASM'* compute on, differently, say $\underline{x}$ and $\underline{x}'$, then the direct translation of a proof map is a function, which maps a tuple of values for $\underline{x}'$ to a tuple of values for $\underline{x}$. Since we found no need for the connection between $\underline{x}$ and $\underline{x}'$ to be a function, we allow it to be an arbitrary *relation*, which we describe by a (DL-)formula *INV*($\underline{x}$,$\underline{x}'$), with free variables $\underline{x}$ and $\underline{x}'$. We call this formula a ***coupling invariant.***

Induction over the number of rule applications is replaced by induction on the number of iterations of the while loop (for details of the formalization, see (Schellhorn and Ahrendt, 1997)), and it can be proved, that the following three goals are sufficient to prove goal (1).

$$\text{CompAssum}(\underline{\text{in}},\underline{\text{in}}') \rightarrow \text{INV}(\underline{\text{t}},\underline{\text{t}}') \quad (2)$$

$$\text{INV}(\underline{\text{x}},\underline{\text{x}}') \rightarrow \text{stop} = \text{stop}' \wedge \text{out} = \text{out}' \quad (3)$$

$$\begin{aligned} &\text{INV}(\underline{\text{x}},\underline{\text{x}}') \wedge \text{stop}' = \text{run} \wedge \text{stop} = \text{run} \\ &\rightarrow \langle \text{BODY}'(\underline{\text{x}}')\rangle\, \langle \text{BODY}(\underline{\text{x}})\rangle\, \text{INV}(\underline{\text{x}},\underline{\text{x}}') \end{aligned} \quad (4)$$

The first goal states that the coupling invariant holds before execution of the two while loops, corresponding to PO1. The second goal says that the coupling invariant implies that both loops stop at the same time with the same answer substitution (PO2). These two goals are usually rather trivial. The complexity of verification is buried in finding an invariant INV such that the last goal (4)is provable. This last goal corresponds to PO3 and formalizes commutativity of Fig. 4. The proof of (4) splits into one case for each corresponding pair of rules.

5. The WAM Analysis of Börger and Rosenzweig

To make this chapter as self-contained as possible, the following section introduces some of the Abstract State Machines of our case study. We will closely follow [Sections 1, 2] from (Börger and Rosenzweig, 1995) and deviate only in some notations. This will set up the verification task and enable us to discuss the problems we encountered in solving it (see Sect. 6). We will use the abbreviation *i*/*j* to mean the refinement from ASM*i* to ASM*j*.

5.1. *The First Interpreter: Search Trees*

The two most important data structures needed to represent a *Prolog computation state* are the *sequence* of Prolog literals still to be executed and the current *substitution*. This state is modified by

1. unifying the first literal of the sequence, called *act* (activator), with the *head* of a clause
2. replacing *act* by the *body* of that clause
3. applying the unifying substitution to the resulting sequence and
4. composing the unifying substitution with the 'old' substitution.

If this leads to failure, alternative clauses have to be chosen by backtracking. Due to this the interpreter has to keep a record of the former computation states and the corresponding clause choice alternatives. This history is represented by a tree of *nodes*, connected from leaves to the root by a function *father*. Information on alternative clauses, which may be tried at a node *n*, is stored as a list *cands(n)* of candidate nodes. Each node in this list refers via a function *cll* to a clause line in the Prolog program. The initial *cands*-list is constructed with the help of a function *procdef*, which is assumed to return the program lines containing the candidate clauses for a given literal. The current computation state is carried by a distinguished node, the *currnode*.

To handle the *cut*, an extension of the state representation is required. A *cut* updates the *father* of the current node to the *father* of that computation state whose *act* caused the introduction of the considered *cut*. For this we have to 'remember' where a *cut* has been introduced. An uniform solution is to attach the *father* of the (old) *currnode* to each clause body being introduced to the literal sequence. This attachment divides the sequence of literals into subsequents, called goals, each decorated by one node. The resulting (decorated goal) sequence *decglseq* looks as follows

$$decglseq = [\langle \underbrace{[\overbrace{g_{1,1}}^{act}, g_{1,2}, \ldots, g_{1,k_1}], \overbrace{n_1}^{cut\,pt}}_{goal} \rangle, \ldots, \langle [g_{m,1}, \ldots, g_{m,k_m}], n_m \rangle]$$

$$cont = [\langle \quad [g_{1,2}, \ldots, g_{1,k_1}], n_1 \rangle, \ldots, \langle [g_{m,1}, \ldots, g_{m,k_m}], n_m \rangle]$$

The continuation *cont*, which is the *decglseq* without *act*, will later on help to describe the construction of a new *decglseq*.

To introduce the rules of the ASM we will now consider the evaluation of the query `?- p.` on the following Prolog program.

```
1 p :- fail.        3 q.
2 p :- q,!.         4 p.
```

which is stored as the value of a constant *db* (database) in the initial algebra of the ASM. Line numbers are shown explicitly in the program for explanatory purposes.

The query `?- p.` is stored as the *decglseq* of node *A* in the initial search tree depicted in Fig. 6. The two nodes (labeled $\perp$ and *A*) form the initial domain of a dynamic universe *node*, which is extended by the rules of the ASM. Tree structure is stored in a function *father : node* $\rightarrow$ *node*, indicated by the arrow in Fig. 6, so we have *father(A)* $= \perp$ (the father of $\perp$ is undefined). Root node $\perp$ serves only as a marker when to finish search and does not carry information. The initial *currnode* is *A*, as indicated by the double circle. Computed substitutions (attached to the nodes with the *sub* function) are not shown in the figures, since they do not matter in the example.

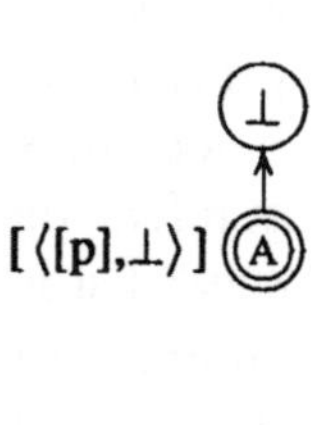

Fig. 6.

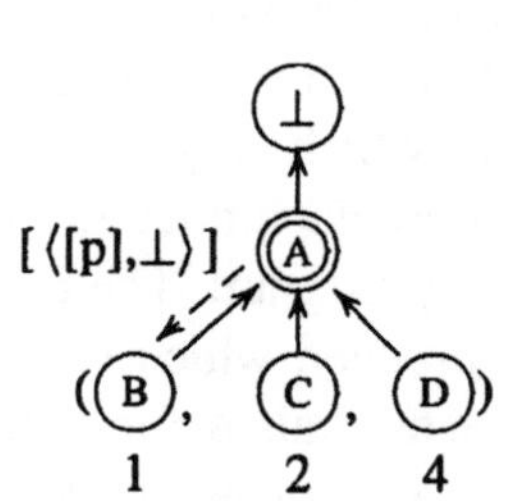

Fig. 7.

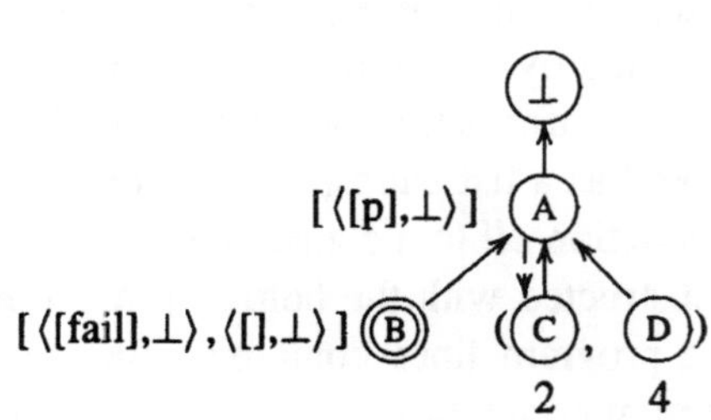

Fig. 8.

The ASM run is controlled by two program variables (i.e. 0-ary dynamic functions) *mode* and *stop*. The value of *mode* switches between *call* and *select*, while the value of *stop* remains *run* until it finally changes to *halt*. This stops the evaluation by falsifying all rule guards.

In *call* mode, which is the initial mode, the candidate nodes are computed (for a guard which involves *act*, we assume that *decglseq* ≠ *[]*, *goal* ≠ *[]*).

call rule

```
IF   stop = run
   ∧ is_user_defined(act)
   ∧ mode = call
THEN
   LET [cll_1,...,cll_n] =procdef(act,db)
   EXTEND node
   BY tmp_1,...,tmp_n
   WITH father(tmp_i) := currnode
          cll(tmp_i) := cll_i
          cands := [tmp_1,...,tmp_n]
   ENDEXTEND
   mode := select

where
backtrack ≡
IF father = ⊥
THEN stop := halt
        subst := failure
ELSE currnode := father
        mode := select
```

select rule

```
IF   stop = run
   ∧ is_user_defined(act)
   ∧ mode = select
THEN
   IF cands = []
   THEN backtrack
   ELSE
   LET clau =
            rename(clause(cll(first(cands))),vi)
   LET mgu = unify(act,head(clau))
   IF mgu = failure
   THEN cands := rest(cands)
   ELSE
   currnode := first(cands)
   decglseq(first(cands)) :=
      apply(mgu,[⟨body(clau), father⟩|cont])
   sub(first(cands)) := sub ∘ mgu
   cands := rest(cands)
   vi := vi +1
   mode := call
```

In this rule, the computed candidates for currnode, *cands(currnode)*, are abbreviated with *cands*, and we will use similar abbreviations for the functions *father*, *decglseq* and *sub* in the following rules.

The EXTEND construct, by expanding the universe *node*, allocates one node for every clause whose head 'may unify' with the literal *act*. This list of clause lines is computed by *procdef(act,db)* and is assumed to contain at least those clauses, whose head unify with the activator, and at most those with the same leading predicate symbol as *act*. The result of the rule application is depicted in Fig. 7.

The *cands* list (of node *A*) is indicated by a dashed arrow to its first element and brackets around the elements. The clause lines corresponding to the candidates are attached to the new nodes via the function *cll*, as shown by numbers below the nodes. The change of the *mode* variable activates the select rule. This rule causes backtracking if there are no (more) alternatives to select. Otherwise, by repeated application, it removes all candidates whose heads do not unify with *act*. When the first candidate is reached, for which a most general unifier ***mgu*** exists (variable index *vi* is used to rename the implicitly universal quantified clause variables to new instances), this node becomes *currnode*. A new *decglseq* is computed by replacing the activator of the old *decglseq* with the body of the 'selected clause'. As a cutpoint the *father* of the old *currnode* is attached to this new goal. The ***mgu*** is applied to the resulting *decglseq* and composed (with ∘) with the old substitution *sub*.

The result of applying the select rule in our example is shown in Fig. 8. Now mode is *call* again, but since the activator *fail* is not user defined, instead of the call rule the fail rule fires.

fail rule
IF stop = run ∧ act = fail THEN **backtrack**

It sets *currnode* to *A* again. Note that node *B* is not formally deallocated (i.e. it remains in the *node* universe). Again in *select* mode, the next candidate node of *A*, node *C*, is selected and its *decglseq* is computed as [⟨[q,!],⊥⟩,⟨[],⊥⟩]. Then the call rule allocates one new candidate node *E* for the only appropriate clause q. After selection of node *E* the ASM arrives at the state shown in Fig. 9.

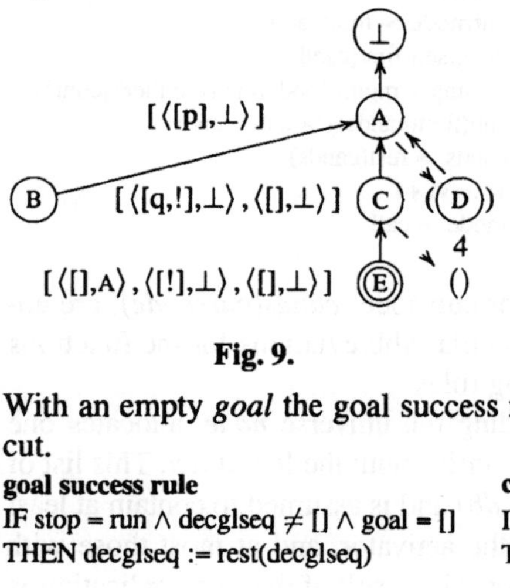

Fig. 9.

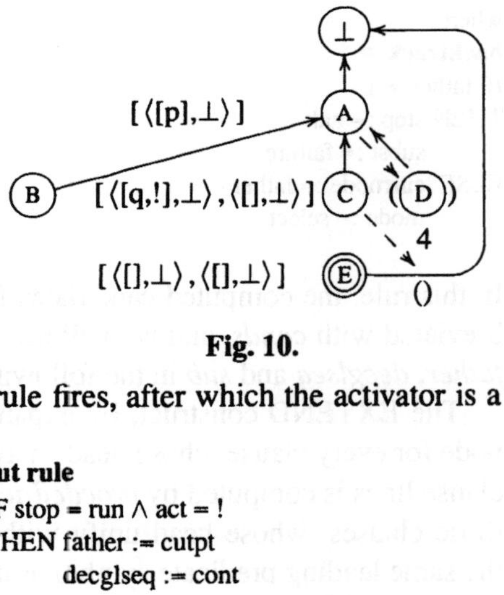

Fig. 10.

With an empty *goal* the goal success rule fires, after which the activator is a cut.

goal success rule
IF stop = run ∧ decglseq ≠ [] ∧ goal = []
THEN decglseq := rest(decglseq)

cut rule
IF stop = run ∧ act = !
THEN father := cutpt
decglseq := cont

The cut succeeds (is removed) and the *father* function is updated to the cut-point decorating the goal where the cut appears (see Fig. 10). This action is the only purpose for decorating goals with nodes.

Finally, with another two applications of the goal success rule, *decglseq(E)* becomes empty. Since this means that the initial query is completely solved, the ASM sets the answer substitution *subst* to *sub(currnode)* (here, of course, the empty substitution).

final success rule

IF stop = run ∧ decglseq(currnode) = [] THEN stop := halt
subst := sub

Since *stop* is no longer *run*, no more rule is applicable and the ASM halts.

5.2. *The Second Interpreter: Stacks of Choicepoints*

Here we summarize the first refinement of the ASM described above towards the Warren Abstract Machine (WAM), following (Börger and Rosenzweig, 1995), [Section 1.2]. There are three main differences between the first and the second ASM.

First, function *father* is renamed to *b*. This change indicates that *b* now points *b*ackwards in a chain of nodes, which form a *stack*.

Second, the new ASM (ASM2) provides the registers *cllreg, decglseqreg, breg* and *subreg* corresponding to *cll, decglseq, father* and *sub* applied to the *currnode*. Thereby it avoids allocation of *currnode*.

Third, instead of providing a list of candidate nodes, ASM2 attaches the *first* candidate directly via the *cll*-function. This is possible if we assume that clauses whose head starts with the same predicate are stored in successive clause lines followed by a special marker *nil*. The ("compiled") representation of our example Prolog program for ASM2 thus has to look like

```
1 p :- fail.        3 p.      5 q.
2 p :- q,!.         4 nil     6 nil
```

A new *procdef'* function is needed, such that *procdef' (act,db)* now yields the first clause line whose head may unify with the activator *act*. For *act* = *p* we get *procdef' (p,db)* = *1*, the first line of a clause with head *p*. The connection to the old *procdef* function is stated in the following *compiler assumption* about function *compile*, which is used as an axiom in correctness proofs. Let *db'* = *compile(db)* in

mapcl(procdef(act,db),db)
= mapcl(clls(procdef'(act,db'),db'),db') (5)

Here *clls* collects successive line numbers, until a *nil* is found and *mapcl* selects the clause at each line number. Instead of allocating a candidate list, ASM2 simply assigns *procdef' (act,db)* to *cllreg*. Incrementing *cllreg* then corresponds to removing a candidate from *cands*. If the clause at *cllreg* becomes *nil*, no more candidates are available. Allocation of a new node is now done only in *select* mode, when a new candidate clause is visited. The changed rules are

call rule

```
IF   stop = run
   ∧ is_user_defined(act)
   ∧ mode = call
THEN cllreg := procdef'(act,db')
     mode := select
```

where
backtrack ≡

```
IF breg = bottom
THEN stop := halt
     subst := failure
ELSE decglseqreg := decglseq(breg)
     subreg := sub(breg)
     breg := b(breg)
     cllreg := cll(breg)
     mode := select
```

select rule

```
IF   stop = run
   ∧ is_user_defined(act)
   ∧ mode = select
THEN
   IF clause(cllreg) = nil
   THEN backtrack
   ELSE
   LET clau = rename(clause(cllreg),vi)
   LET mgu = unify(act, head(clau))
   IF mgu = failure
   THEN cllreg := cllreg +1
   ELSE
   EXTEND node BY tmp
   WITH breg := tmp
        b(tmp) := breg
        decglseq(tmp) := decglseqreg
        sub(tmp) := subreg
        cll(tmp) := cllreg +1
   ENDEXTEND
   decglseqreg :=
      apply(mgu, [ ⟨body(clau),ctreg⟩ |cont])
   subreg := subreg ∘ mgu
   vi := vi +1
   mode := call
```

In the other rules of the previous ASM, only function *father* is renamed to *b*, and the abbreviations *decglseq(currnode)*, *father(currnode)* and *sub(currnode)* are replaced by *decglseqreg*, *breg* and *subreg*.

In our example, ASM2 goes through the states shown in Fig. 11 and 12, which correspond to those from Fig. 8 and 9 for the first ASM. Dashed arrows now point to the *cll* of a node. Since the former values attached to the *currnode* are now stored in registers, allocation of nodes corresponding to *B* and *D* is avoided. On the other hand, when node *A* is visited by backtracking (by executing the fail instruction in the state shown in Fig. 6), its choicepoint is moved to registers, and the following select instruction allocates the new, similar choicepoint *A'*.

In ASM2, the nodes which may be visited in the future are always reachable from *breg* via the *b* function. They form a stack, but note that there may still be abandoned nodes in the *node* universe, which are no longer reachable (here *A*). This causes one of the problems in the verification of the refinement of ASM1 to ASM2. The tuple of values *decglseq(n)*, *sub(n)*, *cll(n)* and *b(n)* attached to a stack node *n* is usually called a *choicepoint*.

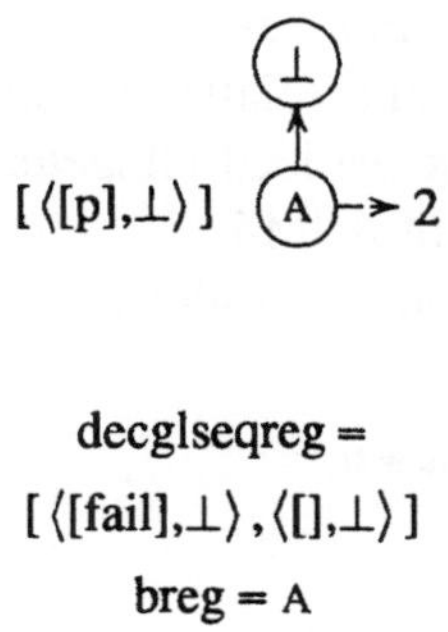

Fig. 11.

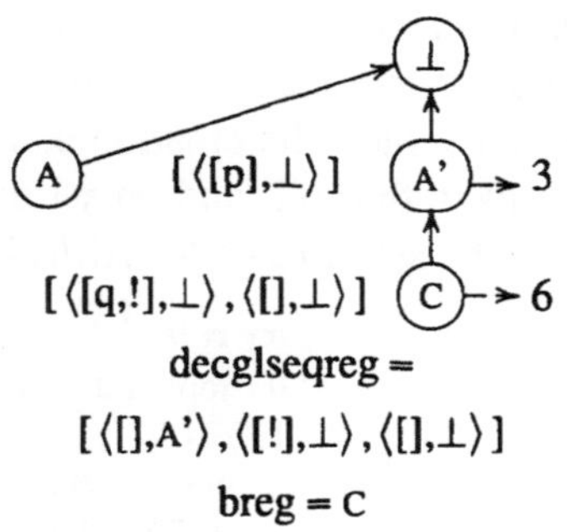

Fig. 12.

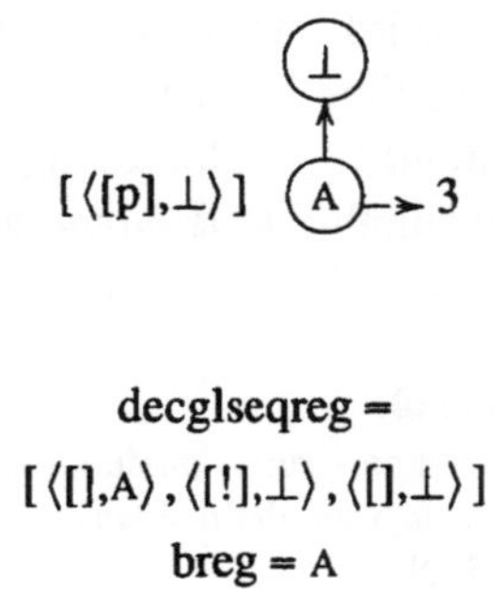

Fig. 13.

5.3. *The Third and Fourth Interpreter: Optimizations*

Although the second interpreter allocates fewer nodes than the first, there are two possibilities for improvements, which are exploited in ASMs 3 and 4.

The first one is again visible in our example. When the first clause for activator p is tried, select rule allocates a new node *A*, and set the values *decglseq(A)*, *sub(A)* and *cll(A)* of the new choicepoint.

Since the first alternative does not lead to a solution, the interpreter executes a *backtrack* instruction, which removes the node *A* from the stack. Thereby the whole choicepoint becomes inaccessible. The subsequent select rule for the second alternative then pushes the new choicepoint *A'* on the stack, which is exactly the same as the one for the first alternative, except that *cll(A')* has been incremented.

The optimization done in ASM3 avoids deallocation and reallocation of choicepoints. Instead it *reuses* the existing choicepoint. The optimization is achieved by replacing the removal of a choicepoint in the else-branch of backtracking with the assignment *mode := retry*, which activates a new rule, retry rule. This rule combines the effects of the else-branch of *backtrack* and *select*. It is executed instead of select rule for every alternative except the first. Its action is to remove a choicepoint (i.e. to set *breg* to *b(breg)*) only on execution of the last alternative. Otherwise it reuses the old choicepoint as required by incrementing *cll(breg)*.

The old select rule, which allocates a new choicepoint is now only called for the first alternative clause, is renamed to try rule. To avoid code duplication the common parts of try and retry rule are moved to a new enter rule.

The second place for improvement is addressed in interpreter 4. It is the allocation of choicepoints with empty lists of alternatives. Such a *useless* choicepoint is created e.g. for queries with just one alternative in the try rule of the third interpreter (resp. in select rule of the second). An example node is C in Fig. 7. Such a choicepoint is useless, since it will immediately be discarded by backtracking when visited. Its creation can be avoided altogether by suitable *look ahead* guards in the call- and retry rule of ASM4.

With both optimizations, the set of rules for ASM4 looks as follows

```
call rule
IF   stop = run ∧ mode = call
   ∧ is_user_defined(act)
THEN
    IF clause(procdef'(act,db')) = nil
    THEN backtrack
    ELSE
    cllreg := procdef'(act,db')
    ctreg := breg
       /* look ahead guard */
    IF clause(procdef'(act,db')+1, db')
       ≠ nil
    THEN mode := try
    ELSE mode := enter

enter rule
IF stop = run ∧ mode = Enter
THEN
    LET clau = rename(clause(cllreg),vi)
    LET mgu = unify(act, hd(clau))
    IF mgu = nil
    THEN backtrack
    ELSE
    decglseqreg :=
        apply(mgu,
              [<bdy(clau),ctreg> |cont])
    subreg := subreg ∘ mgu
    vi := vi +1
    mode := Call

goal success rule
IF stop = run ∧ goal = []
THEN decglseqreg :=
           rest(decglseqreg)

final success rule
IF   stop = run ∧ decglseqreg = []
THEN stop := halt
       subst := subreg

try rule
IF stop = run ∧ mode = try
THEN
    mode := enter
    EXTEND node BY tmp
    WITH breg := tmp
           b(tmp) := breg
           decglseq(tmp) := decglseqreg
           sub(tmp) := subreg
           cll(tmp) := cllreg +1
    ENDEXTEND

retry rule
IF stop = run ∧ mode = retry
THEN
    decglseqreg := decglseq(breg)
    subreg := sub(breg)
    cllreg := cll(breg)
    ctreg := b(breg)
    mode := Enter
       /* look ahead guard */
    IF clause(cll(breg) +1) ≠ nil
    THEN cll(breg) := cll(breg) +1
    ELSE breg := b(breg)

cut rule
IF stop = run ∧ act = !
THEN father := cutpt
         decglseqreg := cont

fail rule
IF stop = run ∧ act = fail
THEN backtrack

where
backtrack ≡
IF breg = ⊥
THEN stop := halt
        subst := failure
ELSE mode := retry
```

An additional register *ctreg* is now set in call and retry rule, to have the right *cutpt* available in enter rule. By the two optimizations, the state of ASM4 corresponding to the one of ASM2 from Fig. 12 is now given by Fig. 13.

5.4. *The Fifth Interpreter: Compilation of Backtracking Structure*

The first three refinement steps can be viewed as an optimization of the first ASM, which do not change the representation of the Prolog program. In contrast, the refinement from ASM4 to ASM5 compiles the predicate structure of Prolog. For the first time instructions are introduced, which will also be present in the final WAM.

The basic idea behind this refinement step is to move control of rule selection from the *mode* variable to the compiled code. In addition to the clauses, ASM5 code contains instructions, which control the stack manipulation as the mode did before.

The former register for the current clause line, *cllreg*, is therefore renamed to a program counter *pcreg*. Consequently, the continuation addresses for backtracking, attached to each choicepoint *n* by *cll(n)*, are now represented by *pc(n)*. Selection of a clause at a clause line (function *clause*) becomes selection of the code (clause or instruction) at an address (function *code*). Checks for the value of *mode* are replaced by checks on the type of the code stored at *pcreg*. As an example, the following clauses for a predicate p in a Prolog program

```
p(X)    :- body1.    p(g(X)) :- body3.
p(f(X)) :- body2.    p(g(X)) :- body4.
```

are translated to the code fragment (labels L1 – L4 are symbolic addresses)

```
L1: try_me_else(L2)
    p(X)    :- body1.
L2: retry_me_else(L3)
    p(f(X)) :- body2.
L3: retry_me_else(L4)
    p(g(X)) :- body3.
L4: trust_me
    p(g(X)) :- body4.
```

On a query ?- p(X), call rule of ASM5 (now called when *pcreg* is at a special *start* address) sets *pcreg* to address L1.

call rule
```
IF   stop = run
∧ is_user_defined(act)
∧ pcreg = start
THEN
   ctreg := breg
   IF code(procdef5(act,db5)) = nil
   THEN backtrack
   ELSE pcreg := procdef5(act,db5)
```

```
where
backtrack ≡
IF breg = ⊥
THEN stop := halt;
        subst := failure
ELSE pcreg := pc(breg)
```

Execution of the `try_me_else(L2)` instruction at address `L1` has a similar effect as the try rule of ASM4 had. The continuation address for alternative clauses stored in *pc(n)* now is `L2`. By incrementing *pcreg* the next clause considered is the one at `L1 +1`. Similarly, execution of a clause at `L2 +1` or `L4 +1` executes a rule with the same effect as enter rule of ASM4.

try_me rule
```
IF   stop = run
   ∧    code(pcreg,db5)
     = try_me_else(N)
THEN
   EXTEND node BY tmp
   WITH breg := tmp
          b(tmp) := breg
          decglseq(tmp) :=
             decglseqreg
          sub(tmp) := subreg
          pc(tmp) := N
   ENDEXTEND
   pcreg := pcreg +1
```

enter rule
```
IF   stop = run
   ∧ is_clause(code(pcreg,db5))
THEN
   LET clau = rename(code(pcreg),vi)
   LET mgu = unify(act, hd(clau))
   IF mgu = nil
   THEN backtrack
   ELSE
   decglseqreg :=
      apply(mgu,[<bdy(clau),ctreg> |cont])
   subreg := subreg ∘ mgu
   vi := vi +1
   pcreg := start
```

When *pcreg* = `L3` or *pcreg* = `L5` rules are executed that correspond to the then- and the else-branch of retry rule of ASM4.

retry_me rule
```
IF   stop = run
   ∧ code(pcreg,db5) = retry_me_else(N)
THEN
   decglseqreg := decglseq(breg)
   subreg := sub(breg)
   ctreg := b(breg)
   pc(breg) := N
   pcreg := pcreg +1
```

trust_me rule
```
IF   stop = run
   ∧ code(pcreg,db5) = trust_me
THEN
   decglseqreg := decglseq(breg)
   subreg := sub(breg)
   breg := b(breg)
   pcreg := pcreg +1
```

In general, the list of clauses for one predicate given in the original program is compiled to a code fragment stored in the memory of ASM5. The fragment starts with a `try_me_else` instruction and consist of the list of clauses separated by `retry_me_else` instructions, except the last, which is separated by a `trust_me`. Such a code fragment is called a *linear chain.* The requirement, that all code fragments must be linear chains is reflected in the compiler assumption for 4/5. Formally we have

$$\begin{aligned}&\text{mapcl}(\text{clls}(\text{procdef}'(\text{act},\text{db}'),\text{db}'),\text{db}')\\&=\text{chain}(\text{procdef}_5(\text{act},\text{db}_5),\text{db}_5)\end{aligned} \tag{6}$$

where *procdef'* and *db'* are the *procdef*-function and the Prolog program from ASMs 2, 3 and 4. $procdef_5$ is the new *procdef*-function for ASM5 and db_5 is the compiled Prolog program. The partial function *chain* terminates, if the code fragment stored at $procdef_5(act,db_5)$ is a linear chain, and delivers the clauses contained in it. A look at (Börger and Rosenzweig, 1995) shows a problem with the formal definition[1]. It does not guarantee that `try_me_else`, `retry_me_else`, `trust_me` instructions in the compiled Prolog program are only used in that order, as is stated in the text. A correct formalization of linear chains as a recursive program is (*cons* adds an element to the front of a list, instr $\equiv$ code(Ptr,db_5) and next_instr $\equiv$ code(Ptr +1,db_5))

chain(Ptr,db_5) =
  if instr = try_me_else(N) then chain-try-me(Ptr,db_5)
  if is_clause(instr) then [instr] else
  if instr = nil then [] else **undef**

chain-try-me(Ptr,db_5) =
  if instr = try_me_else(N) $\wedge$ is_clause(next_instr)
  then cons(next_instr,chain-retry-me(N,db_5)) else **undef**

chain-retry-me(Ptr,db_5) =
  if instr = retry_me_else(N) $\wedge$ is_clause(next_instr)
  then cons(next_instr,chain-retry-me(N,db_5)) else
  if instr = trust_me $\wedge$ is_clause(next_instr)
  then [next_instr] else **undef**

Non-termination of this program can be caused either by a cyclic pointer structure or by an attempt to compute **undef**. Therefore termination of *chain* is required in compiler assumption (6) above.

5.5. *The Sixth and Seventh Interpreter: Switching*

Until ASM5, the problem how to select clauses whose head 'may unify' with an activator has been delayed by using an abstract (underspecified) *procdef*-function. The problem is addressed in the two refinement steps from interpreter 5 to 7. ASM6 groups clauses together by introducing nested chains (at

[1] The definition is given for nested chains (our ASM6), while we consider only linear chains in ASM5, but the problem is the same.

positions, where linear chains contain clauses, a nested chain may contain another chain), and ASM7 introduces switching instructions. These restrict the possible implementations of the *procdef*-function. E.g. for an activator *act* = `p(X,g(Y))`, switching instructions may select the clauses with leading predicate symbol `p`, whose second argument is either a variable or starts with function symbol `g`. A detailed description of switching instructions is beyond the scope of this chapter, we refer the reader to (Börger and Rosenzweig, 1995) and (Aït-Kaci, 1991).

6. Verification

This section describes our work on the verification of the refinement steps from ASM1 to ASM7, which we have proven correct so far.

The following subsections will discuss the problems we found in the verification of each of the individual refinements. Before, let us give some impression on the verification process in general. As was discussed in Sect. 4, the critical point for a successful formal proof always is to find a coupling invariant INV($\underline{x}$,$\underline{x}$'), such that the diagram of Fig. 4 from Sect. 4.3 commutes for every corresponding pair of rules.

Verification of the refinements revealed, that it is impossible to state INV in a first proof attempt or to find all properties listed in INV in a pencil-and-paper proof. Depending on the complexity of the refinement step, between 6 and 17 iterations were needed to find an invariant, for which the proof goes through.

Our general experience was that every time one finds INV to be insufficient and therefore adds new properties, this again causes unprovable goals. To discharge these new goals INV has to be improved again, leading to an evolutionary process of improving INV by verification attempts. The good support KIV offers for this process (correctness management, the possibilities to inspect and modify proof trees and the reuse of proofs) were very important for minimizing the overhead of the iterations.

Sect. 6.1 gives some impression of the verification problems of the first refinement. Verification of the optimizations 2/3 and 3/4 is shortly discussed in Sect. 6.2.

Sect. 6.3 is concerned with the verification of the first proper compilation step 4/5. We will try to give an impression of the evolutionary development of the coupling invariant for this refinement, by stating the initial coupling invariant, and by indicating, what changes were necessary during iterated proof attempts. We will also show, how an error was uncovered, which is present in the rules of all ASMs from ASM3 on. The error was found during the verification of 4/5, since 4/5 was done before 2/3 and 3/4. The reason was that we wanted to find out if the different type of refinement (compilation, not optimization) would pose different new problems.

The equivalence proofs for the refinements from ASM5 to ASM7 were the most complex verification problems we have tackled so far in this case study. Verification uncovered some problems in the compiler assumptions as well as a missing backtracking-clause in the switching instructions. For more details on this verification the interested reader is referred to (Schellhorn and Ahrendt, 1997).

The final subsection 6.4 gives some statistics.

6.1. *From Trees to Stacks: Verifying the First Refinement*

Verification of the first refinement started from the 9 properties as given in (Börger and Rosenzweig, 1995), and a proof sketch given in (Schmitt, 1994), which added 3 more and made some minor corrections. These properties just formalize our description (see 5.2) of the changes that were made in the first refinement. Since ASM1 and ASM2 allocate different sets of nodes, a mapping *F* from nodes of ASM2 to corresponding nodes of ASM1 is used (which e.g. maps *breg* to *currnode*), just as in (Börger and Rosenzweig, 1995).

As already (Schmitt, 1994) pointed out, *F* cannot be given statically, but has to be defined by induction on the number of rule applications. Therefore we use an existentially quantified dynamic function for this purpose (note, that without our representation of dynamic functions as data structures, the quantification would not be first-order).

During verification we had to learn that the initially given properties were far from being sufficient for a formal proof. Another 20 properties had to be added in 12 iterated proof attempts. Some major ones were the characterization of the *stack* nodes of ASM2 (those nodes, which are reachable from *breg* via the *b*-function), the injectivity of the mapping *F* for those stack nodes, and a precise characterization of the search tree structure (all candidate nodes must be different and disjoint from the stack nodes, all cutpoints stored in decorated goal sequences must be stack nodes etc.).

A detailed discussion on how we found these new properties during proof attempts in two months of work is given in (Schellhorn and Ahrendt, 1997).

6.2. *Verifying the Optimizations*

Verification of the refinements from ASM2 to ASM4 is easier (done in 3 weeks) than the verification of 1/2 and 4/5 done before, since coupling invariants sufficient for a correctness proof are much easier to find. Although some additional properties to the ones given in (Börger and Rosenzweig, 1995) are necessary, (e.g. the nodes mentioned in them must be stack nodes), they provide a good starting point.

The main problem in the equivalence proof of 2/3 is how to generalize the proof technique as described in sect. 4.3 to cases, where m rule applications of ASM2 correspond to n rule applications of ASM3 (there are cases with m:n = 2:3, 1:2). A general solution, which guarantees that correctness and completeness can be shown in *one* proof, will be described in (Schellhorn, 1998).

In the verification of the refinement from 3/4 we (like (Börger and Rosenzweig, 1995) too) use a coupling invariant, which relates computation states that have the same "useful" choicepoints with non-empty clause list.

Unfortunately, with this coupling invariant proof obligation PO3 from section 4.3 (goal (3)), which states that both interpreters must terminate at the same time, is not provable. ASM4 may already have stopped with result *failure*, while ASM3 still has to backtrack from useless choicepoints. This situation of *non-simultaneous termination* has to be considered carefully in the coupling invariant. An additional argument is needed to guarantee that for two states of ASM3 and ASM4 related by the coupling invariant, removing useless choicepoints by executing rules of ASM3 will keep the invariant and eventually lead to a state where *stop* = *stop'* holds again. A generalized version of PO3 then becomes provable with this argument.

6.3. *Verifying Compilation of Backtracking Structure*

At a first glance, the refinement 4/5 seems to be easy to verify, since the mapping between rules is 1:1 (one rule of ASM4 corresponds to one of ASM5), and both interpreters allocate the same set of nodes, so no function F (like in 1/2) is required.

Nevertheless, proving the refinement of 4/5 poses some new problems, since this refinement is the first, in which a compiler assumption plays an important role. To demonstrate the complexity (we needed 9 iterations), we will try to give an impression of the development process necessary to find a correct coupling invariant (see also (Ahrendt, 1995)). Unfortunately, we cannot avoid to confront the reader with a lot of details, which were uncovered during the verification. Only the consideration of these details leads to the detection of all hidden assumptions, which are necessary to ultimately *guarantee* the absence of errors in the refinement. And indeed, the attempt to verify the refinement revealed an error in one of rules shown in Sect. 5.4.

Let us start with our first attempt to define a coupling invariant. According to the description in Sect. 5.4, most of the corresponding registers from ASM4 and ASM5 should be identical. This gives the following properties (using the convention, that variables of ASM5 are written with a prime).

1. vi = vi'
2. stop = stop'
3. subreg = subreg'
4. breg = breg'
5. decglseqreg = decglseqreg'
6. ctreg = ctreg'
7. sub ^ n = sub' ^ n
8. b ^ n = b' ^ n
9. decglseq ^ n = decglseq' ^ n

Application of dynamic functions is written with infix ^, the notation used in KIV (see Sect. 4.1). The node *n* in Properties 7–9 must be universally quantified over the set *ns* of currently allocated nodes, except the root node ⊥, and we adopt this convention whenever a node *n* is mentioned in the following. The universal quantification causes a first problem. All relevant nodes must be guaranteed to be in this set (which is the same for both interpreters). We additionally need

10. breg ∈ ns
11. ctreg ∈ ns
12. decglseqreg $\in_{ctp}$ ns
13. decglseq ^ n $\in_{ctp}$ ns
14. b ^ n ∈ ns
15. ns = ns'

where *dg* $\in_{ctp}$ *ns* is defined to mean "all cutpoints contained in *dg* (a decorated goal sequence) are contained in *ns* (a set of nodes)".

Now it remains to define the relation between the clause lines stored in *cllreg* and *cll ^ n* in ASM4 and those stored in *pcreg* and *pc ^ n* in ASM5.

Starting with *cllreg* and *pcreg*, a comparison of the rule guards shows, that the *type* of the code, *pcreg* points to in ASM5, corresponds to the value of *mode* in ASM4. Therefore the definition will be by case distinction on the *mode*. Moreover, the invariant must guarantee that the *content* of the code(-sequent), *pcreg* points to, must always be equal to the clause(-list), *cllreg* points to.

mode = call: In this case, the content of *cllreg* is irrelevant (it will be overwritten by the call rule), and the value of *pcreg* is start.

16. mode = call → pcreg = start

mode = enter: Here, *pcreg* must point to a clause, in particular to the same as *cllreg* does.

17. mode = enter → clause(cllreg,db') = code(pcreg,db_5)

mode = try: Two properties must be guaranteed. First, *pcreg* must point to a try_me_else instruction, and second the clauses contained in the following instructions must be the same as those following *cllreg*. The second property is formalized by

18. mode = try → mapcl(clls(cllreg,db')) = chain-try-me(pcreg,db_5)

But note that the first property also follows from 18, since the definedness of chain-try-me implies that *pcreg* points to a try_me_else instruction. Actually, the check for the instruction at *pcreg* to be a try_me_else instruction was introduced just for this reason. It would not have been necessary, if chain-try-me were only used as an auxiliary function for chain in the compiler assumption, since chain contains the same check already. This is a general result for the definition of compiler assumptions. Design auxiliary procedures (or assumptions) in such a way that they are suitable to describe intermediate execution states of the ASM.

mode = retry: In this mode, *pcreg* has been set to *pc ^ breg'*:

19. mode = retry → pcreg = pc ^ breg'

This property is sufficient to ensure that *pcreg* points to a *retry_me_else* or *trust_me*-instruction, provided that we specify this property for all stack nodes *pc ^ n*:

20. mapcl(clls(cll ^ n,db')) = chain-retry-me(pc ^ n,db_5)

Again, definedness of *chain-retry-me* guarantees that *pc ^ n* indeed points to a retry_me_else instruction.

This completes the description of the initial invariant. What we have shown so far is the process of looking for a first version of an invariant, *before* starting the first proof attempt with the system. We demonstrated that doing this is not a trivial task at all. But, in our experience of verifying refinements, each hour invested in a good starting point can save days of verification time.

Now we like to describe the process of completing the invariant by iterated proof attempts with KIV until it suffices for the inductive proof. We give a rough overview of this *search* rather then describing the logical deduction, explaining how hidden assumptions were detected (if the proof needed them explicitly) and how proving these new formulas leads to new gaps an so on. We take this proof-historical point of view to emphasize the evolutionary nature of solving the given problem.

With the current invariant we tried to prove that the diagram shown in Fig. 4 (Sect. 4.3) commutes for each corresponding pair of rules. The first case we tried was the one with the two cut rules (since this was one of the most problematic proofs in the verification of 1/2). After some minutes of proving, we arrived at the subgoal, in which property 19 had to be proved to hold after rule application. The property holds trivially, since the cut rule of ASM4 is always called in *call* mode, and does not change it. But our invariant was too weak to allow a proof of this property. Taking into account, that fail rule is similar, we added (note again the convention, that guards of rules, which mention *act*, implicitly contain the conjunct decglseqreg $\neq$ [] $\wedge$ goal $\neq$ [])

21. decglseqreg=[] $\vee$ goal=[] $\vee$ $\neg$ is_used_defined(act) $\rightarrow$ mode = call

to the coupling invariant. With the new invariant the case for the cut rules succeeds. The case for the retry rules revealed another problem caused by the formula we just introduced. In retry rule *decglseqreg* is loaded with *decglseq ^ breg* from the stack. If this decorated goal sequence were empty, we would have mode = retry, violating property 21. To resolve this problem, we have to ensure, that the activator of every decorated goal sequence stored on the stack always is user defined (we abbreviate the goal and the activator of decglseq ^ n with goal ^ n and act ^ n).

22. decglseq ^ n $\neq$ [] $\wedge$ goal ^ n $\neq$ [] $\wedge$ is_user_defined(act ^ n)

A further attempt to prove the case of retry rules uncovers a new problem. We cannot guarantee, that retry rule of ASM4 is executed with breg $\neq \perp$ (which would load the registers with the undefined values at the root node). Therefore we strengthen property 19 to

19a. mode = retry $\rightarrow$ pcreg = pc ^ breg' $\wedge$ breg $\neq \perp$

With this invariant, the proof for most of the cases goes through. One of the last rules we considered was the fail rule (assuming this proof should be one of the easy ones ...). But a proof attempt finally gives the goal

$\neg$ (decglseqreg = [] $\vee$ goal = [] $\vee$ $\neg$ is_user_defined(act))

which is obviously not provable. Analysis of the proof branch shows, that this goal arose from trying to prove

decglseqreg=[] $\vee$ goal=[] $\vee$ $\neg$ is_user_defined(act)
$\rightarrow$ retry=call

which is itself an instance of property 20 after fail rule. This means, that our assumption, that fail rule would be called only in call mode (like cut rule) was wrong. But if it is wrong, then there is something wrong not only with our invariant, but also with fail rule!

To see the problem, look again at the fail rule from ASM4, as it was shown in Sect. 5.3. The obvious *intention* of the rule is that retry rule should be executed afterwards.

Now it seems to be obvious that the only rule applicable after execution of the fail rule is indeed retry rule. But our correctness proofs reveals that fail rule does not invalidate its own guard, so it may be executed again, leading to an infinite loop. The rule system is therefore indeterministic (or following the newer terminology of (Gurevich, 1995), inconsistent), and does no longer correctly implement a Prolog interpreter.

Although the error is easy to correct (the conjunct *mode* = *call* must be added to the guard of fail rule), we think this is a typical error that is very difficult to find even by intensive inspection (and, of course, we *had* to inspect the code thoroughly before we could make an attempt to define a coupling invariant). A reader will always unconsciously resolve the indeterminism in the intended way. Nevertheless, an implementation is blind for intentions, and will possibly resolve the conflict in the wrong way (and ours did!).

Correction of the fail rule still gives *act* = fail, and *mode* = retry after its application. Therefore property 21 must be changed to

21a. (decglseqreg=[] ∨ goal=[] ∨ act = !) → mode = call

Unfortunately, still one detail is incorrect in our coupling invariant. In the case of the proof, which considers the two try rules, we find that we cannot prove that the choicepoint pushed on the stack is user defined. We have to add

23. mode = try → is_user_defined(act)

Finally, the equivalence proof of 4/5 succeeds with the invariant

vi = vi' ∧ stop = stop' ∧ subreg = subreg' ∧ breg = breg' ∧ ctreg = ctreg'
∧ decglseqreg = decglseqreg' ∧ breg ∈ ns ∧ ctreg ∈ ns ∧ decglseqreg $\in_{ctp}$ ns
∧ ns = ns' ∧ (mode = call → pcreg = start)
∧ (mode = enter → clause(cllreg,db') = clause(pcreg))
∧ (mode = try
  → is_user_defined(act) ∧ mapcl(clls(cllreg,db')) = chain-try-me(pcreg,db_5))
∧ (mode = retry → pcreg = pc ^breg'∧ breg ≠ ⊥)
∧ ((decglseqreg = [] ∨ goal = [] ∨ act = !) → mode = call)
∧ (∀ n. n ∈ ns ∧ n ≠ ⊥
  → sub ^n = sub' ^n ∧ b ^n = b' ^n
  ∧ decglseq ^n = decglseq' ^n ∧ b ^n ∈ ns
  ∧ decglseq ^n ≠ [] ∧ goal ^n ≠ []
  ∧ is_used_defined(act ^n) ∧ decglseq ^n $\in_{ctp}$ ns
  ∧ mapcl(clls(cll ^n,db')) = chain-retry-me(pc ^n,db_5))

6.4. *Statistics*

The following table gives the number of KIV proof steps, the number of interactively given proof steps and the number of theorems for the final version of each refinement proof. Also the time it took to specify and verify each refinement, the number of iterations needed to find the final invariant and its size (in lines) are given.

The size of the interpreters starts with 120 lines of (PASCAL-)code and reaches 240 lines for ASM7. The algebraic specifications of all datatypes is composed of about 90 subspecifications with altogether 454 axioms. A first version of the specifications and the interpreters was written within a day. Only minor corrections were necessary. 580 first-order lemmas are used, 249 of them were from the library. The remaining 331 were added on demand during the proofs of the main equivalence theorems. Their proofs needed 441 interactions and 1462 proof steps.

	1/2	2/3	3/4	4/5	5/6	5/7
Proof Steps	1475	4425	3988	2936	6190	20085
Interactions	246	450	580	237	666	1970
Theorems	16	32	31	32	53	72
Iterations	12	8	5	9	8	17
Verif. time	2 months	2 weeks	1 week	1 month	2 weeks	2 months
Size of INV	20	25	25	14	53	97

As the statistic shows, the proofs for 5/7 were extraordinary complex. *All* of the problems we encountered in the refinements of ASM1 to ASM5 are present in the equivalence proof of ASM5 and ASM7: stack nodes must be characterized (like in 1/2), rules do not correspond 1:1 (like in 2/3), the interpreters may terminate non-simultaneously (like in 3/4), and the compiler assumption must be carefully translated to assumptions about intermediate interpreter states (like in 4/5). Also the verification of 5/6 showed that splitting verification of 5/7 into two parts complicates the work instead of simplifying it. Therefore we constructed the equivalence proof between ASM5 and ASM7 directly.

The statistic also shows a drastic increase of productivity. The main reason (besides increasing familiarity with the topic) are improvements in the KIV system, which were done from the experience we had gained from verifying refinements 1/2 and 4/5. A lot of heuristics (most notably the ones for quantifier instantiation, unfolding recursive procedures and for loops) were improved, as well as the efficiency of the simplifier (see Chapter II.1.1).

7. Related Work

Work on compiler verification in general (or even more general: data refinement and other refinement relations) is so numerous that we will not even attempt to give an overview.

From the work on formal system-supported verification of compilers we exemplarily want to mention the work with NQTHM on the formal verification of a compiler for an imperative language ((Moore, 1988), (Young, 1988)). This work is based on the notion of "interpreter equivalence" which is quite similar to our notion of equivalence of ASMs. It also contains a lot of references to related work.

Of specific work on the formal verification of a Prolog compiler we are aware only of the parallel work of C. Pusch in Munich. She also verified some refinement steps with the Isabelle system (Pusch, 1996).

The formalism used in Isabelle are inductively defined relations on the tuple of variables, which correspond to the semantics of our imperative programs as relations over their values. Pattern matching notation and polymorphism as used in functional languages allow to write the rules of an ASM in a more concise notation than our notation as PASCAL programs.

In contrast to our approach, which starts from a Prolog semantics based on search trees and tries to model the ASM approach as faithfully as possible, verification in Isabelle started from an operational Prolog semantics which is already based on stacks. Stacks were modeled as simple lists, which allows to avoid the characterization of stack nodes.

Instead of our refinement from ASM1 to ASM2 two other refinement steps were verified. Refinements 3 and 4 as verified in Isabelle are the same as our refinements from ASM2 to ASM4. In Isabelle, the Prolog-construct *fail* was not considered, therefore the error we found in the interpreters 3 and 4 was not present in the case study.

The verification effort for the four refinement steps as given in (Pusch, 1996) was 7 person months and 3500 interactions. These numbers are about two times the numbers we got for the verification of the three refinements to reach ASM4. We suspect that this is largely due to the use of an asymmetric proof technique using proof maps, which requires two separate large proofs for correctness and completeness for each refinement step instead of one symmetric proof.

8. Conclusion

We have presented a framework for the formal verification of the Prolog to WAM compilation as given in (Börger and Rosenzweig, 1995). The framework is based on the translation of sequential Abstract State Machines to imperative programs over algebraic specifications. With this translation correctness and completeness of the refinement between two ASMs is expressible as program equivalence in Dynamic Logic.

We introduced a proof technique based on coupling invariants, which corresponds to the use of proof maps over ASMs. We have found that the correct coupling invariants, which are needed to show correctness and completeness of refinement steps, are far too complex to be stated correctly in a first attempt. The incremental development of a correct version takes much more time than the verification of the correct solution. Therefore, besides the pure power of the theorem prover, the 'proof engineering' support offered by the verification system (explicit proof trees, correctness management, reuse of proofs etc.) is crucial for the feasibility of the case study.

Verification showed that (Börger and Rosenzweig, 1995) is indeed an excellent analysis of the compilation problem from Prolog to WAM. Nevertheless an unintended indeterminism in one of the ASMs had to be removed (6.3), and minor corrections were necessary on the formalization of the compiler assumptions. These results show that, to *guarantee* compiler correctness, mathematical analysis should be followed by formal verification.

Let us conclude with an outlook on the continuing work on this case study. The next two of the remaining 6 refinement steps are concerned with the compilation of single clauses ([Section 3] in (Börger and Rosenzweig, 1995)). Their correctness should be easy to show and require no new proof techniques. New problems will have to be overcome to verify that Prolog-Terms can be represented by pointer structures (the final [Section 4] in (Börger and Rosenzweig, 1995)). Finally it would remain to verify a compiler built on the basis of the compiler assumptions.

Although we are currently only about half the way from Prolog to the WAM, verification of the first levels has confirmed our belief that verification of the WAM is a feasible, but challenging task.

ACKNOWLEDGEMENTS

We thank our colleagues Wolfgang Reif, Kurt Stenzel and Matthias Ott for their valuable comments on drafts of this chapter.

REFERENCES

Ahrendt, W. (1995). Von PROLOG zur WAM — Verifikation der Prozedurübersetzung mit KIV. Diplomarbeit, Fakultät für Informatik, Universität Karlsruhe. (in German).

Aït-Kaci, H. (1991). *Warren's Abstract Machine. A Tutorial Reconstruction.* MIT Press.

Börger, E. and Rosenzweig, D. (1995). The WAM—definition and compiler correctness. In Beierle, C. and Plümer, L., editors, *Logic Programming: Formal Methods and Practical Applications*, volume 11 of *Studies in Computer Science and Artificial Intelligence*. North-Holland, Amsterdam.

Goldblatt, R. (1982). *Axiomatising the Logic of Computer Programming*. Springer LNCS 130.

Gurevich, M. (1995). Evolving algebras 1993: Lipari guide. In Börger, E., editor, *Specification and Validation Methods*. Oxford University Press.

Harel, D. (1984). Dynamic logic. In Gabbay, D. and Guenther, F., editors, *Handbook of Philosophical Logic*, volume 2, pages 496–604. Reidel.

Moore, J. (1988). PITON: A Verified Assembly Level Language. Technical report 22, Computational Logic Inc. available at the URL: http://www.cli.com.

Pusch, C. (1996). Verification of Compiler Correctness for the WAM. In *Proc. of the 1996 Int. Conf. on Theorem Proving in Higher Order Logics*, LNCS 1125.

Reif, W. (1995). The KIV-approach to Software Verification. In Broy, M. and Jähnichen, S., editors, *KORSO: Methods, Languages, and Tools for the Construction of Correct Software – Final Report*. Springer LNCS 1009.

Reif, W., Schellhorn, G., and Stenzel, K. (1995). Interactive Correctness Proofs for Software Modules Using KIV. In *Tenth Annual Conference on Computer Assurance*, IEEE press. NIST, Gaithersburg (MD), USA.

Reif, W., Schellhorn, G., and Stenzel, K. (1997). Proving System Correctness with KIV 3.0. In *14th International Conference on Automated Deduction. Proceedings*, pages 69 – 72. Townsville, Australia, Springer LNCS 1249.

Reif, W., Schellhorn, G., Stenzel, K., and Balser, M. (1998). Structured Specification and Interactive Proofs with KIV. In Bibel, W. and Schmitt, P., editors, *Automated Deduction — A Basis for Applications*, volume I, 3. Kluwer. to appear.

Reif, W. and Stenzel, K. (1995). Reuse of Proofs in Software Verification. In Köhler, J., editor, *Workshop on Formal Approaches to the Reuse of Plans, Proofs, and Programs*. Montreal, Quebec.

Schellhorn, G. (1998). A general proof technique for ASM refinements. to appear as a Technical Report of the University of Ulm.

Schellhorn, G. and Ahrendt, W. (1997). Reasoning about Abstract State Machines: The WAM Case Study. *Journal of Universal Computer Science (J.UCS)*. available at the URL: http://hyperg.iicm.tu-graz.ac.at/jucs/.

Schmitt, P. H. (1994). Proving WAM compiler correctness. Interner Bericht 33/94, Universität Karlsruhe, Fakultät für Informatik.

Sterling, L. and Shapiro, E. (1986). *The Art of Prolog*. MIT Press.

Young, W. D. (1988). A Verified Code Generator for a Subset of Gypsy. Technical report 33, Computational Logic Inc. available at the URL: http://www.cli.com.

INGO DAHN AND JOHANN SCHUMANN

CHAPTER 8

USING AUTOMATED THEOREM PROVERS IN VERIFICATION OF PROTOCOLS

1. INTRODUCTION

Protocols play an important role for any kind of applications where two or more participants (systems, users, services) have to communicate with each other. In particular with the growing importance of distributed computer systems and services (networking, distributed operating systems), protocols are becoming more and more important. As a standardized basis for protocols, the ISO/OSI stack of protocol levels is usually used. Given the transport medium on the lowest level, upper levels (e.g., transport layer, application layer) provide additional services to handle and secure data transmission.

Typically, *communication protocols* (e.g., Alternating Bit-, Sliding Window-, Stenning-protocol) are located on level 2 of the OSI/ISO stack: communication protocols have to ensure secure transmission of data, although the medium may be unreliable. This means that no data packages must be lost or modified. On the other hand, *authentication protocols* are located on a higher level on the OSI/ISO model. They assume that the communication between the partners is physically reliable. These protocols have to ensure that data can be transmitted safely, i.e., that no intruder can overhear data, provide false data or disturb the protocol.

All protocols must be highly reliable. In particular, in business applications or aircraft/spacecraft applications lives and millions of dollars are at stake. Experience shows that many errors usually occur in designing a protocol. In order to ensure reliability, formal methods must be used. Only during their application, it is possible to obtain proofs and thus guarantee the absence of errors.

Proving properties for protocols has usually been done manually or with highly interactive tools. Thus, the verification of a protocol turned out to be a time-consuming and error-prone procedure. In this chapter, we present the results of three case studies which show that for the purpose of protocol verification, automated theorem provers for first order predicate logic can be used in a convincing way.

This chapter proceeds as follows: the first section covers the verification of communication protocols. After a short introduction into general issues, the

W. Bibel, P. H. Schmitt (eds.), Automated Deduction. A basis for applications. Vol. III

verification of the Alternating Bit Protocol using the tool ILF is presented. ILF has an interactive theorem prover which allows us the export of proof tasks to automated theorem provers. A second case study shows the results of experiments made with a FOCUS specification of the Stenning protocol. Both case studies illustrate the use of automated provers in different situations of protocol design. While the first concerns the proof that a given protocol has certain properties, the second originated from a situation where the task was to design a protocol meeting given conditions. Section 3 presents an approach to automated verification of authentication protocols using the BAN-logic. After an introduction into the BAN-logic, we describe the transformation of BAN-formulas into first order logic and their execution with the automated theorem prover SETHEO.

A discussion and perspectives of using automated theorem provers in the verification of protocols conclude this chapter. The appendix contains selected parts of the formal treatment of the Kerberos authentication protocol.

2. Verification of Communication Protocols

2.1. *Introduction and Related Work*

Communication protocols are tools to ensure reliable communication over unreliable communication channels. They are realized in distributed systems having (at least) the following components:

- a sender,
- a receiver,
- a message channel to transmit information from the sender to the receiver and
- an acknowledge channel to transmit information from the receiver to the sender.

Each of these components has its own structure. At least they will have a storage to hold pieces of information.

Initially, the message storage of the sender contains the finite sequence of messages to be transmitted and that of the receiver is empty. When the sender puts a message to the message channel, it adds some control information. The channel may transmit the message and the control information correctly or it might scramble or even lose them. The receiver gets messages and control information from the message channel and analyzes them. Then it decides whether or not to trust the information, and puts an appropriate control information into the acknowledge channel. The acknowledge channel may be as unreliable as the message channel. After having read the acknowledge information, the sender decides whether it has to retransmit information or if it can continue to transmit messages which have not yet been sent.

The specification of a communication protocol describes the relevant properties of the aforementioned components. From this description, safety and liveness of the protocol have to be proved. By *safety*, we mean that the right data are transmitted in the correct order, i.e. that the sequence of received messages is always an initial segment of the sequence of messages to be transmitted. *Liveness* ensures that the system does not get blocked so that eventually all messages will be transmitted.

Correctness of communication protocols can be established by a variety of means. If the number of messages and the number of steps are finite and bound by a number which can be fixed in advance, model checking methods can be used successfully. (Bezem et al., 1995) verify a variant of the Alternating Bit Protocol in the COQ system, using a specification as an algebra of communicating processes. (Shankar and Lam, 1992) give correctness proofs for stepwise refined protocol specification using ad-hoc rules of inference. (Larsen and Milner, 1992) proves correctness of the Alternating Bit Protocol using bisimulation. (Huggins, 1995) verifies the Kermit Protocol and the Alternating Bit Protocol informally, starting from a specification by means of abstract state machines. (Nipkow and Slind, 1995) give a formal verification of a similar protocol based on a description using automata.

In the following subsections, we shall report on the interactive verification of safety and liveness properties for the Alternating Bit Protocol and for the Stenning Protocol, both supported by the use of automated theorem provers. The first case study is done with an interactive theorem prover which automatically does the necessary preprocessing for the automated provers. This is a self contained verification, starting from a given specification in the language Z.

In the second case, we describe what can be gained when proof problems are formulated in a way which is optimal for a special theorem prover. Here, we analyze a verification previously done by a human expert and show that the occurring proof problems can be solved automatically.

2.2. *Verification of the Alternating Bit Protocol using ILF*

In this subsection, we discuss the use of automated theorem provers in an interactive environment for the verification of communication protocols. Using a verification of the Alternating Bit Protocol as an example, we shall follow the steps from a specification given in Z up to the generation of proof problems for first order automated provers. The verification has been actually performed by Th. Baar (Baar, 1997) within the ILF system using an interactive prover developed by the first author. This interactive prover made use of the automated theorem provers DISCOUNT (Denzinger et al., 1997) and SETHEO (Letz et al., 1992). The latter prover was developed by the second author.

Communication protocols have been verified in the past with several interactive provers. In connection with these experiments it has been also described, how to translate constructs of the Z specification language into higher order logic. However, currently most automated provers use unsorted first order logic. Moreover, they have specific ways to react to certain syntactic constructs. This has to be taken into account when input files for automated provers are to be generated from Z specifications.

We shall introduce basic features of the Z specification language and of the Alternating Bit Protocol. Then, we describe the translation of the Z specification of the Alternating Bit Protocol into a formal theory in a sorted first order logic. We consider the resulting proof problems and the challenges they pose to an interactive theorem prover.

Passing from an interactive prover to an automated prover requires a translation of the proof problems into unsorted first order logic by encoding type information. We describe a method to achieve this in the given field of application.

The Alternating Bit Protocol and Z

The Alternating Bit Protocol, first proposed in (Bartlett et al., 1969), is very simple, but its main idea is also used in other more powerful protocols. The sender tags each message to be transmitted with an additional bit. The receiver returns the tags to the sender through the acknowledge channel. He will consider a message as correct only if the tag contains the expected bit value. Similarly, the sender will proceed to send the next message (with a flipped tag bit) only if he received the last used tag bit from the acknowledge channel.

The protocol is safe only under some additional assumptions. For example the channels may loose messages, but they may not change them or change the order in which they appear. In order to prove the liveness of the protocol, some fairness assumptions must be met. For example it must be assumed that the channels do not lose all messages. All these conditions have to be described in the specification of the protocol.

The specification language Z is widely used for the specification of distributed systems and protocols. It is based on set theory and sorted first order predicate logic. A library of well-known mathematical theorems — called Mathematical Toolkit — is an integral part of Z.

Translation of Z Specifications to Formal Theories

Z can be considered as a set of tools that facilitate the formulation of complex theories. The first of these tools is the type system. Starting from user defined basic types and from the type of natural numbers, more complex types are formed by two type constructors only: Power set and Cartesian product. Moreover, some other syntactic constructs in Z notation are printed like types, though Z type checkers do not evaluate them.

For example the Z notation suggests that it is possible to quantify variables for functions which take arguments of a type t_1 and yield values of a type t_2. However, such a quantifier will in fact range over subsets of the Cartesian product of t_1 and t_2 and will be relativized by a condition that selects graphs of functions from arbitrary binary relations. For the user, this is just a matter of notation. For formal verification it makes an essential difference.

For the user writing a formal specification the Z type checker is a tool to discover errors in specifications. Type checking in Z is deterministic and static. The main problem of theorem provers is to restrict the proof search. Now, this becomes another important purpose of the type system.

When a prover realizes that a variable cannot be instantiated with a term because of a type clash, it can rule out a possibility of proof construction. Hence, an extensive type system can lead to a considerable restriction of the search space.

To achieve this, an appropriate type checker must be called whenever a variable is to be instantiated with another variable or non-variable term, i. e. type checking becomes dynamic. Unification is a basic operation in theorem provers. Hence, type checking must not be be too expensive.

Especially, type systems are inappropriate when they admit the definition of types by explicit conditions, like in PVS. The verification of these conditions requires more inferences - this can in fact extend the provers search space. Instead, such types have to be introduced together with function declarations which give purely algorithmic and syntactic criteria to check whether a term has a specific type.

Then the restriction of a type by a condition leads — in extension of the Z type system — to a hierarchy of types. Such a type hierarchy can also reduce the number of proof obligations. Instead of proving separately that each object of type t_1, t_2 respectively has a certain property, it suffices to prove this property once for all objects of a common super-type containing t_1 and t_2. Especially, basic mathematical knowledge is often available for most general classes of objects.

Z permits generic types, e. g. the type of all binary relations with arguments of type α, β where α and β are type variables. Hence, polymorphic types are necessary for a language where Z constructs can be interpreted directly.

It is well-known in automated theorem proving that equality deserves special attention. In Z, the equality predicate requires arguments of equal types. Semantically this means that each type has its own equality relation. But automated theorem provers are not prepared to handle a variety of equalities.

Theorem provers without explicit treatment of equality, like SETHEO, can be given several binary relations with axiomatizations of the basic properties of equality or a single ternary equality relation which has the actual type as an additional parameter. Thus they can access Z equality.

But, for these provers, the substitution of equal terms deep inside some complex constructs, will require a long chain of deductions using substitutivity axioms. Hence, such provers will not perform well on equality problems. Since equality plays an important role in the axiomatization of communication protocols, the use of such provers has to be combined with the use of equational provers. In the experiments described below, the DISCOUNT system has been used to treat pure equality problems. This enforced the use of a single untyped equality relation.

The translation of a specification given in Z into the first order clause logic used by automated theorem provers is done in two steps. First, the Z specification is manually rewritten as a theory in a polymorphic sorted first order logic. The ILF input language used as a target language for this step has a more expressive type system than Z. Thus, Z phrases written like types could be translated into ILF types, even if they were not types in Z. For example ILF permits directly variables which can be instantiated with functions only.

The semantic of Z is static. Z specifications describe regular runs of systems. These runs are sequences of states, i.e. functions from the set of natural numbers into sets of states. While this is implicit in Z, it must be made explicit for the formal verification.

The basic construct of Z is the *schema*. It consists of a set of declarations of symbols and a set of related axioms. Schemas are named. They can be joined by aggregating their names. In this way, a hierarchy of theories is generated which reflects the structure of a complex system with several components in a natural way.

Schemas can be copied with decorations, i.e. all basic symbols of a schema are modified in a standard way, say by attaching a dash. This provides a convenient way to describe the differences in the states of a system before and after an action has occurred.

Z schemas are similar to descriptions of classes of mathematical structures, consisting of several component structures. The ILF input language permits quantification over such structures. However, the schemas of the protocol verification turn out to be inappropriate for the *verification* of communication protocols, though they are very convenient for their *specification*.

Z schemas combine all properties of a certain component of the system, especially its behavior under all relevant actions. In contrast, most verification proof problems require knowledge on the behavior of all system components when a specific action occurs. This suggests to resolve Z schemas into their basic components and to re-structure the theory in an action-oriented way.

This translation procedure leads to rather complex formulas. For example there is an axiom describing the value of the first member of the message storage of the sender at time $i+1$ in terms of values of other equally complex components.

It is error prone for the user to handle these complex terms. Therefore, ILF provides support for user defined shorthands. These are also used by the system to present complex machine-generated terms to the user. The user may introduce such shorthands at any time, even during the interactive proof editing process. Note that this has also the effect, that a formula may be presented by the system in form which is more compact than that entered by the user.

Safety and liveness of communication protocols concern properties of traces. In the formal theory describing a communication protocol, a trace is a sequence of system states. Sequences are functions with the set of natural numbers as its domain. They are treated as objects.

The theory obtained from the Z specification must be augmented by formulas with general mathematical contents — e. g. theorems on functions — which can be derived from the Mathematical Toolkit. They describe relevant properties of sequences, lists and other auxiliary concepts.

An Interactive Theorem Prover for Protocol Verification

The Alternating Bit Protocol was verified with the interactive theorem prover PROOFPAD. The verification can be performed simply by entering formulas to be proved. The PROOFPAD will extract the proof problems from the user input. They are analyzed and transformed automatically into problems suitable for the automated theorem provers DISCOUNT and SETHEO. Then the theorem provers are launched in the local network and, depending on the results of their work, the user is informed whether the gaps in his proof could be completed. If the provers did not succeed, the user has to insert further formulas or to restrict the set of axioms to be used for a specific proof problem. More details on the PROOFPAD can be found in Chapter II.4.14.

Each domain has its specific theorem proving techniques. For the PROOFPAD, they can be encoded as proof tactics.

For the verification of properties of sequential systems, an important method is to prove properties of runs by induction where the induction variable describes the time. A proof tactic can easily split such a proof problem into a problem on the initial states and a problem on state changes. Then, no further induction is required and non-inductive theorem provers can be used.

Unfolding of definitions is also a standard technique that can be performed by a poof tactic. When it replaced derived concepts in the proof problems on initial states of the Alternating Bit Protocol, the newly generated problems are simple enough to be solved by the available automated provers.

The induction step is automatically split into cases according to the possible combinations of preconditions that activate changes in the systems. In each of these single cases, a uniform proof tactic models the effect of the actions by applying substitutions suggested by the respective axioms. This tactic makes use of appropriate naming conventions for the axioms, derived from the Z specification. Its application removes the necessity to deal with changing time.

In complex terms many sub-terms can be replaced by other terms which are potentially equal. Hence, the reduction of term complexity by renaming techniques is an essential tool to facilitate the work for automated provers. Each case has its specific complex terms which have to be abbreviated by renaming.

By analyzing the concepts occurring in a problem, tactics can also restrict the set of axioms that should be used in proof attempts. E.g. in the induction step, axioms describing the initial state should be neglected. Note that this is based on a structuring of the theory according to the sequential behavior of the system, where the structure of the original Z specification only reflects its composition from sender, receiver and channels.

Clearly, all these tactics are domain specific but not specific for the verification of a particular protocol. They are helpful whenever the behavior of a system is specified by describing initial states and changes in discrete time steps. This is the case when Z or abstract state machines are used as specification tools.

The safety of the Alternating Bit Protocol is proved by a single call to a domain specific tactic. This tactic initiates the induction proof, simplifies the occurring terms by renaming, isolates the relevant cases, and generates proof problems that can be solved automatically.

The proof of the liveness of the protocol required 5 such tactics. (Nipkow and Slind, 1995) reports on the verification of a similar protocol using 480 tactic applications in the interactive theorem prover Isabelle.

In order to make the proof problems accessible to automated theorem provers like SETHEO and DISCOUNT, the type information contained in the formulas has to be translated into a form that is intelligible for the automated theorem provers used. This is achieved by encapsulating terms u into terms

$h(\tau, u)$ where τ is a term denoting the minimal value type of u. In this way, type checking is encoded into unification.

Type parameters were used like object parameters. Since all type constructors required for the verification of the Alternating Bit Protocol took arbitrary Z types as parameters, there was no need for additional encoding of types for type parameters.

DISCOUNT is based on term rewriting. Therefore, it turned out to be more efficient — for this prover — to encode type information only at the occurrences of variables. Then additional equations permitted DISCOUNT to insert the necessary type information when a unification was to be made. This leads to much smaller terms.

Experience

The experiments confirmed, that current automated theorem provers can bridge the gap between informal proofs in a specific domain and proofs in a formal logic calculus. The user had to supply all essential ideas of the proof. But the task of formal verification could be performed by the machine using automated provers and proof tactics realizing standard domain specific inference techniques. Thus, the user could use arguments from protocol verification instead of formal logic.

Although the automated provers did not find very complex proofs in this study, ILFs ability to present these proofs in a human readable format was a surprisingly valuable tool. It happened several times, that the user proposed an intermediate step in the proof which was in contradiction with other properties. From this contradiction, automated provers in the background justified subsequent lines in a formally correct way. Some of these user errors were discovered only when the user inspected the automatically generated proof. Since proofs in protocol verification can contain large terms, still expanded by encoding type information, a machine generated human readable presentation of these proofs is indispensable.

The available theorem provers are not yet optimal for the occurring proof problems. Absence of type handling made it necessary to encode types in order to enable the possible reduction of search space for the provers. However, the provers were not able to to treat the sub-terms used for the encoding of type information in a special way. This leads to inefficient behavior, especially unnecessary attempts to use equality reasoning with these terms.

Interactive editing of a large proof leads to the generation of numerous related proof problems. These problems will share many axioms. Parameter settings successful for one problem are likely to be successful for other problems too. But current provers are unable to utilize previous experience. Specifically, this leads to duplicated work in the prover's preprocessing phase.

The verification of the Alternating Bit Protocol required the use of 12 theorems on lists. These theorems did not contain any of the concepts from the specification and have not been proved in this experiment. They have to be taken from an attached general mathematical library or proved interactively from the Z Mathematical Toolkit by an appropriate specialist — i.e., by a mathematician.

2.3. *Verification of a* FOCUS *specification using SETHEO*

This case study describes experiments with the automated theorem prover SETHEO. The prover has been applied to proof tasks which arise during the formal development of a specification for the *Stenning* communication protocol. Its development and verification in FOCUS on the level of traces is described in "C. Dendorfer, R. Weber (1992a): *Development and Implementation of a Communication Protocol – An Exercise in* FOCUS".

During the formal development of a specification, the correctness of each step has to be proven. Many of these proofs are quite simple — when done by hand they require only few lines. Nevertheless, all these proofs have to be carried out on a highly formal level. This tends to be very time-consuming and error-prone, since a large number of proof tasks occur even in small specifications.

This case study has been made in order to evaluate how the automated theorem prover Setheo (Letz et al., 1992) can be applied to such proof tasks.

Focus

The typical development of a protocol specification in FOCUS (Dederichs et al., 1993) starts from an abstract description of the services provided by the upper layer and those provided by the lower layer (here: level of transport medium). FOCUS covers all design steps of the development from a non-constructive global specification down to an executable program. In our case study, we focus on the first steps, namely the refinement of the *global* requirements specification (a trace specification) to a *modular* requirements specification (also a trace specification). This refinement process is necessary, since the global requirements of the upper layer cannot be separated directly into local requirements of the transmitter, the receiver, and the transport medium.

Trace logic is used for these specifications. A trace specification states requirements for histories ("traces") of a distributed system as a sequence of *actions*. The description is in a way similar to linear time temporal logic (c.f. (Pneuli, 1986)). The protocol is defined by its actions (e.g., $snd(d), d \in$ UDU – send a data unit). Then, a trace is a sequence of such actions. The requirements of the protocol are given as a set of *liveness* and *safety* properties using predicate logic formulae, specifying "allowed" traces.

The task of a protocol verification is then to show that liveness and safety properties are not violated during any refinement step.

For the following, no detailed knowledge about the FOCUS method is required. We only give a list of required operators working on traces. For detailed definitions see (Dendorfer and Weber, 1992a). For traces t, u, v, and actions a, b, c

- $u \cdot t$ or $a \cdot t$ denotes concatenation of traces.
- $t \sqsubseteq u$ means "trace t is a prefix of u".
- $\#t$ denotes length of t. If the trace is infinite, $\#t = \infty$.
- $\langle a_0, \ldots, a_n \rangle$ denotes the trace consisting of actions $a_0, \ldots, a_n$.
- $t[k]$ denotes the k-th element of t (strong definition).
- $a©t$ denotes the filtered trace t that contains only actions a. For example, $Snd©t$ results in a trace, containing *snd*-actions only.
- a *in* t holds, if and only if action a occurs in trace t.
- $\langle d, k \rangle$ is a pair, consisting of a piece of data d and an integer k.

The Proof Tasks

The given proof tasks originate from a case study about the specification of the *Stenning protocol* (Stenning, 1976). The Stenning Protocol ensures reliable communication of upper layer data units (UDU)[1] using an unreliable transport medium ("lower layer"). This layer can loose packages ("lower layer data units", LDU) or permute the sequence of packages. Reliable communication is accomplished by adding a unique sequence number to each UDU and by introducing acknowledge messages: each package is sent repeatedly, until an acknowledge message with the correct sequence number is received by the transmitter.

In our FOCUS specification, the actions which define the Stenning protocol are: $snd(d)$ and $rec(d)$ (send/receive an upper layer data unit, $d \in UDU$), $get_R(x)$ and $put_T(x)$ (receive/send a lower layer data unit, $x \in LDU$), and $get_T(ack)$ and $put_R(ack)$ to receive (and send respectively) an acknowledge message (over the transport medium)[2]. The FOCUS specification of the Stenning protocol starts with services which are provided by the upper layer and the lower layer. Those provided on the upper layer ("reliable communication") are specified in the liveness property UL and the safety property US.

---

[1] These data units are seen as black boxes and are not further specified.

[2] The set of all actions of a given kind is defined as $Snd := \{snd(d) | d \in UDU\}$. *Rec*, Get_R, Put_T are defined in a similar way. Furthermore, a *clock* action "$\surd$" is also defined.

All properties have been directly taken from (Dendorfer and Weber, 1992a) and are shown for reference in Table I and Table II. UL means that the number of packages (UDU's) sent is equal to the number of received packages, i.e., no package has been lost, and US expresses that nothing "wrong" has been received. The properties LL^1, LL^2 and LS^1, LS^2 specify the behavior of the transport medium (in both directions): $LS^{1,2}$ states that if a data package (LDU) is received on one side, it must have been put previously onto the medium on the other side. However, it does not say that any sent package must be also received; losses are possible. $LL^{1,2}$ specifies that the medium is not broken forever (if we have sent an infinite number of packages, then an infinite number of packages is received after an infinite number of clock-cycles).

Table I. Liveness Properties

$$
\begin{aligned}
UL(t) &\equiv \#(Rec©t) = \#(Snd©t) \\
LL^1(t) &\equiv (\#(\mathrm{put}_T(e)©t) = \infty \Rightarrow \#(\mathrm{get}_R(e)©t) = \infty) \wedge \#(\surd©t) = \infty \\
LL^2(t) &\equiv (\#(\mathrm{put}_R(e)©t) = \infty \Rightarrow \#(\mathrm{get}_T(e)©t) = \infty) \wedge \#(\surd©t) = \infty \\
L_1(t) &\equiv data(Snd©t) \sqsupseteq \langle d_0,\ldots,d_n\rangle \Rightarrow \forall k \leq n : \mathrm{put}_T(\langle k,d_k\rangle)\ in\ t \\
L_2(t) &\equiv \mathrm{put}_T(\langle k,d\rangle)\ in\ t \Rightarrow \mathrm{get}_R(\langle k,d\rangle)\ in\ t \\
L_3(t) &\equiv (\forall k \leq n : \mathrm{get}_R(\langle k,d_k\rangle)\ in\ t) \Rightarrow data(Rec©t) \sqsupseteq \langle d_0,\ldots,d_n\rangle \\
L_4(t) &\equiv \mathrm{put}_T(\langle k,d\rangle)\ in\ t \Rightarrow \mathrm{get}_T(ack(k))\ in\ t \\
L_5(t) &\equiv \#(Snd©t) > k \wedge \neg A(t,k) \wedge \#(\surd©t) = \infty \Rightarrow \\
&\quad\ \exists j,d : \neg A(t,j) \wedge \#(\mathrm{put}_T(\langle j,d\rangle)©t) = \infty \\
&\quad\ where A(t,k) \equiv \exists s : s \cdot \mathrm{get}_T(ack(k)) \sqsubseteq t \wedge \#(Snd©s) > k \\
L_6(t) &\equiv \#(\mathrm{get}_R(\langle k,d\rangle)©t) = \infty \Rightarrow \#(\mathrm{put}_R(ack(k))©t) = \infty
\end{aligned}
$$

The refinement of the specification with the goal of a modular requirements specification is performed in several steps. All requirements which are not local to the receiver or the transmitter, or which are not provided by the medium, have to be refined. In (Dendorfer and Weber, 1992a) this is accomplished by introduction of sequence numbers ($Step_1$), introduction of acknowledge messages ($Step_2$), and complete localization ($Step_3$). In the original paper (Dendorfer and Weber, 1992a), these proof obligations gave rise to several propositions and lemmata.

Table II. Safety Properties

$$
\begin{aligned}
US(t) &\equiv \forall s \sqsubseteq t: data(Rec©s) \sqsubseteq data(Snd©s) \\
LS^1(t) &\equiv \forall s \sqsubseteq t: \mathrm{get}_R(e)\ in\ s \Rightarrow \mathrm{put}_T(e)\ in\ s \\
LS^2(t) &\equiv \forall s \sqsubseteq t: \mathrm{get}_T(e)\ in\ s \Rightarrow \mathrm{put}_R(e)\ in\ s \\
S_1(t) &\equiv \forall s \sqsubseteq t: data(Rec©s) = \langle d_0, \ldots, d_n\rangle \Rightarrow \forall k \leq n: \mathrm{get}_R(\langle k, d_k\rangle)\ in\ s \\
S_2(t) &\equiv \forall s \sqsubseteq t: \mathrm{put}_T(\langle k, d\rangle)\ in\ s \Rightarrow (Snd©s)[k] = snd(d) \\
S_3(t) &\equiv \forall s \sqsubseteq t: \mathrm{put}_R(ack(k))\ in\ s \Rightarrow \exists d: \mathrm{get}_R(\langle k, d\rangle)\ in\ s
\end{aligned}
$$

Table III. Definition of the refinement steps

$$
\begin{aligned}
Step_0(t) &\equiv LS^{1,2}(t) \wedge LL^{1,2}(t) \wedge UL(t) \wedge US(t) \\
Step_1(t) &\equiv LS^{1,2}(t) \wedge LL^{1,2}(t) \wedge S_1(t) \wedge S_2(t) \wedge L_1(t) \wedge L_2(t) \wedge L_3(t) \\
Step_2(t) &\equiv LS^{1,2}(t) \wedge LL^{1,2}(t) \wedge S_1(t) \wedge S_2(t) \wedge S_3(t) \wedge L_1(t) \wedge L_3(t) \wedge L_4(t) \\
Step_3(t) &\equiv LS^{1,2}(t) \wedge LL^{1,2}(t) \wedge S_1(t) \wedge S_2(t) \wedge S_3(t) \wedge L_3(t) \wedge L_5(t) \wedge L_6(t)
\end{aligned}
$$

Each step (see Table III) consists of a conjunction of the liveness and safety properties as shown in Table I and Table II. Then, a proof task is of the form: $Step_i(t) \Rightarrow Step_{i-1}(t)$. Due to space restrictions, we cannot describe the details of the refinements. For details see (Dendorfer and Weber, 1992b) or (Schumann, 1995). In several cases, more than one proof task arises from one proposition, e.g., by splitting "– " into the "⇐" and the "⇒" case. Furthermore, we processed two additional lemmata (Lemma 3.4 and Lemma 3.5) which have been defined and used in (Dendorfer and Weber, 1992a) for the proof of Prop. 3.3 and Prop. 3.4, respectively. In total, we obtained 10 proof tasks.

Formalization for SETHEO

The proof tasks described above are the basis for our case study. Hence, each proof task must be transformed into a representation suitable for Setheo. This transformation is accomplished in 4 steps:

1. transformation into first order predicate logic (FOL): all axioms, theorems, and lemmata must be represented in first order predicate logic.
2. transformation of the notation: the entire formula must be represented in a syntax which is readable by Setheo.
3. axiomatization of the underlying theory: for all operators, axioms describing all properties must be added.
4. transformation into clausal normal form (CNF): the formula must be transformed into a set of clauses, and quantifiers must be removed by Skolemization.

Since all formulae in this case study already were available in first order logic, the first step could be skipped. The syntactic transformation (2) mainly involved a syntactic transformation into prefix notation of all operators and deciding which operator (or function symbol) should be represented as a predicate symbol and which symbols are written as (syntactic) function symbols. For better legibility, we have selected equality (=), and all relational binary operators (i.e., $\leq, in, >, \sqsubseteq$) to be represented as predicate symbols of arity 2 ("equational representation"). This means that an expression $a = b$ is written as `equal(a,b)`. All other symbols occurring in a formula are transformed into prefix function symbols or syntactic constants.

Finding the appropriate set of axioms (3) for all operations and relations is an extremely difficult task for any application of an automated theorem prover. A number of important decisions must be made here: which *kind* of axiomatization is to be used for which *subset* of operations and relations, and *which* axioms (and lemmata) are to be added to the formula.

The aim of this study has been to *find* proofs for the given proof tasks, using Setheo. Therefore, we started with a small set of "common" axioms (for equality) and added more axioms and primitive lemmata as needed. Since these lemmata are quite obvious and follow directly from the definition of the operators, they were not proven within our case study. Hence, we will call them *high-level axioms*. However, these axioms needed for our proof tasks are often rather weak (e.g. $\forall t, a : \#(a © t) = \infty \Rightarrow a \; in \; t$, i.e., if an action occurs infinitely often in a trace, it occurs at least once in that trace). Those axioms are too weak to be useful in a general context. Finally, all formulas must be converted into clausal normal form (4). For this task, we used a standard algorithm (see e.g., (Loveland, 1978)).

Table IV. Experimental results for all proof tasks

task	definition	cl	T_{run} [s]	R	length of proof total	Ext	Red
1	$Step_1(t) \Rightarrow US(t)$	23	0.2	–	10	9	1
2	$Step_1(t) \Rightarrow UL(t)$ (strong)	22	0.2	–	6	5	1
3	$Step_2(t) \Rightarrow Step_1(t)$	24	0.6	f	18	18	0
4	task 3 without Lemma (8)	23	7.7	Δ	20	20	0
5	$Step_3(t) \Rightarrow Step_2(t)$	25	17.0	Δ	20	20	0
6	–"–	40	8.0	–	20	20	0
7	–"–	37	0.4	–	6	6	0
8	Lemma for task 3	24	0.2	–	11	11	0
9	Lemma for task 5,6,7	26	0.1	–	3	3	0
10	–"–	24	0.1	–	10	10	0

Experiments and Results

All 10 proof tasks have been processed by SETHEO after the preparation steps described above had been performed. Table IV gives an overview of the results. For each proof task, we show its definition, and the number of clauses after transformation into clausal form (cl). Run-times for SETHEO (V3.2) are in seconds and have been obtained on a SUN Sparc10. SETHEO was started with its default parameters, and iterative deepening over the depth of the tableau (A-literal depth). Only in those cases, where Setheo did not find the proof within a few seconds, two additional techniques have been used to tackle the problem: enabling the additional "fold-up" inference rule (Letz et al., 1994), and using the preprocessor DELTA. DELTA (Schumann, 1994) generates selected unit clauses in a bottom-up way during the preprocessing phase. These are added to the formula before the top-down search with Setheo starts. Table IV shows the necessary resources R ("–" for standard options, f for fold-up, and Δ for DELTA) and the length of the proof (measured in Model Elimination extension and reduction inference steps).

The figures in this table show that, after transformation and with an appropriate set of axioms, SETHEO can find the requested (non-trivial) proofs automatically within short run-times. All proofs have been post-processed by ILF-SETHEO (Wolf and Schumann, 1997) in order to obtain human-readable proofs in LaTeX type-setting.

Assessment
The run-times shown in Table IV are in the range of a few seconds. This time, however, does *not* reflect the overall time needed to carry out the case study. An estimation of the time needed for the entire case study is difficult to give because the author did not log the sessions. In total, this work was completed within a few days. Major time-consuming tasks were (in decreasing order): setting up an environment ("infra-structure") to keep all formulae and results consistent, debugging formulae (finding out why things went wrong and correcting the formulae (mostly mis-prints)), finding and formulating the appropriate axioms, and checking the proofs for correctness (in order to detect possible flaws in the axioms or theorems). However, the entire case study could be carried out *without* detailed knowledge of FOCUS and the problem domain. In the following, we will summarize the experiences made during each step of formalization and execution of the proof tasks by Setheo.

The step of finding the right axioms and lemmata for the actual proof task is probably the most difficult problem in applying an automated theorem prover to this kind of application. In our case study, the most naïve approach has been taken: besides standard axioms for equality, obvious axioms defining the operators and lemmata ("high-level axioms") have been added where necessary. Here, emphasis was put on a set of consistent axioms which was *sufficient* to find the proofs, not one which completely defines the operations and relations. A full set of axioms would span a search space too large to be handled within reasonable run-times. Here, mechanisms for preselecting axioms (see also Chapter III.2.9 in this volume) would be required. High-level axioms describe obvious features of the operators (e.g., getting the *n*-th value of a trace) without the necessity of using the low-level definition of the operators. Therefore, high-level axioms help to avoid many simple cases of *induction* which would require extra preprocessing steps since SETHEO alone cannot handle induction. This approach is also reflected in the original manual proofs in (Dendorfer and Weber, 1992a) which do not use induction.

In general, formulas, arising from Focus specifications are implicitly sorted: each object carries a unique sort (e.g., UDU, LDU, trace,...). In this case study, we could drop this sort information, since all formulas and axioms have been designed sort-consistent, i.e., no incorrect proofs can occur when the sorts are neglected. For further case-studies, however, compiling sorts into the terms will be required and helpful. Such a compiler (e.g., within ILF, described above, or the one developed within Protein (Baumgartner and Furbach, 1994)) takes the formula and the sort specifications and generates a new set of clauses, leaving the check for sort consistency to the standard unification procedure within SETHEO. Besides guaranteeing sort-consistent proofs, such a tool is very helpful for finding typing errors which consumed much time in this case study.

A theorem prover can be successfully applied only if it can be used without detailed knowledge about the prover's internal details (e.g., optimal parameter setting). Therefore, only standard parameters (3 different sets, see above) for Setheo have been used for the experiments. Further developments, such as parallel competition (e.g., p-SETHEO, see Chapter II.3.10) relieves the user from setting parameters and can automatically select good parameter settings without user interaction.

3. VERIFICATION OF AUTHENTICATION PROTOCOLS

3.1. *Authentication Protocols*

One particular kind of protocols are *authentication protocols*. In contrast to the communication protocols discussed above, the goal of an authentication protocol is that "*the principals should be entitled to believe that they are communicating with each other, and not with intruders.*" [3] Principals can be persons, other computers, or even services.

Therefore, authentication protocols are used in many security relevant areas. Typical applications include tele-banking, tele-shopping or login-procedures in distributed systems. Since most applications are in commercial environments where millions of dollars are at stake, security of such protocols must be *guaranteed.* Flaws in an authentication protocol are hard to detect. Therefore, many methods and tools for the analysis and verification of authentication protocols have been studied and developed (for an overview see e.g. (Geiger, 1995; Meadows, 1994)). In order to find weak points (or rather their absence) in such a protocol, typical intruder scenarios can be simulated (c.f. the interrogator system (Millen et al., 1987)). Methods for formal verification have been developed since about 1989. The most well-known approach in that area is presented in (Burrows et al., 1989). The authors have proposed a logic of *belief,* called BAN logic (named after the initials of the authors). This logic is capable of expressing the behavior of a protocol and allows specification of safety and security properties. With the help of a logic calculus these properties can be shown formally.

Several problems with the BAN logic (e.g., with respect to modeling intruders and specifying confidentiality properties) and a missing denotational semantics has led to many extensions and modifications of this logic, e.g. GNY (Gong et al., 1990), SVO (Syverson and van Oorschot, 1994), or AUTLOG (Kessler and Wedel, 1994).

[3] (Burrows et al., 1989), pg. 1.

Several approaches with computer-supported verification of authentication protocols using one of these logics have been reported: AUTLOG (Kessler and Wedel, 1994) uses a PROLOG system with forward reasoning, whereas Brackin (1996) uses the higher-order theorem prover HOL to perform reasoning with the BAN logic. Craigen and Saaltink (1996) transform the BAN logic into first-order logic using an approach similar to our own. The prover EVES is then used to perform reasoning. In order to improve the level of automatic processing, a specific forward reasoner has been added.

Quite a different approach has been proposed by Kindred and Wing (1996). Here, protocol runs and intruder scenarios are modeled as finite-state systems which are then explored with symbolic model checking. However, this approach cannot reflect — due to its finiteness — all the possible ways an intrusion can occur. This difficulty could be overcome with a CSP-based approach which is able to model a potentially infinite number of interleaving protocol runs. (Paulson, 1997) uses Isabelle to interactively show security and safety properties of authentication protocols.

We apply the automated theorem prover SETHEO to *automatically* verify theorems formulated in the BAN logic.

3.2. *The BAN Logic*

The BAN logic (Burrows et al., 1989) is a multi-sorted modal logic. The basic sorts of objects are: ***principals***, ***encryption keys***, and ***formulas***. Principals are denoted by A, B, P, S and pA, pB, pS; formulas as X, Y. Encryption keys (shared between two principals) are written as K_{ab}, K_{as}, K_{ba}. Public keys are K_a, K_b, K_c, whereas $K_a^{-1}, \ldots$ denote the corresponding secret keys. Nonces (e.g., time-stamps) are T, T_a, T_b. The main operator of that logic is belief: P **believes** X (or $P| \equiv X$) means that principal P believes a statement X. This also means that P acts as if X was true. This operator and other BAN operators (see Table V) may be connected via "$\wedge$" to construct new formulas. For a complete list of operators and a detailed description see (Burrows et al., 1989) or (Burrows et al., 1990).

The BAN logic as presented in (Burrows et al., 1989) is defined via a list of logical postulates (or inference rules). The most important inference rules are shown in Table VI. The *message-meaning* rule (MM_{sk}, Table VI) concerns the interpretation of a message. As soon as a principal P believes that it communicates safely with some other principal Q using encrypted messages (with shared key K), and such a message arrived, we may infer that P believes that Q has actually sent the message X. For public keys or shared secret keys similar inference rules are defined.

Table V. Operators and function symbols of the BAN logic

Shorthand	Description	Shorthand	Description
$P \mid\equiv X$	**P believes X**	$\{X\}_K$	X encrypted with key K
$P \triangleleft X$	**P sees X**	$P \overset{K}{\leftrightarrow} Q$	comm. via shared key K
$P \mid\sim X$	**P said X**	$P \overset{K}{\rightleftharpoons} Q$	comm. via secret key K
$P \mid\Rightarrow X$	**P controls X**	$\overset{K}{\mapsto} Q$	comm. via public key K
$\#X$	**fresh**(X)	$\langle X\rangle_Y$	msg combined of X and Y
K^{-1}	inverse of K	$\{M_1,\ldots,M_n\}$	msg consists of $M_1,\ldots,M_n$

Table VI. BAN inference rules

MM_{sk}	Message Meaning	$\dfrac{P \mid\equiv P \overset{K}{\leftrightarrow} Q \qquad P \triangleleft \{X\}_K}{P \mid\equiv Q \mid\sim X}$
NV	Nonce-verification	$\dfrac{P \mid\equiv \#X \qquad P \mid\equiv Q \mid\sim X}{P \mid\equiv Q \mid\equiv X}$
JD	Jurisdiction	$\dfrac{P \mid\equiv Q \mid\Rightarrow X \qquad P \mid\equiv Q \mid\equiv X}{P \mid\equiv X}$
SY	Symmetry 1,2	$\dfrac{P \mid\equiv Q \overset{K}{\leftrightarrow} R}{P \mid\equiv R \overset{K}{\leftrightarrow} Q} \qquad \dfrac{P \mid\equiv Q \overset{K}{\rightleftharpoons} R}{P \mid\equiv R \overset{K}{\rightleftharpoons} Q}$

Nonce-verification: Nonces are specific messages generated by a principal. Their main purpose is to guarantee the freshness of a message. Therefore, often time-stamps are used as nonces. The nonce-verification rule (NV) states that a message is "recent", i.e., that the sender (Q) still believes in it when the corresponding nonce X is fresh ($\#X$). The *jurisdiction* rule states that if P believes that Q has jurisdiction (control) over X, then P trusts Q on the truth of X[4]. Two inference rules (which are not defined in (Burrows et al., 1989), but used there) define the symmetry of communication (SY) via shared keys and shared secrets.

Several inference rules of the BAN logic concern the "packing" and "unpacking" of messages which consist of more than one atomic message. Fur-

[4] (Burrows et al., 1989), pg. 3.

thermore, rules exist that a principal can decipher messages, for which the appropriate keys are known. For a list of these inference rules see (Burrows et al., 1989).

In general, an authentication protocol is specified as a finite sequence of messages which are being transmitted between the principals, usually written as (principal P sends a message to principal Q):

$$P \rightarrow Q : \text{message}$$

For analysis with BAN logic, all messages of the protocol must be *idealized.* This means that uncrypted parts of a message are not included, such that all messages are of the form $\{X\}_K$ or sequences thereof. As an example, consider the message $A \rightarrow B : \{A, K_{ab}\}_{K_{bs}}$. This message means that A sends its own identification plus the shared key K_{ab}, encrypted with the key K_{bs} to principal B. It is idealized as[5]

$$A \stackrel{id}{\rightarrow} B : \{A \stackrel{K_{ab}}{\leftrightarrow} B\}_{K_{bs}}$$

If a message $P \stackrel{id}{\rightarrow} Q : X$ has been received, the formula $Q \triangleleft X$ (Q **sees** X) holds. Thus, a list of such formulas (one for each message) comprises the input specification of a protocol in BAN.

The theorems we want to prove in general depend on the protocol we are analyzing. However, there are default requirements for the services which an authentication protocol should provide. These are of such a form that the principals (A and B) believe that they properly communicate via a shared key K: $A| \equiv A \stackrel{K}{\leftrightarrow} B$ and $B| \equiv A \stackrel{K}{\leftrightarrow} B$, and the principals believe the beliefs of their respective partners: $B| \equiv A| \equiv A \stackrel{K}{\leftrightarrow} B, A| \equiv B| \equiv A \stackrel{K}{\leftrightarrow} B$.

3.3. *Transformation for SETHEO*

The BAN logic is defined by an alphabet of atoms (principals and keys), functions and operators, and inference rules. One way to transform such a logic into first order predicate logic (FOL) is "simulation": On a meta-level, all formulas of BAN are represented as first-order *terms*, and the inference rules are specified as first-order formulas. Since all BAN formulas are ground, and quantifiers (universal quantifiers if any) only occur on the outermost level, this transformation is safe. Such an approach has been successfully used e.g.,

[5] Here, the assumption is implicitly made that if A possesses the key K_{ab}, then it communicates with B in a secure way. For a discussion of this process see Section 3.6.

for preparing single-axiom propositional calculi for automated theorem proving (McCune and Wos, 1991). Hence, for the BAN logic we obtain the following transformation. A unary FOL predicate $holds(X)$ is defined which has the meaning: $holds(X)$[6] is true, if and only if the BAN-formula X is *derivable* from other BAN formulas using the BAN inference rules. For example, the BAN-formula $A|\equiv \#T_s$ (A believes the freshness of nonce T_s) is transformed into: $holds(A|\equiv \#T_s)$.

Additionally, each BAN inference rule is transformed into a FOL formula. Given an inference rule of the form

$$\frac{F_1 \quad \dots \quad F_n}{G}$$

we obtain: $holds(F_1) \wedge \dots \wedge holds(F_n) \rightarrow holds(G)$.

Variables which occur in the formulas of the inference rule become all-quantified in the transformation. E.g., the nonce-verification rule NV from Table VI is written as:

$$\forall P, Q, X : holds(P|\equiv \#X) \wedge holds(P|\equiv Q|\sim X) \rightarrow holds(P|\equiv Q|\equiv X)$$

Then, in our approach, the entire input formula for SETHEO is of the form

$$A_1 \wedge \dots \wedge A_n \wedge M_1 \wedge \dots \wedge M_m \wedge R_1 \wedge \dots \wedge R_k \rightarrow T$$

where the A_i are the transformed assumptions, R_i are the transformed inference rules, M_i are the transformed messages of the idealized protocol, and T is the theorem to be shown. Then, this formula can easily be transformed into clausal normal form obtaining a set of Horn clauses.

Of course, our transformation has to make sure that the sorts of all objects of the BAN logic (principal, message, formula) are handled correctly. The transformation described above would yield formulas of a multi-sorted FOL. Due to the structure of the inference rules, however, this sort information can be neglected: given correctly-sorted BAN formulas, all derivable formulas are automatically of the correct sort[7].

Within the BAN logic, messages can consist of a single atomic message, or a (finite) sequence of messages enclosed in braces $\{\}$. A notation of such sequences of variable length as well as predicates for accessing, breaking-up and constructing sequences have to be defined in first-order logic as *auxiliary*

---

[6] In shorthand, we write $\vdash X$.

[7] Explicit handling of sorts, however, greatly facilitates user input. With its help, many syntactical errors can be detected easily.

predicates. For these purposes, we have defined an additional function symbol *cons* with arity two, and a symbol *nil*, representing the empty sequence. Then, a message $\{m_1, m_2, \ldots, m_n\}$ is represented as the term $cons(m_1, cons(m_2, \ldots cons(m_n, nil) \ldots)$. Then, auxiliary predicates to select *one* message out of such a combined message M ("$\iota(M)$") and to select a subsequence $S \sqsubseteq M$ have been defined by the appropriate axioms.

3.4. *A Prototypical Tool*

Based on this transformation, we have designed and implemented a simple prototypical tool (Wagner, 1997). This tool accepts the specification of the idealized protocol, the assumptions and the conjectures in a simple specification language. From these, the tool generates all required first order proof obligations. Each proof, found by SETHEO is automatically converted into a human-readable form using the tool ILF-SETHEO (Wolf and Schumann, 1997). If all proof obligations could be shown (controlled by UNIX's make), a detailed LaTeX proof protocol, containing the specification and the proofs is generated. At no point, any knowledge about the Model Elimination Calculus, SETHEO, or the transformation described above, is required. A complete example (concerning the Kerberos protocol) is shown in the appendix.

3.5. *Experiments and Results*

Using our tool, we experimented with well-known protocols found in the literature, namely the Kerberos protocol, the Secure RPC-handshake, the Needham-Schroeder, the Needham-Schroeder with public keys, the Otway-Rees, the Wide-mouthed frog, the Yahalom, and the CCITT-509 protocol. For each of the experiments, the idealized protocols, assumptions and the theorems to be proven have been taken directly from the literature.

Table VII shows the results of the experiments. For each protocol, we show the number of messages of the protocol (#msgs), the number of proof tasks to be shown, and total run-time to process *all* proof tasks ΣT_{tot} obtained on a SUN Sparc II. This run time consists of the time needed to compile the formula and to carry out the search. Since post-processing by ILF-SETHEO involves sending e-mail, we did not add this time. Under normal network conditions, however, it only takes a few seconds, before the LaTeX documents arrive.

SETHEO could show all required proof tasks within a few seconds of run-time. Whereas most proofs are relatively short and straight-forward, the proofs of several theorems for the Yahalom and the Otway-Rees protocols are rather lengthy and require considerable run-times.

Table VII. Experimental results

Protocol	#msgs	#tasks	ΣT_{tot}[s]
Andrew Secure RPC Handshake	4	6	2.71
Needham-Schroeder	5	5	4.35
Needham-Schroeder/w public keys	7	6	3.15
Otway-Rees	4	6	5.85
Wide-mouthed frog	2	2	2.23
Yahalom	4	12	61.78
CCITT-X509	3	2	1.00

3.6. *Assessment*

As demonstrated in the previous section, all proof tasks arising from the verification of well-known authentication protocols with BAN logic could be shown fully automatically within short run-times. Our approach does not require any knowledge about the underlying theorem prover (SETHEO) or the model elimination calculus. Specifications and generated proofs are presented in a problem-oriented way.

However, verification of a protocol with BAN-logic —although performed automatically— leaves scope for much improvement: the BAN is a rather weak logic, which means that many important issues (e.g., confidentiality) cannot be expressed within this framework. The idealization of protocols makes many hidden assumptions[8], is difficult, and would require additional formal reasoning. Furthermore, the original BAN-logic has no clean denotational semantics.

Therefore, our BAN-based tool cannot be used "stand-alone" for full protocol verification. Other methods and tools are required to capture all possible aspects of modeling intruders. Ideally, multiple methods should be combined for handling different phases and levels of protocol verification. In particular for early stages of protocol development our tool will be helpful, since it allows for a fast and easy specification (i.e., fast finding of more obvious bugs) and produces short and understandable proofs.

[8] For example, this formalization assumes that each principal is trustworthy, i.e., it does not broadcast its secret keys.

4. Conclusions

We have demonstrated that the deductive power of automated theorem provers has reached a level where they can be used successfully as a tool in the verification of protocols. We have seen that relevant concepts in this domain can be formulated in a straightforward way for input into existing first order provers. This concerns communication protocols as well as authentication protocols.

We do not claim that protocol verification can be completely automated in the near future. But the use of automated theorem provers can make formal verification feasible with relatively little conceptual overhead for the verifying human, compared with manual verification procedures. Furthermore, more reliable results can be obtained.

The use of automated theorem provers lets the focus of attention of the human shift from the formal logic calculus to the conceptual level. He is less concerned with the construction of proofs in a formal logic since this can be left to the provers. This has the remarkable consequence that formal verification can be made accessible for users which are less trained in formal logic. Their main concern will be the formal specification of the protocols, of the conditions to be verified and the guidance of the verification process. All this has to be done in a way that supports the needs of the automated provers.

In order to facilitate this, theorem provers must be encapsulated in an environment with a convenient user interface. We have shown that existing provers can provide output in a format that is familiar to experts in protocol verification. This has to be augmented by similar prover input possibilities.

In our case studies, the initial specifications have been translated manually into formal theories but there is no obstacle to automating this process. This automated preprocessing could include the transformation of proof problems with domain specific tools. The target of such a transformation must be to reduce the search space of the provers. This includes preference for certain terms as well as the selection of the relevant parts of the knowledge base. It requires expertise in both the possibilities provided by the special features of protocol specifications and the demands of modern theorem provers.

Moreover, a convenient user interface is needed. It should support input in a dedicated specification language or with graphic tools. (Saaltink, 1995) and (Eastaughffe et al., 1997) describe promising approaches in both directions.

Combining all these features of existing systems will not eliminate the inherent complexity of formal protocol verification - but it can make it manageable for more users.

REFERENCES

Baar, T.: 1997, 'Formaler Beweis der Korrektheitseigenschaften einer Z-Spezifikation des Alternating-Bit-Protokolls unter Benutzung des Werkzeuges ILF'. Berlin, diploma paper, Humboldt University, Berlin. in German.

Bartlett, K. A., R. A. Scantlebury, and P. T. Wilkinson: 1969, 'A Note on Reliable Full-Duplex Transmission over Half-Duplex Links'. *Communications of the ACM* **12**, 260–261.

Baumgartner, P. and U. Furbach: 1994, 'PROTEIN: A *PRO*ver with a *T*heory *E*xtension *I*nterface'. In: *12th International Conference on Automated Deduction*, Vol. 814 of *LNAI*. Springer.

Bezem, M., R. Bol, and J. F. Groote: 1995, 'Formalizing Process Algebraic Verification in the Calculus of Constructions'. Technical Report 2/95, Eindhoven Technical University.

Brackin, S. H.: 1996, 'A HOL Extension of GNY for Automatically Analyzing Cryptographic Protocols'. In: *Proc. IEEE Computer Security Foundations Workshop IX*. IEEE.

Burrows, M., M. Abadi, and R. Needham: 1989, 'A Logic of Authentication'. In: *ACM Operating Systems Review 23(5) / Proceedings of the Twelfth ACM Symposium on Operating Systems Principles*.

Burrows, M., M. Abadi, and R. Needham: 1990, 'A Logic of Authentication'. *ACM Transactions on Computer Systems* **8**(1), 18–36.

Craigen, D. and M. Saaltink: 1996, 'Using EVES to Analyze Authentication Protocols'. Technical Report TR-96-5508-05, ORA Canada.

Dederichs, F., C. Dendorfer, and R. Weber: 1993, 'FOCUS: A Formal Design Method for Distributed Systems'. In: A. Bode and M. D. Cin (eds.): *Parallel Computer Architectures*. pp. 190–202, Springer.

Dendorfer, C. and R. Weber: 1992a, 'Development and Implementation of a Communication Protocol – An exercise in FOCUS'. SFB-bericht Nr. 342/4/92 A, Technische Universität München, Institut für Informatik.

Dendorfer, C. and R. Weber: 1992b, 'From Service Specification to Protocol Entity Implementation – An Exercise in Formal Protocol Development'. In: R. Linn and M. Ü. Uyar (eds.): *Protocol, Specification, Testing and Verification XII*, Vol. C-8 of *IFIP Transactions*. pp. 163–177.

Denzinger, J., M. Kronenburg, and S. Schulz: 1997, 'DISCOUNT. A Distributed and Learning Equational Prover'. *Journal of Automated Reasoning* **18**, 189–198.

Eastaughffe, K. A., M. A. Ozols, and A. Cant: 1997, 'Proof Tactics for a Theory of State Machines in a Graphical Environment'. In: *Conference on Automated Deduction (CADE) 14*, Vol. 1249 of *LNAI*. pp. 366–379, Springer.

Geiger, J.: 1995, 'Formale Methoden zur Verifikation kryptographischer Protokolle'. Fortgeschrittenenpraktikum, Technische Universität München. in German.

Gong, L., R. Needham, and R. Yahalom: 1990, 'Reasoning about belief in cryptographic protocols'. In: *Proc. of IEEE Symposium on Security and Privacy, Oakland, Ca.,USA*. pp. 234–248, IEEE.

Huggins, J. K.: 1995, 'Kermit Protocol - Formal Specification and Verification'. In: E. Börger (ed.): *Specification and Validation Methods.* Oxford Univ. Press, pp. 247–293.

Kessler, V. and G. Wedel: 1994, 'AUTLOG — An Advanced Logic of Authentication'. In: *Proc. IEEE Computer Security Foundations Workshop IV*. pp. 90–99, IEEE.

Kindred, D. and J. Wing: 1996, 'Fast, Automatic Checking of Security Protocols'. In: *2nd USENIX Workshop on Electronic Commerce*. pp. 41–52.

Larsen, K. G. and R. Milner: 1992, 'A Compositional Protocol Verification Using Relativized Bisimulation'. *Information and Computation* **99**, 80–108.

Letz, R., K. Mayr, and C. Goller: 1994, 'Controlled Integration of the Cut Rule into Connection Tableau Calculi'. *Journal Automated Reasoning (JAR)* **13**, 297–337.

Letz, R., J. Schumann, S. Bayerl, and W. Bibel: 1992, 'SETHEO: A High-Performance Theorem Prover'. *Journal of Automated Reasoning* **8**(2), 183–212.

Loveland, D. W.: 1978, *Automated Theorem Proving: a Logical Basis.* North–Holland.

McCune, W. W. and L. Wos: 1991, 'Experiments in Automated Deduction with Condensed Detachment'. Technical report, Argonne National Laboratory.

Meadows, C. A.: 1994, 'Formal verification of Cryptographic Protocols: A Survey'. In: *Proc. AsiaCrypt.*

Millen, J., S. Clark, and S. Freedman: 1987, 'The interrogator protocol security analysis'. *IEEE Trans. Software Engineering* **SE-13**(2).

Nipkow, T. and K. Slind: 1995, 'I/O Automata in Isabelle/HOL'. In: *Types for Proofs and Programs.* Springer LNCS vol. 996, pp. 101–119.

Paulson, L.: 1997, 'Proving Properties of Security Protocols by Induction'. In: *Proc. 10th Comp. Security Workshop*. pp. 70–83, IEEE.

Pneuli, A.: 1986, 'Applications of temporal logic to the specification and verification of reactive systems'. In: J. W. de Bakker (ed.): *Current Trends in Concurrency*, Vol. 224 of *LNCS*. pp. 510–584, Springer.

Saaltink, M.: 1995, 'The Z/EVES System'. Technical report, ORA Canada.

Schumann, J.: 1994, 'DELTA — A Bottom-up Preprocessor for Top-Down Theorem Provers, System Abstract'. In: *Proc. of Conference on Automated Deduction CADE 12*, Vol. 814 of *LNAI*. pp. 774–777, Springer.

Schumann, J.: 1995, 'Using SETHEO for Verifying the Development of a Communication Protocol in FOCUS – A Case Study –'. In: P. Baumgartner, R. Hähnle, and J. Posegga (eds.): *Proc. of Workshop Analytic Tableaux and Related Methods*, Vol. 918 of *LNAI*. pp. 338–352, Springer.

Shankar, A. U. and S. S. Lam: 1992, 'A Stepwise Refinement Heuristic for Protocol Construction'. *ACM Transactions on Programming Langanguages and Systems* **14**, 417–461.

Stenning, V.: 1976, 'A data transfer protocol'. *Computer Networks* **1**, 98–110.

Syverson, P. F. and P. van Oorschot: 1994, 'On Unifying Some Cryptographic Protocol Logics'. In: *Proc. of the IEEE Comp. Soc. Sympos. on Research in Security and Privacy*. pp. 14–28.

Wagner, K.: 1997, 'PIL: Ein SETHEO-basiertes Werkzeug zur Analyse kryptographischer Protokolle'. Fortgeschrittenenpraktikum, Technische Universität München, Institut für Informatik. in German.

Wolf, A. and J. Schumann: 1997, 'ILF-SETHEO: Processing Model Elimination Proofs for Natural Language Output'. In: *Conference on Automated Deduction (CADE) 14*, Vol. 1249 of *LNAI*. pp. 61–64, Springer.

A. Example: the Kerberos Protocol

In this appendix we present the formalisation and analysis of a well-known protocol, the Kerberos protocol.

The Kerberos protocol (developed at MIT within the Athena project) establishes a shared key between two principals *pA*, *pB* with the help of an authentication server *pS*. Its 4 messages and their idealization are shown in Table VIII. With the first message, *pA* contacts the server *pS*, sending its own name and that of the proposed communication partner *pB*. Since all components of this message are uncrypted this message is not included in the idealization of the protocol. *pS* provides *pA* in the second message with the session key K_{ab}, and a certificate conveying the session key and *pA*'s identity (encrypted by the key K_{bs}). The entire message is encrypted by *pA*'s key K_{as}. Then, the latter part of the message is sent from *pA* to *pB*, along with a time-stamp (nonce T_a), issued by *pA*. Finally, *pB* acknowledges the establishment of a secure communication by incrementing the time-stamp T_a and sending it back to *pA* (encrypted now by the mutual session key K_{ab}).

For the analysis of this protocol, we must give a set of additional *assumptions*: statements about shared keys, jurisdiction, and the freshness of the nonces. These assumptions (from (Burrows et al., 1989)) are shown in Table IX[9].

The security properties which have to be shown, and which comprise our proof tasks $K_1, \ldots, K_4$ for the SETHEO experiments are:

$$K_1: \quad pA|\equiv pA \overset{K_{ab}}{\leftrightarrow} pB \qquad K_3: \quad pA|\equiv pB|\equiv pA \overset{K_{ab}}{\leftrightarrow} pB$$
$$K_2: \quad pB|\equiv pA \overset{K_{ab}}{\leftrightarrow} pB \qquad K_4: \quad pB|\equiv pA|\equiv pA \overset{K_{ab}}{\leftrightarrow} pB$$

In the following, we show the (slightly abbreviated) input specification for the Kerberos protocol in our specification language PIL (Wagner, 1997),

9 The experiments with SETHEO revealed that one assumption essential for the proofs, $pA|\equiv \#T_a$, is not described in the literature. This fact has also been detected (independently) in (Craigen and Saaltink, 1996). With this additional assumption, proof task $K_\triangle$ can be shown.

Table VIII. The Kerberos Protocol: original and idealized messages

msg #	Kerberos protocol	idealized Kerberos protocol
1	$pA \rightarrow pS : pA, pB$	—
2	$pS \rightarrow pA : \{T_s, L, K_{ab}, pB,$	$pS \stackrel{id}{\rightarrow} pA : \{T_s, pA \stackrel{K_{ab}}{\leftrightarrow} pB,$
	$\{T_s, L, K_{ab}, pA\}_{K_{bs}}\}_{K_{as}}$	$\{T_s, pA \stackrel{K_{ab}}{\leftrightarrow} pB\}_{K_{bs}}\}_{K_{as}}$
3	$pA \rightarrow pB : \{T_s, L, K_{ab}, pA\}_{K_{bs}},$	$pA \stackrel{id}{\rightarrow} pB : \{T_s, pA \stackrel{K_{ab}}{\leftrightarrow} pB\}_{K_{bs}},$
	$\{pA, T_a\}_{K_{ab}}$	$\{T_a, pA \stackrel{K_{ab}}{\leftrightarrow} pB\}_{K_{ab}}$
4	$pB \rightarrow pA : \{T_a + 1\}_{K_{ab}}$	$pB \stackrel{id}{\rightarrow} pA : \{T_a, pA \stackrel{K_{ab}}{\leftrightarrow} pB\}_{K_{ab}}$

Table IX. The Kerberos Protocol: assumptions for the analysis

$pA, pB \mid\equiv pA \stackrel{K_{as}}{\leftrightarrow} pS$	$pA \mid\equiv (pS \mid\Rightarrow pA \stackrel{K}{\leftrightarrow} pB)$	$pA \mid\equiv \#T_s$
$pB \mid\equiv pB \stackrel{K_{bs}}{\leftrightarrow} pS$	$pB \mid\equiv (pS \mid\Rightarrow pA \stackrel{K}{\leftrightarrow} pB)$	$pB \mid\equiv \#T_s$
$pS \mid\equiv pA \stackrel{K_{as}}{\leftrightarrow} pS$		$pB \mid\equiv \#T_a$
$pS \mid\equiv pB \stackrel{K_{bs}}{\leftrightarrow} pS$		$pA \mid\equiv \#T_a$ (†)
$pS \mid\equiv pA \stackrel{K_{ab}}{\leftrightarrow} pB$		

and the proof for K_3, as generated by SETHEO. This piece of LaTeX-code has been included exactly as it produced by ILF-SETHEO (Wolf and Schumann, 1997). *No* manual changes (except one line break) had to be applied.

A.1. *Proof generated by ILF-SETHEO*

THEOREM A.1. *query.*

Proof (by SETHEO). We show directly that

(8.1) *query.*

Because of *message_3*

(8.2) $\vdash B \triangleleft (\{\{\{T_S, A \stackrel{K_{A,B}}{\leftrightarrow} B\}\}_{K_{B,S}}, \{\{T_A, A \stackrel{K_{A,B}}{\leftrightarrow} B\}\}_{K_{A,B}}\}).$

Because of *query_3*

(8.3) $query : - \vdash B \mid\equiv A \mid\equiv A \stackrel{K_{A,B}}{\leftrightarrow} B.$

Table X. Specification of the Kerberos Protocol

```
Objects:                    // declaration of all objects
   principal A,B,S;
   sharedkey K_A_b, K_a_s, K_b_s;
   variable sharedkey K;
   statement T_a, T_s;

Assumptions:                // assumptions (abbreviated)
   (A,S) believes sharedkey K_a_s;
   (B,S) believes sharedkey (B,S,K_b_s);
   ...

Idealized Protocol:
   message 2: S -> A  {T_s, sharedkey (A,B,K_a_b),
          {T_s, sharedkey(A,B,K_a_b)}(K_b_s)}(K_a_s);
   message 3: A -> B  {T_s, sharedkey K_a_b}(K_b_s),
          {T_a,sharedkey K_a_b}(K_a_b);
   message 4: A <- B  {T_a, sharedkey K_a_b}(K_a_b);

Conjectures:                // conjectures (4 tasks)
   after message 2: A believes sharedkey K_a_b;
   after message 3: B believes sharedkey K_a_b;
                    B believes A believes sharedkey K_a_b;
   after message 4: A believes B believes sharedkey K_a_b;
```

Because of *break_up_message*

(8.4) $\vdash A|\equiv B|\equiv C : - C \sqsubseteq D \wedge \vdash A|\equiv B|\equiv D.$

Because of *nonce_verification*

(8.5) $\vdash A|\equiv B|\equiv C : - \vdash A|\equiv B|\sim C \wedge \vdash A|\equiv \#C.$

Because of *freshness* $\vdash A|\equiv \#B : - \vdash A|\equiv \#C \wedge C = \iota(B)$. Because of *assumption_11* $\vdash B|\equiv \#T_A$. Therefore

(8.6) $\vdash B|\equiv \#(\{T_A, A \overset{K_{A,B}}{\leftrightarrow} B\}).$

Because of *message_meaning*

(8.7) $\vdash A|\equiv B|\sim C : - \vdash A \triangleleft \{C\}_D \wedge \vdash A|\equiv B \overset{D}{\leftrightarrow} A.$

Because of *conjecture_2*

(8.8) $\vdash B| \equiv A \overset{K_{A,B}}{\leftrightarrow} B.$

Because of *sees_components* $\vdash A \triangleleft B : - \vdash A \triangleleft C \wedge B = \iota(C)$. Hence by (8.2) $\vdash B \triangleleft \{\{T_A, A \overset{K_{A,B}}{\leftrightarrow} B\}\}_{K_{A,B}}$. Hence by (8.7) and by (8.8) $\vdash B| \equiv A|\!\sim (\{T_A, A \overset{K_{A,B}}{\leftrightarrow} B\})$. Hence by (8.5) and by (8.6) $\vdash B| \equiv A| \equiv (\{T_A, A \overset{K_{A,B}}{\leftrightarrow} B\})$. Hence by (8.4) $\neg$ *query*. Hence by (8.3) *query*. Thus we have completed the proof of (8.1).

q.e.d.

WOLFGANG REIF AND GERHARD SCHELLHORN

CHAPTER 9
THEOREM PROVING IN LARGE THEORIES

1. Introduction

The motivation behind this chapter is the question: how can formal software verification benefit from automated first-order theorem proving? To answer the question we used the software verification tool KIV (Reif et al., 1995), (Reif, 1995), (Reif et al., 1997) as a test environment. We did comparative experiments with five automated theorem provers as dedicated subsystems for the non-inductive first-order goals that showed up during proofs of specification- and program properties in KIV. The five provers were Otter (Wos et al., 1992), Protein (Baumgartner and Furbach, 1994), SETHEO (Goller et al., 1994), Spass (Weidenbach et al., 1996) and $_3T\!A\!P$ (Beckert et al., 1996). The experiments were performed in the context of a joint project between the research groups of $_3T\!A\!P$ and KIV at the Universities of Karlsruhe and Ulm on the conceptual integration of interactive and automated theorem proving. In this project we use KIV and $_3T\!A\!P$ as an experimental platform (for other results of this project, see Chapter II.1.4). Therefore, $_3T\!A\!P$ could be called from inside KIV, whereas for the other provers the problems had to be transferred by hand.

As a general result of the experiments we found that the automated provers did not satisfy the expectations induced by the good results some of them normally get in standard benchmarks from the TPTP library (Sutcliffe et al., 1994). In our experiments either the success rate was rather low or the required time was unexpectedly high.

One major reason for this behavior is the large number of axioms in typical software specifications. Whereas typical TPTP problems only have a few axioms, typical software specifications in KIV comprise up to several hundred axioms. The provers were misled or even got lost in the search space although the majority of the axioms were irrelevant to the proofs under consideration. We found that in the context of large software specifications it is important to find out the relevant axiom set (for the proof of a theorem) or at least a good approximation to it.

In this chapter we present a reduction technique that takes a large theory and a goal, and computes a reduced axiom set by filtering out as many irrelevant axioms as possible. The proof search then is performed with the re-

W. Bibel, P. H. Schmitt (eds.), Automated Deduction. A basis for applications. Vol. III

duced axiom set. The reduction is independent from the actual prover and the calculus. Furthermore, it can be applied recursively during a proof to the subgoals and sublemmas of the original theorem. They often have progressively smaller sets of relevant axioms than the theorem itself.

To evaluate the reduction technique we repeated the original experiments once more, but now with the reduced axiom sets. In the largest example, the original theory had 500 axioms. The reduced axiom sets for the test examples had around 20 axioms. With the reduction the provers were able to prove more theorems than before. Furthermore, for those theorems that could already be proved without the axiom reduction we got considerably shorter proof times. As a side effect it turned out that the susceptibility of the five provers to the problem of large axiom sets is quite different. Furthermore, the experiments shed some light on the question what makes one large theory more harmful than another.

In the next section we illustrate the problem with an example. In Sect. 3 we present the reduction criteria and the assumptions about the specification structure. The reduction procedure is explained in Sect. 4. Sect. 5 reports on the experimental results, and Sect. 6 draws some conclusions.

2. An Example

The example is a specification of a single datatype and not of a whole software system. But it is sufficient to demonstrate the basic reduction criteria, and large enough to cause some problems for the automated theorem provers. The example deals with the data type of finite enumerations. These are bijections from a finite subset of data elements to an initial segment of the natural numbers. Examples are mappings which associate unique keys to database entries, or enumerate the nodes in a graph. Actually, the specification of enumerations was part of a larger KIV case study on the verification of depth-first search on graphs.

Enumerations can be constructed by $\emptyset$ (the empty bijection), and by en $\oplus$ d which adds a data element *d* (with fresh code number) to the enumeration *en*. Adding an element twice has no effect. The size function # *en* returns the number of elements recorded in the enumeration, and the predicate $d \in en$ tests for membership. The selector *num_of(d, en)* returns the number of the element *d* inemphen, and *el_of(n, en)* gives back the element number *n* in *en*. Both operations are unspecified if $d \notin en$ or if $n \geq \#\, en$, respectively. Finally, the operation $en \ominus d$ removes the element *d* and its code number from the enumeration (if there is such an element). In addition all code numbers greater than *num_of(d, en)* are decremented by one in order to avoid gaps in the range

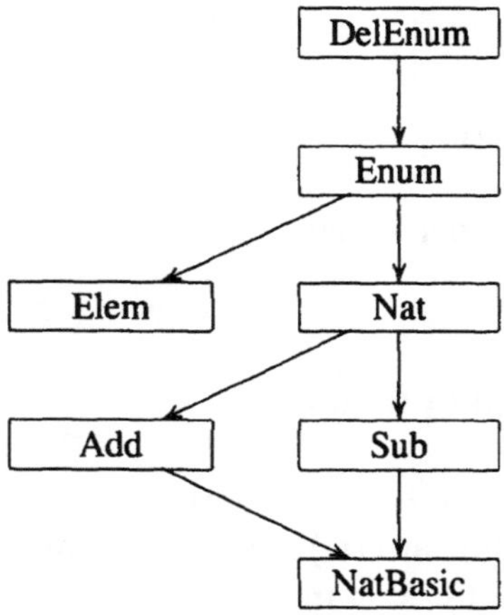

Figure 1. Structure of the example specification

of $en \ominus d$. Enumerations can be implemented in various ways, depending on what operations are critical for efficiency. Possible implementations are duplicate free lists, arrays, or hashtables (to speed up *num_of*).

2.1. *The formal specification*

With a little experience the above description of finite enumerations can be directly translated into a bunch of axioms. However, from the software engineering point of view an amorphous list of formulas is a bad representation. In the software design process, the specification is preceded by a careful problem analysis identifying decomposable subproblems and their interrelations. With the standard algebraic specification language used in KIV, this information can be made explicit in the specification structure. This is not only good software engineering practice but also very helpful for automated deduction.

The formal specification of finite enumerations is shown in Figs. 1 – 3. It identifies seven specification modules. *DelEnum* is the toplevel specification describing the overall functionality of finite enumerations over the parameter *Elem*. It imports the subspecification *Enum* and enriches it by $\ominus$ and its axioms (called the Δ of the enrichment). *Enum* specifies the remaining operations for finite enumerations. The constraint "enum **generated by** $\emptyset$, $\oplus$" gives a structural induction principle.

The specification is formulated relative to a standard specification of natural numbers (actually taken from the KIV library): *NatBasic* is the freely generated fragment of the natural numbers with 0 and successor succ, with

NatBasic =
data specification
nat = 0 | succ (pred : nat);
variables m, n: nat;
order predicates . < . : nat × nat;
end data specification

Generated axioms:

nat **freely generated by** 0, succ;
pred(succ(n)) = n,
m + 0 = m, m + succ(n) = succ(m + n),
¬ m < 0, 0 < succ(n), succ(m) < succ(n) ↔ m < n,
¬ m < m, m < n ∧ n < k → m < k

Add =
enrich Nat **with**
functions . + . : nat × nat → nat ;
axioms

n + 0 = n, m + succ(n) = succ(m + n)

end enrich

Sub =
enrich NatBasic **with**
functions . − . : nat × nat → nat;
axioms

m − 0 = m, m − succ(n) = pred(m − n)

end enrich

Nat = Add + Sub

Figure 2. The example specification (part 1)

the additional predecessor function pred and the ordering <. It is written like an ML (Milner et al., 1989) datatype declaration for which the axioms can be generated automatically. The enrichments *Add* and *Sub* introduce addition and subtraction by recursive definitions. *Nat* is the union of *Add* and *Sub*. The formal semantics of the specification language is described in (Wirsing, 1990), (Reif, 1992).

The specification *DelEnum* (inclusive of all subspecifications) has 17 axioms that are directly given in Fig. 3. 8 axioms are generated automatically for the data specification *NatBasic* in Fig. 2. Furthermore, the specification *Nat*

```
Elem =
specification sorts elem; variables d: elem;
end specification
---------------------------------------------
Enum =
generic specification
parameter Elem using Nat
sorts enum;
constants ∅ : enum;
functions
   . ⊕ .    : enum × elem → enum ;
   num_of : elem × enum → nat   ;
   el_of    : nat × enum   → elem ;
   # .      : enum         → nat   ;
predicates . ∈ . : elem × enum;
variables en: enum;
axioms
   enum generated by ∅, ⊕;
   d ∈ en → en ⊕ d = en, ¬ d ∈ ∅, d1 ∈ en ⊕ d ↔ d = d1 ∨ d1 ∈ en,
   # ∅ = 0, ¬ d ∈ en → #(en ⊕ d) = succ(# en),
   ¬ d ∈ en → num_of(d, en ⊕ d) = # en,
   d ≠ d1 → num_of(d, en ⊕ d1) = num_of(d, en),
   d ∈ en → el_of(num_of(d, en), en) = d
end generic specification
---------------------------------------------
DelEnum =
enrich Enum with
functions . ⊖ . : enum × elem → enum;
axioms
   ¬ d ∈ en → en ⊖ d = en,
   d ∈ en → #(en ⊖ d) = pred(# en),
   d1 ∈ en ⊖ d ↔ d ≠ d1 ∧ d1 ∈ en,
      d ∈ en ∧ d1 ∈ en ∧ num_of(d1, en) < num_of(d, en)
   → num_of(d1, en ⊖ d) = num_of(d1, en),
      d ∈ en ∧ d1 ∈ en ∧ num_of(d, en) < num_of(d1, en)
   → num_of(d1,en ⊖ d) = pred(num_of(d1, en))
end enrich
```

Figure 3. The example specification (part 2)

from the library is associated with 77 additional standard lemmas to improve arithmetic reasoning. These are persistent over the lifetime of the specification, and have been proved long time ago once and for all. Generally, the reuse of a formal library specification may include axioms that are relevant to one application but irrelevant to another. For *Enum* we get altogether 102 axioms.

As test examples we selected the 52 proof obligations for *DelEnum*, that showed up during a KIV case study on depth-first search in graphs. In a number of large case studies we found that in order to prove theorem n, it is a good idea to add all the n-1 previously proved theorems as lemmas to the theory. Although this enlarges the theory again, the effect is positive. With the redundant 77 lemmas of *Nat* and the discipline to add all previously proved test examples to the theory, the success rate of the five provers was doubled.

2.2. *Proving a goal*

Now we want to prove the theorem

$$\text{th-45: } \#\,\text{en} = \text{succ}(0) \rightarrow \emptyset \oplus \text{el\_of}(0, \text{en}) = \text{en}$$

from our example theory. In the test suite, th-45 was the 45th theorem. Therefore we had at this stage 146 (= 102 + 44) axioms in the theory. Potentially all of them could be used in the proof. Actually an interactive proof in KIV used only 12 of them (6 of the previously proved theorems, 4 axioms from the *Enum*-specification and two axioms from *NatBasic*).

Of course, it is impossible by syntactic consideration to find out in advance *exactly* which axioms will be relevant for a proof. But already the structure of the specification allows us to reduce the set of relevant axioms and lemmas considerably, as we will now show informally for the example. The reduction is useful for hierarchy persistent specifications. A precise definition of the reduction criteria is given in Sect. 3.

First, we select the minimal subspecification, whose signature comprises that of the theorem. In our case this is the specification *Enum*, since the operation $\ominus$ defined in *DelEnum* does not occur in the theorem. Thereby 5 axioms and 17 lemmas can be removed from the theory.

Second, the arithmetic operations + and − are not required. They are neither used to define any operation in the *Enum* specification nor do they occur in the theorem (this does not mean that *Add* and *Sub* are superfluous. They could be used for example in a superspecification of *DelEnum* in a larger case study). Removing all axioms and theorems for + and − saves additional 4 axioms and 57 lemmas which no longer must be passed to the automated theorem prover.

Third, with the removal of $\ominus$ we have eliminated all axioms, which make use of the predecessor function. Since the theorem does not use it, and *NatBasic* could be split into a specification, which defines 0,succ and an enrichment for pred, we can also remove all axioms and lemmas which use the predecessor function (this is not a special case, it is true for all selectors of datatype specifications). Again this removes 2 axioms and 12 lemmas from the theory.

Finally, with a similar argument as for the predecessor function, the 4 axioms and 6 lemmas for the $<$ predicate can be dropped.

After four reductions we are left with 38 axioms (10 axioms from the specification, 28 additional lemmas), which may be relevant to the proof of the theorem. The set of relevant axioms was reduced by a factor of almost 4. Although, this is not optimal (only twelve of them are actually needed) it makes a big difference for automated theorem provers. E.g., Otter was not able to prove the theorem with the full set of axioms within 5 minutes. With the reduced set of axioms the time to prove the theorem was about 11 seconds (a summary of the results for all provers will be given in Sect. 5).

3. REDUCTION CRITERIA

Specifications in KIV are built up from elementary first-order theories with the usual operations known in algebraic specification: union, enrichment, parameterization, actualization and renaming. Their semantics is the class of all models (loose semantics). Reachability constraints like "nat generated by 0, succ" or "list generated by nil, cons" restrict the semantics to term-generated models. The constraints are reflected by induction principles in the calculus for theorem proving used in KIV. We treat them as (higher-order) axioms. The structure of a specification is visualized as a specification graph, like the one we have seen in Fig. 1.

Structuring operations are not used arbitrarily in formal specifications of software systems. Enrichments "ESPEC = **enrich** SPEC **by** Δ", where Δ consists of a signature sig(Δ) and axioms ax(Δ) to be added, are supposed to have the property of *hierarchy persistency*. This property says that every model of SPEC can be extended to a model of ESPEC (therefore such enrichments are sometimes also called model conservative). Hierarchy persistency of an enrichment implies safe reduction of the set of necessary axioms to prove a theorem φ: Every theorem φ that holds in (all models of) ESPEC and uses only the signature defined in SPEC, already holds in SPEC. Without loss of generality we can also assume sig(Δ) $\neq \emptyset$ in the following, since otherwise the axioms in ax(Δ) would already be theorems over SPEC.

Hierarchy persistency cannot be checked syntactically, but there are sufficient criteria: e.g. definitional extensions and recursive definitions over free datatypes are easily proved to be hierarchy persistent. More generally, if the functions and predicates defined in the Δ of an enrichment **enrich** SPEC **with** Δ can be implemented by abstract programs, which use only the operations defined in SPEC, terminate on any input and satisfy the given axioms, the enrichment is hierarchy persistent. E.g. function $\ominus$ defined in *DelEnum* could be implemented by a program which given an enumeration en and an element d to delete, would construct the result by inserting all elements of en except d in a new empty enumeration (using a loop from 0 to # en).

The criterion of hierarchy persistency for enrichments can be used to derive criteria for the safe reduction of axioms for all other types of specifications: A generic specification like *Enum* must be a hierarchy persistent enrichment of the parameter. Nothing is required for renamings, and actualized specifications must be removed from the specification graph by actually computing the instantiated specification (this may duplicate some specification structure).

For a union specification $\text{SPEC}_1 + \text{SPEC}_2$ the criterion required is that the operations common to SPEC_1 and SPEC_2 are defined in a set of subspecifications $\text{SP}_1, \ldots \text{SP}_n$ of both SPEC_1 and SPEC_2, which is shared in the specification graph and that both the enrichments of $\text{SP}_1 + \ldots + \text{SP}_n$ to SPEC_1 and to SPEC_2 are hierarchy persistent. In this case the union is said to *have explicit sharing*, and the proof of a theorem φ with $\text{sig}(\varphi) \subseteq \text{sig}(\text{SPEC}_i)$ requires only the axioms of SPEC_i (i=1,2).

Structured specifications in which every enrichment is hierarchy persistent and every union has explicit sharing are called *modular*. An example is the specification from the last section given in Fig. 1 and Fig. 2. All enrichments are hierarchy persistent, and the union specification *Nat* has explicit sharing: the common part of *Add* and *Sub* is the specification *NatBasic* shared by both specifications.

Modular specifications are the usual case in software development[1]. Often even the structure of an implementation by a system of software modules follows the structure of the specification (see (Reif, 1995) for details). The specification is then called an *architectural specification* ((Mosses, 1996)).

For a modular specification we can now prove the following criteria for the reduction of axioms:

- *minimality criterion.* To prove a theorem one never needs more axioms than those of the minimal subspecification MSPEC whose signature cov-

[1] The situation is different in mathematics: consider e.g. the enrichment of rings to fields, which is not hierarchy persistent.

ers the signature of the theorem. This criterion was applied to the example theorem in Sect. 2 to eliminate the axioms of *DelEnum*. In practice, this criterion usually does not lead to much reduction, since usually theorems are formulated over the minimal subspecification.

– *structure criterion.* If the enrichment "'**enrich** SPEC **by** Δ"' is a subspecification of the minimal specification MSPEC, such that the operations from Δ are neither used in specifications above the enrichment nor in the theorem, the axioms from Δ can be dropped. This criterion was applied to eliminate the axioms for + and −.
– *specification criterion.* For an elementary specification the criteria described above can also be used to determine, if an operation defined in it is a hierarchy persistent enrichment of the rest. This criterion was applied to split the datatype specification *NatBasic* into a basic specification *NatB* and enrichments for the predecessor function and the less predicate.
– *recursion criterion.* The three previous criteria can be applied recursively, until the set of axioms gets stable. This criterion was used to eliminate the axioms for the predecessor function.

By applying the four criteria on a modular specification, a theory can be determined, over which the theorem can be *guaranteed* to hold, if it holds at all. Even if there are some non hierarchy persistent enrichments in a structured specification, the axiom reduction is still a very good heuristic. For actually proving theorems, we use the automated theorem provers as "oracles", which relieve us, if successful, of some interactions in KIV. The approach is correct, but not complete, since we are dealing with inductive theories. Reachability constraints such as "Nat generated by 0,succ" are disregarded by the automated theorem provers. In rare cases (we found one example in the three case studies we considered), an inductively provable theorem may even require to be generalized to a theorem, which requires a larger signature.

4. THE REDUCTION ALGORITHM

With the reduction criteria from the previous section, a set of relevant axioms RelAx(SP,φ) for a structured specification SP and a theorem φ can be computed. Of course, "relevant" axioms are a superset of those actually needed. In practical applications, there will be a lot of theorems $\varphi_1, \varphi_2, \ldots$ which have to be proved over the same specification SP. Therefore, it is a bad idea to start the computation of relevant axioms and lemmas for every theorem anew. Instead, an algorithm can be developed which divides computation of RelAx(SP,φ) in two phases. In the first phase the specification structure is compiled to a table T := Comp(SP). For every axiom ψ the table T records a set T(ψ) of Δ's from the enrichments of SP. Basic specifications are thereby viewed as enrichments of the empty specification $\emptyset$. T(ψ) roughly describes the minimal necessary part of SP over which ψ is a syntactically correct formula. The first phase is independent of the theorem to prove, and can be computed when the specification and its structure are known.

In the second phase the information stored in the table T is evaluated to give the relevant axioms Eval(T,φ) for a theorem φ to be deduced from SP. RelAx(SP,φ) := Eval(Comp(SP), φ). Comp does the main computation, Eval is a simple iteration function over all axioms.

The two parts of the algorithm are based on the observation that the semantics of a modular specification depends on the Δs of enrichments, but not on the initial structure of the specification graph. Any *syntactically correct* specification graph formed with the same Δ's defines a semantically equivalent modular specification. This observation (proved by induction on the number of Δ's) is the rational behind the following theorem.

THEOREM 1. *The minimal specification required to prove a theorem φ with signature sig(φ) can be computed by the following three steps:*

1. *Apply the specification criterion from Sect. 3 to break up Δs into smaller pieces.*
2. *Compute the set $\mathcal{S}_0$(sig(φ)) of all Δ's for which sig(Δ) and sig(φ) have a non-empty intersection.*
3. *Compute the minimal superset $\mathcal{S}$(sig(φ)) of $\mathcal{S}_0$(sig(φ)) such that it represents a syntactically correct specification. This is the minimal specification required to prove φ.*

The theorem is proved by rearranging the specification structure, such that the Δs from $\mathcal{S}$(sig(φ)) form a subspecification. Because of modularity, the axioms from this subspecification then are sufficient to prove φ.

As an application of Theorem 1 consider the goal

$$\text{th-19: } n < \#\,en \rightarrow (\text{num\_of}(\text{el\_of}(n,en), en) = n$$

which is dual to the last axiom

$$d \in en \rightarrow \text{el\_of}(\text{num\_of}(d, en), en) = d$$

from specification *Enum*. The first step in Theorem 1 splits *NatBasic* into a basic specification *NatB* and two enrichments with less predicate (specification *Less*) and predecessor function (*Pred*). For the second step we find that th-19 contains symbols defined in the signature of Δ_{Enum} (the Δ of the enrichment *Enum*) and Δ_{Less}, so $\mathcal{S}_0$(sig(th-19)) = $\{\Delta_{Enum}, \Delta_{Less}\}$. $\mathcal{S}_0$(sig(th-19)) does not represent a syntactically correct specification since the axioms of Δ_{Enum} and Δ_{Less} contain symbols from Δ_{NatB}. Therefore we include Δ_{NatB} in $\mathcal{S}$(sig(th-19)). This is the only addition needed, since **enrich (enrich (enrich $\emptyset$ with** Δ_{NatB}) **with** Δ_{Less}) **with** Δ_{Enum} is a syntactically correct specification. Theorem 1 now guarantees that a proof of th-19 requires the axioms of Δ_{Enum}, Δ_{Less} and Δ_{NatB} only.

For our example theorem th-45 from Sect. 2, things are easier. We have $\mathcal{S}_0$(sig(th-45)) = $\{\Delta_{Enum}\}$ and $\mathcal{S}$(sig(th-45)) = $\{\Delta_{Enum},\Delta_{NatB}\}$ as was stated informally in Sect. 2.

Now we are ready to sketch the two phases of the algorithm, which computes the relevant axioms and lemmas for the proof of a theorem: The first phase only relies on the specification SP. Comp(SP) computes

- the modified specification SP' according to the first step in Theorem 1.
- $\mathcal{S}(\{f\})$ with respect to SP' for every symbol (sort, function and predicate) of SP'
- $\mathcal{S}(\text{sig}(\psi))$ for every axiom ψ. $\mathcal{S}(\text{sig}(\psi))$ is the union of all $\mathcal{S}(\{f\})$ with f $\in \text{sig}(\psi)$

The result of Comp(SP) is a table T which associates $T(\psi) := \mathcal{S}(\text{sig}(\psi))$ with every axiom ψ.

In the second phase φ, the theorem to prove comes in. Eval(T,φ) first computes $\mathcal{S}(\text{sig}(\varphi))$ (using the precomputed information on $\mathcal{S}(\{f\})$ for every symbol f). The result is the set of relevant axioms Eval(T,φ) := $\{\psi\colon T(\psi) \subseteq \mathcal{S}(\text{sig}(\varphi))\}$. After a successful proof of theorem φ, we can add it to the theory, and update T assigning $\mathcal{S}(\text{sig}(\varphi))$ to φ.

This algorithm can still be improved. One optimization is to remove any Δ from $T(\psi)$, which is in $\mathcal{S}(\text{sig}(\Delta'))$ for another $\Delta' \in T(\psi)$.

A second optimization is based on the observation that an inference step between an axiom ψ and a theorem φ_0 is only useful, if φ_0 and ψ share some signature symbols (in many calculi, no other steps are even possible). Therefore RelAx(SP,φ_0) can be computed incrementally as follows: Instead of computing the full set $\mathcal{S}(\text{sig}(\varphi_0))$ we use the signature of φ_0 to compute a first approximation

$$\text{RelAx}_0(\text{SP},\varphi_0) := \{\psi : \text{sig}(\varphi_0) \cap \text{sig}(\Delta) \neq \emptyset \text{ for every } \Delta \in T(\psi)\}$$

of the relevant axioms. RelAx$_0$(SP,φ_0) are those axioms ψ for which $T(\psi)$ is already a subset of $\mathcal{S}_0(\text{sig}(\varphi_0))$. Adding these axioms as preconditions to the goal to prove, i.e. setting $\varphi_1 := Cl_\forall(\bigwedge RelAx_0(SP,\varphi_0)) \rightarrow \varphi_0$, where $Cl_\forall$ is universal closure, we can either compute a second approximation

$$RelAx_1(SP,\varphi_1) := \{\psi : sig(\varphi_1) \cap sig(\Delta) \neq \emptyset \text{ for every } \Delta \in T(\psi)\}$$

immediately (iteration will finally yield all relevant axioms), or first try some deduction steps before considering the axioms of the next approximation. Often the theorem will already be provable with the axioms from $RelAx_0$. In fact, for the example theorem th-19, $RelAx_0$ will already include the necessary axioms, consideration of the axioms from *NatB* is avoided.

Iterated computation of relevant axioms is even more beneficial when a proof calculus is used that has a notion of "subgoals" (like sequent- or tableau-calculi). Then the second iteration of computing relevant axioms may use the subgoals produced by the inference steps instead of φ_1. Since these subgoals usually have a smaller signature that φ_1 itself, even fewer axioms will be relevant for them.

5. Experimental Results

To evaluate the results of the axiom reduction we first tried the automated theorem provers Otter, Protein, SETHEO, Spass and $_3T\!A\!P$ on 45 non-inductive (+ 7 inductive) theorems defined over the *DelEnum* specification from Sect. 2. No theorem was invented for this case study, all were drawn from the case study on verifying depth-first search in graphs. All provers were given 2 minutes of proof time on a SPARC 20. Their settings can be found in Appendix A. We used the reduction algorithm including the first optimization (the second optimization, incremental computation of the reduced axiom set, was not possible, since the provers did not give access to subgoals). Axiom reduction is inexpensive: the time to compute the table T from a specification (function *Comp* from the previous section) was below one second, and the relevant axioms for a theorem were computed from this table (with *Eval*) in 0.1 – 0.2 seconds.

The results are summarized in Table I (a full listing of all axioms and theorems, as well as a full account of all proof times, is given in (Schellhorn and Reif, 1997)). The first line in the table gives the number of theorems which could be proved with the full set of axioms, the second line gives the number for the reduced set of axioms.

The numbers show that all provers benefited from the reduction of axioms, but there were enormous differences. Very significant improvements were made by Spass and $_3T\!A\!P$, while the other three provers benefited less. The time necessary to prove theorems was reduced by a factor of three on average.

Table I. Results for the Enum example

	Otter	Protein	SETHEO	Spass	$_3T^AP$
full axiom set	35	30	34	22	9
reduced set	36	32	36	37	21

Table II. Results for the compiler verification example

	Otter	Protein	SETHEO	Spass	$_3T^AP$
full axiom set	48	39	45	4	8
reduced set	48	45	48	45	21

A possible explanation for the good behaviour of Protein, Otter and SETHEO is, that their proof search concentrates on the distinguished goal clause. E.g. if we intentionally take a wrong goal clause, the time SETHEO needs to find a proof often increases from a few seconds to some minutes. Similarly, the best settings we found for Otter use the goal as set of support and binary resolution. In auto-mode, which does not distinguish the goal, only 19 theorems could be proved without and 26 with reduction.

As a second case study we considered 54 simple non-inductive first-order theorems that showed up during the verification of a Prolog compiler in KIV (see Chapter III.2.7). These are formulated over a specification which is built up from a lot of standard datatypes (lists, tuples, pairs etc.) in various different instantiations. Therefore the specification structure contains many different sorts, but the hierarchy of specifications is relatively flat. The theorems are easier than the ones found in *DelEnum* example, but the initial set of axioms is much larger (ca. 400). The axiom reduction is more effective than in the previous example: the reduced set contains in most cases only between 4 (!) and 25 axioms. Again the results varied largely as can be seen in Table II. Otter cannot prove more theorems, but the time to prove them decreases in several examples from over 30 seconds to about 2.

It seems, that Otter's, Protein's and SETHEO's heuristics for using axioms are already strong enough, to avoid all axioms involving sorts (encoded as constants), which do not occur in the theorem. In a flat specification structure, this set is already a good approximation to the set of relevant axioms. To see how the heuristics of Otter and SETHEO (Protein is very similar to

Table III. Results for the graph example

	Otter	SETHEO
full axiom set	24	18
reduced set	31	29

SETHEO, so we did not try it) would behave in general, we finally tried an example with the opposite characteristic: only few sorts, but many operations. The example is from the KIV library of standard specifications. There, a specification *Graph* is defined. The full set of axioms contains over 500 axioms. The 40 theorems considered were theorems on the sets of nodes. Axiom reduction yields below 100 relevant axioms for all these theorems. Table III gives the results, which clearly show the positive effect of axiom reduction.

Let us finally make a technical remark. One criterion for the effective use of automated theorem provers in the verification of first-order theorems from software verification is that they separate expensive preprocessing of axioms from the actual proof attempts. This avoids preprocessing the axioms over and over again for several theorems over the same specification. The code generation for SETHEO already takes a minute for the test examples of the Prolog compiler with the full set of axioms. This is quite unsatisfactory. The same problem exists for Protein's code generation. For Otter's transformation of formulas to clauses the problem is less severe since generating clauses is done in about 5 seconds. Spass already separates clause generation from actual proof attempts. Here, $_3T\!AP$ offers the best solution, in that it may keep an arbitrary number of preprocessed sets of axioms, which can be referred to by name.

6. Conclusion

We have investigated the performance of automated theorem provers in formal software verification. To this purpose we used the software verification tool KIV as a test environment, and did comparative experiments with the five automated theorem provers Otter, Protein, SETHEO, Spass and $_3T\!AP$. In this application the challenge for the automated provers is the large number of (up to several hundred) axioms in typical software specifications. The major problem is to find out the few axioms that are really relevant to the proof of a given theorem. We have presented a reduction technique for this problem which eliminates for a given theory and a theorem as many irrelevant axioms from the theory as possible. The technique relies on good software specification practice to reflect design decisions in the specification structure. It is safe for hierarchy persistent specifications. The property of hierarchy persistency is proved separately by KIV. Reduction can be applied recursively also to the subgoals of a goal, and it is independent from the particular prover and calculus. The incremental version of the reduction algorithm computes the reduced axiom set stepwise, using the entire set only when needed.

From the experiments we learned that all five provers benefited from the axiom reduction. They could prove more theorems than before, and were faster for those that could be proved already without reduction. However, the susceptibility of the four provers to the problem of large axiom sets was quite different. Furthermore, the experiments shed some light on the question, to what extent an interactive verification system could benefit from an automated prover.

ACKNOWLEDGEMENTS

We thank Kurt Stenzel for reading the draft and making helpful comments. Thanks also to Harald Vogt and Michael Balser for their help with the experiments.

REFERENCES

Ahrendt, W., Beckert, B., Hähnle, R., Menzel, W., Reif, W., Schellhorn, G., and Schmitt, P. (1998). Integration of Automated and Interactive Theorem Proving. In Bibel, W. and Schmitt, P., editors, *Automated Deduction — A Basis for Applications*, volume I, 3. Kluwer. to appear.

Baumgartner, P. and Furbach, U. (1994). Protein: A prover with a theory extension interface. In *Proc. 12th CADE*, LNCS 804. Springer. for the newest version of Protein, see the URL: http://www.uni-koblenz.de/ag-ki/Implementierungen/Protein/.

Beckert, B., Hähnle, R., Oel, P., and Sulzmann, M. (1996). The tableau-based theorem prover $_3T\!AP$, version 4.0. In McRobbie, M., editor, *Proc. 13th CADE, New Brunswick/NJ, USA*, LNCS 1104, pages 303–307. Springer. for the newest version of $_3T\!AP$, see the URL: http://i12www.ira.uka.de/~threetap/.

Goller, C., Letz, R., Mayr, K., and Schumann, J. (1994). SETHEO v3.2: Recent developments – system abstract. In Bundy, A., editor, *12th Int. Conf. on Automated Deduction, CADE-12*, Springer LNCS 814. Nancy, France. for the newest version of SETHEO, see the URL: http://wwwjessen.informatik.tu-muenchen.de-/forschung/reasoning/setheo.html.

Milner, R., Tofte, M., and Harper, R. (1989). *The Definition of Standard ML*. MIT Press, Cambridge, MA.

Mosses, P. D. (1996). CoFI : The common framework initiative for algebraic specification. In Ehrig, H., v. Henke, F., Meseguer, J., and Wirsing, M., editors, *Specification and Semantics*. Dagstuhl-Seminar-Report 151. http://www.brics.dk/Projects/CoFI.

Reif, W. (1992). Correctness of Full First-Order Specifications. In *4th Conference on Software Engineering and Knowledge Engineering. Proceedings*. Capri, Italy, IEEE press.

Reif, W. (1995). The KIV-approach to Software Verification. In Broy, M. and Jähnichen, S., editors, *KORSO: Methods, Languages, and Tools for the Construction of Correct Software – Final Report*. Springer LNCS 1009.

Reif, W., Schellhorn, G., and Stenzel, K. (1995). Interactive Correctness Proofs for Software Modules Using KIV. In *Tenth Annual Conference on Computer Assurance*, IEEE press. NIST, Gaithersburg (MD), USA.

Reif, W., Schellhorn, G., and Stenzel, K. (1997). Proving System Correctness with KIV 3.0. In *14th International Conference on Automated Deduction. Proceedings*, pages 69 – 72. Townsville, Australia, Springer LNCS 1249.

Schellhorn, G. and Ahrendt, W. (1998). The WAM Case Study: Verifying Compiler Correctness for Prolog with KIV. In Bibel, W. and Schmitt, P., editors, *Automated Deduction — A Basis for Applications*, volume III, 2. Kluwer. to appear.

Schellhorn, G. and Reif, W. (1997). Proving Properties of Finite Enumerations: A Problem Set for Automated Theorem Provers. Ulmer Informatik-Berichte 97-12, Universität Ulm, Fakultät für Informatik.

Sutcliffe, G., Suttner, C., and Yemenis, T. (1994). The TPTP problem library. In Bundy, A., editor, *12th International Conference on Automated Deduction, CADE-12*, Springer LNCS 814. Nancy, France.

Weidenbach, C., Gaede, B., and Rock, G. (1996). SPASS & FLOTTER, version 0.42. In *13th International Conference on Automated Deduction, CADE-13*, Springer LNCS. for the newest version of Spass, see the URL: http://www.mpi-sb.mpg.de/-guide/software/spass.html.

Wirsing, M. (1990). *Algebraic Specification*, volume B of *Handbook of Theoretical Computer Science*, chapter 13, pages 675 – 788. Elsevier.

Wos, L., Overbeek, R., Lusk, E., and Boyle, J. (1992). *Automated Reasoning, Introduction and Applications (2nd ed.)*. McGraw Hill. for the newest version of Otter, see the URL: http://www.mcs.anl.gov/home/mccune/ar/otter/#doc.

A. Appendix. Settings for the Automated Provers

A.1. *Otter*

Otter (version 3.0.4) is an automated theorem prover for unsorted first-order logic with equality. Sorts were encoded as unary functions. Every (sub-)term t of sort s is translated to $s(t)$. Otter was used with the settings of auto-mode, but with the (negated) theorem to prove as the set of support (see p. 552 of (Wos et al., 1992)) and with binary resolution instead of hyper resolution (since the latter is not complete when combined with sos). We tried some other settings, but all resulted in fewer provable theorems.

A.2. *SETHEO*

SETHEO (V 3.3) can handle only clauses from an unsorted logic, and has no built-in equality predicate. Therefore we had to use a standard algorithm for encoding formulas as clauses. The resulting clauses are the same as the ones that Otter generates, except that clauses $\{x \neq t, L_1, \ldots L_l\}$ with $x \notin Vars(t)$ are optimized to $\{L_1[x \leftarrow t], \ldots L_l[x \leftarrow t]\}$. Sorts were encoded as for Otter, the equality predicate was explicitly axiomatized. SETHEO was used with the '-wdr' option, which performed significantly better than the '-dr' option.

A.3. *Protein*

Protein (v 2.12) was given the same input as SETHEO. It was used with the model elimination calculus, which gave the best results.

A.4. *Spass*

Spass (v 0.55) is a theorem prover for unsorted first-order logic with equality. Since unary predicates are treated by special "sort" inference rules we encoded sorts as unary predicates.

A.5. $_3T^AP$

$_3T^AP$ (version 4.0) is a theorem prover for many-sorted first-order logic with equality. There exists an interface to $_3T^AP$ in KIV. Therefore, in contrast to all other provers, $_3T^AP$ could be called directly from KIV. The experiments benefited from the fact, that KIV gives only a subset of the previously proved lemmas to $_3T^AP$.

FRIEDER STOLZENBURG AND BERND THOMAS

CHAPTER 10

ANALYZING RULE SETS FOR THE CALCULATION OF BANKING FEES BY A THEOREM PROVER WITH CONSTRAINTS

1. INTRODUCTION

We show that theorem proving, logic programming and constraint solving can be combined in a straightforward manner. This is shown not only by setting up a theoretical framework, but also by a real world application: the calculation of banking fees. We tackle the problem of deciding whether such a rule set is total and deterministic. Although these questions are undecidable in general, the restricted form of the investigated rule sets makes them decidable and even tractable in practise. Experiences with our system implemented in Prolog as well as some questions one may have with the application are discussed. The success of this application is due to the combination of constraint logic programming and first-order theorem proving, based on the model elimination calculus which is in the focus of the German research program on deduction. Each approach alone is not powerful enough to find the solution in reasonable time. The procedure proposed here can easily be generalized to analyzing arbitrary rule sets.

2. CONSTRAINT MODEL ELIMINATION

One of the most traditional disciplines in research on automated reasoning is theorem proving. In the early days, it concentrated mainly on developing general proof procedures for predicate logic. Because of a shift towards special domain-dependent systems, automated reasoning and logic programming nowadays aim at incorporating specialized and efficient modules which are suited for handling specialized knowledge. Therefore, it seems to be rather natural to combine theorem proving and constraint reasoning.

We will combine the model elimination procedure (Loveland, 1968) and constraint logic programming (Jaffar and Maher, 1994) here. Model elimination is a special tableau method, and constraint reasoning is closely related to theory reasoning. Both topics are discussed in great detail in this enterprise; see chapters I.1.2 and I.1.3 on clausal tableaux and chapter I.2.6 on

W. Bibel, P. H. Schmitt (eds.), Automated Deduction. A basis for applications. Vol. III

theory reasoning. This may be traced back to the work of Bibel (1982) and Stickel (1985). It is interesting to notice that the success of the application to be presented here is due to the combination of both approaches.

2.1. *The PTTP Technique*

Constraint logic programming (CLP) is a field of active research. There, logic programming with Horn clauses is enhanced with an interface to efficient constraint solvers. There are also approaches that combine general, non-Horn theorem proving calculi with constraint processing. For example, Bürckert (1991) introduces the calculus of constrained resolution. However, efficient implementations for these approaches still have to be developed. Here we present a method which is also well-suited for implementation: constraint model elimination (CME). This method has been introduced by Baumgartner and Stolzenburg (1995) and Stolzenburg (1998).

Model elimination (Loveland, 1968) is a linear and goal-directed calculus, thus making it close to the logic programming language Prolog. This allows us to realize model elimination in a straightforward manner on top of existing Prolog systems by using the Prolog technology theorem prover (PTTP) technique proposed by Stickel (1988). The PTTP technique compiles a problem specified in first-order logic into a (constraint) logic program, hence ready to be executed by a Prolog system.

A theorem proving or logic programming system with constraints can be implemented rapidly and easily as will be shown here. For this purpose, the constraint solving mechanisms of the underlying Prolog dialect can be used. So we actually built a PTTP based theorem prover with a very modular architecture that exploits finite domain constraints (van Hentenryck, 1989). Such a cheap and rapid implementation does not seem to be available for constrained resolution and other related approaches.

2.2. *Applying Finite Domain Constraints*

In the last years, researchers in the fields of artificial intelligence, logic programming and deduction systems developed many new ideas and formalisms. They succeeded in showing that their approaches are adequate at least for small examples. But nowadays the following questions often arise:

1. What are the real-world applications of these ideas?
2. Which parts of the formalisms are needed in order to express realistic problems?
3. Are the approaches efficient enough to handle large data sets?

With these questions in mind, we tried to apply our model elimination system Protein (Baumgartner and Furbach, 1994) with finite domain constraints on rule sets for the calculation of banking fees. The underlying problem is as follows: a credit institution sells stocks and shares to its costumers; the fee for them depends on their value, the customer type and other parameters. The rule set is given as a natural language description. However, we will assume that it has already been translated into first-order predicate logic.

As an example, consider the following rule given in natural language: *For every transaction made for an office in Switzerland, a constant fee of 30 SFr is charged.* (Of course, the original rule set was given in German language.) The corresponding first-order logical representation is given as follows, where A stands for the actual deal:

(1) $\forall A\,(country(A,ch) \rightarrow (customer(A,office) \leftrightarrow const(A,30)))$

This formula can be represented by two Prolog rules:

(2) $const(A,30) \leftarrow country(A,ch), customer(A,office).$

(3) $customer(A,office) \leftarrow country(A,ch), const(A,30).$

We translate them into rules with (arithmetical) constraints, i.e. with equations and inequations between parameters. Constraints always occur in rule bodies only. Hence they have to be negated and moved into the body, if they have to occur in the head. This is the case for rule (3), which is translated into (5). Note that $\underline{A}$ stands for the list of input parameters of the actual deal, e.g. $\underline{A} = (Country, Customer)$.

(4) $const(\underline{A},30) \leftarrow Country = ch, Customer = office.$

(5) $\leftarrow Customer \neq office, Country = ch, const(\underline{A},30).$

Schmid (1990) investigates how the following interesting problems concerning such rule sets can be tested: ***Is the rule set total, i.e., does it permit calculation of a fee for every business deal? Is it deterministic, i.e., can only one fee be computed in each case?*** We tried to tackle these questions by means of our theorem proving system with constraints, especially the problem of determinism. We succeeded in finding the answer.

2.3. *Rule Set Verification and Software Engineering*

There is an interesting relationship between our application and automated reasoning in software engineering, which we want to discuss now. In the introduction to their book, Rich and Waters (1986) state that "the ultimate goal of artificial intelligence applied to software engineering is automatic programming". We achieve this goal here to a certain extent, since we get a directly executable specification of a rule set, i.e. a logic program that calculates banking fees according to the given rule set. But, as said above, we want to achieve more than that, namely analyzing these rule sets wrt. critical properties. For this, it seems to be very unlikely to achieve a satisfying state of performance, due to the highly complex and domain-specific problems. One attempt to cut down the problems of automatic programming was to focus on very small and domain-specific problems. In fact, we concentrate here on a special domain: rule sets for banking fees. In addition, we exploit domain-specific knowledge by employing specialized constraint solving modules from the CLP field.

According to Rich and Waters (1986), theorem-proving techniques were used (a) for *deductive synthesis* and (b) for *program verification* in software engineering. Deductive program synthesis is based on the idea, that each step carried out by a constructive theorem prover can be seen as a computational step in a program. Assume that a program takes some input value x and returns an output value y. Furthermore a program specification is given, and the assertions that for a precondition $P(x)$ dependent on the input value x the postcondition $Q(x,y)$ holds. This can be formulated as a theorem $\forall x \exists y\ (P(x) \rightarrow Q(x,y))$ to be proven by the theorem prover, taking the given program specification into account. Let us focus on our approach for the calculation of banking fees and the connection to deductive synthesis. If we consider the rule set to be a program specification and the calculation of a specific transaction under the aspects of determinism and totality as the theorem to be proven, then the derivation or refutation tableau of this proof can be seen as an explanation for the calculation. Experiments with the tool ILF developed to generate natural language proofs from such derivations are very helpful in this context. For more information about ILF, see chapters III.1.1 and III.2.8 and the ILF manual (Dahn et al., 1995).

Program verification makes use of theorem provers in a very similar way. The main goal of program verification is to verify that a program satisfies its formal specification. For this, usually, a set of axioms describing the behavior of the statements, a precondition and a postcondition are required, where the pre- and postconditions are similar to those in program synthesis. The first step of the automatic verification is to construct one large formula by combining the precondition with the behavior of the statements contained in the program. After this formula has been constructed, a theorem prover is used to verify, whether the postcondition follows from this formula. The test for determinism of the rule set can be seen as a verification task, where the postcondition states that at most one fee is calculated for each transaction, and the precondition simply states that the transaction values are within the given domains.

In summary, our practical approach to solve the problems of determinism and totality for a special kind of rule sets is strongly related to basic areas in artificial intelligence and software engineering. Though we focus more on the logic programming and deduction based point of view, the previous mentioned aspects of this work underline the wide area of its research interest.

2.4. *Overview*

The plan for the rest of this chapter is as follows. At first, we will instantiate the CME framework with finite domain constraints in section 3. The reader will be able to see how simple and elegant the system can be implemented. After that, we will present our application of calculating fees and the problems we had to overcome in order to solve the interesting questions in section 4 in more detail. Finally we will give some conclusions in section 5. A preliminary version of the material presented here has been given in Stolzenburg and Thomas (1996).

3. Theorem Proving with Constraints

We now introduce constraint clauses, finite domain constraints (van Hentenryck, 1989), and the CME framework, where ideas from Bürckert (1991) and Jaffar and Maher (1994) are used. Model elimination will be viewed as a tree calculus in this context. The calculus acts with constraint clauses on model elimination tableaux. For further information, the interested reader is referred to Baumgartner and Stolzenburg (1995) and Stolzenburg (1998). Readers not familiar with basic notions of logic programming should consult e.g. Lloyd (1987).

3.1. *Constraint Clauses and their Semantics*

DEFINITION 1. A *constraint clause* is a pair consisting of a disjunction of literals L which is an ordinary clause (over a signature Σ, see below) and a constraint R. It is written as L/R and can be understood as the implication $\forall(R \rightarrow L)$; all free variables are supposed to be universally quantified. The constraints stem from a constraint theory which consists of a set of constraint symbols Δ and a set of constraint models in which these symbols are interpreted. A constraint theory is nothing but a set of first-order structures over the signature Δ.

Constraint theories will now be extended with further predicate symbols. Thus given a constraint theory and a set Σ of predicate symbols—disjoint from Δ—, we augment the constraint models by any possible denotation for these new symbols. A structure $\Im'$ for a constraint clause set C is given by a constraint model $\Im$ of the constraint theory for all symbols in Δ, where in addition all symbols from Σ occurring in left-hand sides L of rules are interpreted by a relation with the respective arity. $\Im'$ is called Σ-*expansion* of $\Im$. The disjointness of the signatures (except for the variables) is crucial for the calculus. This allows a modular architecture of hybrid systems hosting constraint solvers.

DEFINITION 2. If an interpretation $\Im'$ together with a variable assignment α satisfies a literal set L (read as disjunction) we write $\Im'_\alpha \models L$. See Lloyd (1987) for detailed definitions. A constraint clause L/R is satisfied by $\Im'$ (written $\Im' \models L/R$) iff for all variable assignments $\alpha : V \to D$ (where V is the set of the variables in L/R, and D is the domain of the considered constraint theory) with $\Im_\alpha \models R$ it holds $\Im'_\alpha \models L$. We say that a constraint clause set C is satisfied by $\Im'$ iff $\Im' \models L/R$ for every clause $L/R \in C$. The set C is called *satisfiable* iff such an $\Im'$ exists.

3.2. *Finite Domain Constraints*

A *constraint logic program* (also called *constraint clause set* in this context) consists of rules of the form $(L_1 \leftarrow L_2)/R$ where L_1 and L_2 are sets of positive literals (read as disjunction and conjunction of literals, respectively) and R is a constraint. Quite often L_1 is a singleton. A rule is called *disjunctive* iff L_1 contains more than one literal. During a computation when goals are unfolded using constraint clauses, the constraints R in their bodies are collected and tested for consistency. In the sequel, we will consider constraint solvers just as a black box.

In the case of finite domain constraints, we consider the domain of finite sets of constants. Constraint predicate symbols in this context are $=, \neq, \leq, >$ etc. In the case of finite integer domains, operations such as $+$ and $\cdot$ are also useful. All symbols will be interpreted as expected by their usual meaning. Mostly, only linear integer constraints of the form $a_1x_1 + \cdots + a_nx_n \leq b$ are considered where the x_i with $1 \leq i \leq n$ are integer variables and the a_i and b are integer constants. But in general, any relation between two or more variables may be taken into consideration. Without loss of generality, we will assume that all a_i with $1 \leq i \leq n$ are positive. We will specify the domain of a variable x like this: $x :: 0..99$, which means $x \in \{0, 1, \ldots, 99\}$.

The constraint theory consists of only one single model for the finite domain case. This is an important property, because it simplifies our proof procedure. Every constraint theory consisting of only one model enjoys the so-called independence of negated constraints property (Maher, 1993). Because of this, it suffices to consider only one derivation (refutation) for computing answers. This means, the completeness result for CME has a simple structure in this restricted case (see section 3.3). Since we assume that all variables have only finite domains, the only model is even finite. Implementations of finite domain constraints often exploit the implication

$$\text{(6)} \qquad \text{if } \sum_{i=1}^{n} a_i x_i = b \text{ then } b - \sum_{\substack{i=1 \\ i \neq k}}^{n} a_i \max(D_i) \leq a_k x_k \leq b - \sum_{\substack{i=1 \\ i \neq k}}^{n} a_i \min(D_i)$$

for all $1 \leq k \leq n$ where D_i is the domain for the variable x_i, i.e. its value range (van Hentenryck, 1989). Thus, e.g. for $x :: 0..99 \wedge y :: 0..99 \wedge 2x + 5y = 33$, we can infer the domain restrictions $x :: 4..14 \wedge y :: 1..5$ by applying (6) repeatedly. This is left as an exercise to the reader.

However it is not correct to use (6) as the only simplification rule, since not all $(x,y) \in \{4,\ldots,14\} \times \{1,\ldots,5\}$ are solutions, but only $(4,5)$, $(9,3)$ and $(14,1)$. These solutions can be computed by means of the (extended) Euclidean algorithm. Linear integer programming techniques are also applicable here. But for finite domain constraints, usually an algorithm which achieves only local consistency is completed by a backtrack scheme, because this seems to be cheaper than only one monolithic algorithm. In general there is a trade-off between the effectiveness and efficiency of constraint simplification. Hence constraints may be evaluated eagerly or lazily, i.e., the consistency check should be delayed, if it cannot be done efficiently.

3.3. *The Extended Model Elimination Calculus*

Now we want to treat the overall framework in a formal way. This is done in a tableau-like presentation. We have two inference rules in our calculus, namely (constraint) extension and reduction steps. They are used in constraint derivations. Mostly we are interested in derivations ending with an *empty tableau* $\square/R$ where R is a constraint. We will represent the two basic inference rules of the model elimination extended by constraints. For a more detailed description, see Baumgartner and Stolzenburg (1995) and Stolzenburg (1998).

DEFINITION 3. Let C be a constraint clause set. A pair P/R is called *constraint tableau* iff P is a tableau, i.e. a finite set of finite sequences $p = \langle p_1, \ldots, p_n \rangle$ of Σ-literals (which are called paths) and R is a constraint. We will call a sequence of tableaux starting with the initial one L/r—which must be a clause in $\mathcal{C}$ where L is interpreted as a set of paths of length 1—*constraint derivation* in $\mathcal{C}$ with *goal clause* L/r iff in each step one of the following inference rules has been applied:

extension. $$\frac{(P \uplus \{p\})/R}{P'/R'} \text{ if } (L \vee l)/r \in \mathcal{C}$$

where

- $(L \vee l)/r$ is a copy of a clause in $\mathcal{C}$ with fresh variables;
- $R' = R \wedge r \wedge (p_n = \neg l)$ (see explanation below);
- $P' = P \uplus (p \circ L)$ where $\uplus$ denotes (multi-set) union;
- $p \circ L$ denotes the set of paths appending each literal in L to p, i.e. $p \circ L = \{\langle p_1, \ldots, p_n, l\rangle \mid l \in L\}$ provided $p = \langle p_1, \ldots, p_n\rangle$.

reduction. $$\frac{P \uplus \{p\}/R}{P/R'} \text{ if } l \in p$$

where

- $R' = R \wedge (p_n = \neg l)$;
- $p_n = \neg l$ denotes the fact that the last literal in $p = \langle p_1, \ldots, p_n\rangle$ is complementary with l.

The practical combination of model elimination with (finite domain) constraints is quite easy. We had a PTTP-implementation of model elimination already at hand—the Protein system (Baumgartner and Furbach, 1994)—implemented in ECLiPSe-Prolog (ECRC, 1995). Only small modifications were necessary to build the whole system because ECLiPSe features constraint solving mechanisms. The constraint predicates simply can be passed directly to the finite domain constraint solver while the other predicates are treated by our Protein theorem prover. There is a strong interaction between both components. The architecture is modular and simple. We do not need any additional control component. Only one derivation of an empty tableau is necessary in the finite domain case for computing an answer, because the constraint theory consists of only one single model as mentioned before. The next theorem states *soundness and completeness* of CME with finite domain constraints; it is a specialization of the main theorem in Baumgartner and Stolzenburg (1995).

THEOREM 4. *Let $\mathcal{C}$ be a clause set with finite domain constraints attached. $\mathcal{C}$ is unsatisfiable iff there exists one derivation of an empty tableau $\square/R$ such that $\exists R$, i.e., the existential closure of R is valid in the theory of finite domain constraints.*

The definitions stated here slightly differ from those in Stolzenburg (1998), because we establish only refutational completeness of CME here, which is less than answer-completeness. Completeness can be proven by employing the lifting technique to ground proofs. The notion of a ground proof has to be adapted in our context to derivations on ground instances of constraint tableaux. In contrast to Baumgartner and Stolzenburg (1995), the proof of the lifting lemma—relating CME and corresponding ground derivation—is more difficult in the answer-completeness case. Again, the interested reader is referred to Baumgartner et al. (1997) and Stolzenburg (1998) for more details.

4. Banking Applications

At this point we are ready to come back to our application: the calculation of banking fees. Remember the problem: a credit institution sells stocks and shares to its costumers and makes a charge for them. We distinguish different kinds of data in the specification of the problem:

input parameters: value, country, currency, type of customer, etc.;

background facts: Skolem constants or mappings, e.g. conversion table for different currencies;

output value: the customer's fee.

The rule set will be translated into a constraint clause set. In the following we will only refer to the case of finite domain constraints. We are interested in the questions whether the rule set is total and deterministic. At first we want to discuss these problems for unrestricted rule sets and show their undecidability. After that we will introduce some reasonable restrictions which leads to tractability in practice. Finally we will compare two approaches that cope with the restricted rule sets and state some complexity results.

4.1. *Theoretical Problems*

In the sequel we will consider rule sets that compute fees y (the output value) via a binary relation $G(x, y)$ where x comprises all input parameters and background facts. We will assume that the problem specification is always given as a clause set with finite domain constraints. Although the rest of this chapter is mainly concerned with the determinism problem, we will now define two notions, which are interesting in our context.

DEFINITION 5. A binary relation G is called (left) *total* iff for all possible x there exists at least one y with $G(x,y)$. It is called (weakly) *deterministic* iff for all x there exists at most one y with $G(x,y)$. G is *strongly* deterministic iff the rule set allows exactly one way to compute a y for each x. Of course, the latter notion depends on the chosen computation model.

Both kinds of determinism are interesting. In the practical application the strong variant may be desirable, whereas we will prefer the weak one because then we can tackle the problem by means of a Prolog-based theorem proving system quite easily, as we will see next. The following table gives an overview about the complexity of the problems. We will explain it in a while.

problem	semi-decidable	strictly undecidable
G total	yes[1]	yes[2]
G non-total	no	yes
G deterministic	no	yes
G non-deterministic	yes[3]	no

Since predicate logic is undecidable in general, we are able to conclude that the interesting problems here (totality and determinism) are undecidable too. Not both a problem and its complement can be semi-decidable. As we will see, totality may even be strictly undecidable, i.e. not even semi-decidable. Some remarks seem to be necessary to clarify things. They relate to the annotated numbers given in the table.

1. Totality is semi-decidable if we consider arbitrary models. Then we only have to show the theorem $\forall x \exists y\, G(x,y)$. This can be done by means of an ordinary theorem prover.
2. If the elements of x stem from an inductive set, e.g. the Herbrand domain or the natural numbers, then the problem becomes more difficult. This can be seen by the following argument. Logic programs and Turing machines are equivalent in their computational power, and we can reduce our problem to the complement of the halting problem for Turing machines, which is known to be strictly undecidable.
3. Non-determinism turns out to be semi-decidable by the following procedure: enumerate all answers to the query $G(x,y)$ until some x, y, y' are found with $G(x,y) \wedge G(x,y') \wedge y \neq y'$. This works quite well with our theorem proving environment.

$$
\begin{array}{lll}
G(A,K) & \leftarrow & const(A,K). \\
G(A,M) & \leftarrow & \neg const(A,K), min(A,M), value(A,V), \\
 & & V \leq 100 \cdot M. \\
G(A,M) & \leftarrow & \neg const(A,K), \neg country(A,oversea), max(A,M), \\
 & & value(A,V), V \geq 100 \cdot M. \\
G(A,G) & \leftarrow & \neg const(A,K), \neg min(A,Min), \neg max(A,Max), \\
 & & value(A,V), G = V/100. \\
 & & \\
const(A,0) & \leftarrow & country(A,ch), customer(A,office). \\
const(A,30) & \leftarrow & \neg country(A,ch), customer(A,office). \\
const(A,50) & \leftarrow & customer(A,personal). \\
 & & \\
max(A,10000) & \leftarrow & country(A,ch), currency(A,ch). \\
max(A,20000) & \leftarrow & country(A,ch), value(A,V), V > 1000000. \\
 & & \\
min(A,10) & \leftarrow & country(A,ch), type(A,share), value(A,V), V < 100. \\
min(A,10) & \leftarrow & country(A,ch), type(A,oblig), value(A,V), V < 100. \\
min(A,20) & \leftarrow & \neg country(A,ch), value(A,V), V < 200. \\
min(A,50) & \leftarrow & \neg country(A,ch), value(A,V), V \geq 200, V < 500. \\
min(A,80) & \leftarrow & value(A,V), V > 500.
\end{array}
$$

Figure 1. Small rule set.

4.2. *The Treatment of Restricted Rule Sets*

These undecidability results do not sound very encouraging. So it might be a good idea to lower the expressiveness of the rules. But at the same time we want to be able to state all rules for the calculation of banking fees that occur in practice. The following restrictions—see Schmid (1990) for a full list of restrictions—meet both requirements in fact, although they seem to be very strong. The interesting problems then become decidable.

1. All input parameters have finite domains. That allows us to employ finite domain constraints.
2. Recursive loops which cause infinite computations are forbidden.
3. There are no interdependencies, i.e. constraints, among any input parameters (including background facts). Hence only binary constraints between one input parameter and the output value are possible.

Incidentally, the restrictions 1 and 2 alone already suffice for achieving decidability. In fact our system is able to manage rule sets even if none of the restrictions holds, whereas the approach of Schmid (1990) is not. So these restrictions need not to be taken too serious. But our algorithm may not terminate without them.

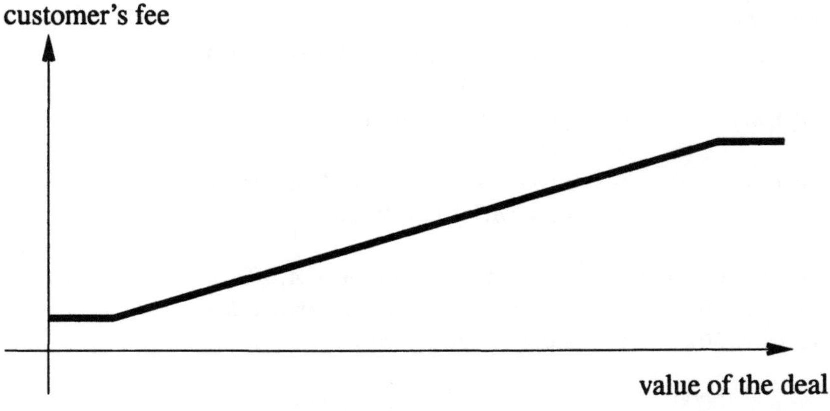

Figure 2. Intended meaning of the rule set.

Let us consider as an example the fragment of a rule set shown in figure 1. It is given in Prolog-like notation. As one can see, negation in the body of rules is allowed. We can interpret it as disjunctive rules, by rewriting e.g. a rule of the form $P \leftarrow \neg Q$ into $P \vee Q$. The variable A serves as an identifier for the deal in question. The first four rules specify different fees depending on special features concerning a transaction A. The most simple calculation e.g. would be a derivation of the first rule of G, followed by an application of the third rule of *const*, which will lead to a calculated fee of 50, if the condition claimed in the *const* rule, namely that the customer of transaction A is a staff member, holds. The intended meaning of this rule set for an "average" customer is shown in figure 2. The reader may look on his own at the rule set for a while. We may distinguish several kinds of predicates:

features: they are binary predicates and specify properties for the parameters of the deal in question, e.g. *country*, *currency* and *value*; these parameters have finite domains which have to be defined elsewhere.

constraints: these predicates, e.g. $W \geq 200, W < 500$ are the arithmetical conditions which may be treated by a finite domain constraint solver; without loss of generality, we will assume that we only have to deal with non-negative integer numbers.

ordinary predicates: they may be computed by further rules in the rule set; examples are G, *const* and *min*.

One solution to our problems is sketched by Schmid (1990). He proposes an implementation in Prolog that uses meta-programming techniques. In this approach, a rule set is processed as follows. Firstly, the program is completely unfolded with its definitions. This procedure terminates because recursive loops are forbidden. Secondly, the semantics of each ordinary predicate is computed as a cross product of the features used in its definition. Of course the constraints must be taken into consideration during this step. In addition, the negation of a predicate may be reduced to the computation of a set complement which is not difficult because of the use of finite domains. Finally we get a set-based representation for the semantics of G (and all other predicates too) which can be investigated further. In order to test (strong) determinism, the intersection of the semantics of alternative rules has to be empty for each predicate.

4.3. *Compiling Rule Sets*

The idea here is to make use of the abilities of our Prolog-based theorem proving systems more directly. Instead of a meta-programming approach we compile the rule set into clauses with finite domain constraints. This leads to a very rapid implementation of the problem exploiting constraint techniques. Our implementation performs a transformation whose main task is to extract the constraints and features to build a constraint clause set description of the rule set. Then the theorem proving system processes it in a top-down manner while the approach of Schmid (1990) works bottom-up. Each predicate call is transformed as described below. The operators with prefix # denote constraint predicates which are processed by the constraint solver.

features: for example, $\neg country(A, ch)$ is rewritten as a constraint condition $Country \,\#\#\, ch$, where ## means $\neq$, i.e. inequality; background facts must ensure that the value of the variable *Country* is taken from the corresponding finite domain.

constraints: for example, $value(A, W) \wedge W < Min * 100$ is replaced by the constraint condition $W \,\#\!<\, Min * 100$.

ordinary predicates: they have to be transformed in a special way; their parameter list consists of a all possible features, more exactly those variables the feature predicates have been transformed into, e.g. *Country*.

We also have to define the domains of the features by background facts and the query to investigate the desired problem, e.g. determinism. The ordinary predicates in the premise are negated and added to the conclusion of the implication, this may lead to a set of disjunctive rules. Figure 3 shows the overall architecture of our application.

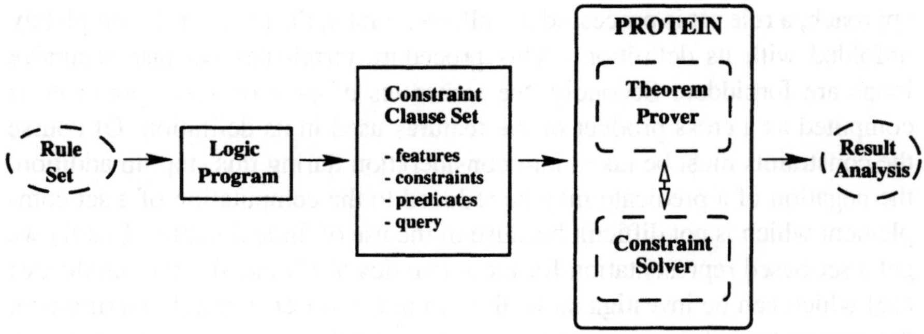

Figure 3. System architecture.

Let us have a closer look at our example after the transformation into a constraint clause set, which is given in figure 4. The parameter list is always abbreviated; the full list is

$$(Min, Max, Value, Country, Type, Customer, Currency, F)$$

where F is the output value, i.e. the fee. Now we just have to define the finite domains of the background facts, and then we are ready to ask a query. This is the last step of the transformation, namely building the query, which is stated in figure 5 (a). The use of constraints helps us to bundle answers and to avoid backtracking. The transformed clause set can now be processed by our Prolog-based theorem prover with constraints giving the answer concerning non-determinism of the rule set, which is shown in figure 5 (b). It says that transactions with the stated features have an ambiguous rate value.

The next table lists the test procedures available in Schmid (1990) (upper half) and our approach (lower half). The totality problem seems to be more difficult than the other one. This is related to the fact that the theoretical complexity of the totality problem is higher than the one of the determinism problem. In order to test for totality, it is planned to collect all answers, which requires that the theorem prover is answer-complete (Baumgartner et al., 1997) for the latter approach.

G non-total	complement of a parameter non-empty
G strongly deterministic	intersection of rule alternatives empty
G is total	collect answers $G(x,y)$ with constraints
G weakly deterministic	query $G(x,y), G(x,y'), y \neq y'$ fails

$G(\ldots,F) \vee \neg const(\ldots,F)$ / *true*.
$G(\ldots,F) \vee const(\ldots,F) \vee \neg min(\ldots,F)$ / *Value* #≤ *Min* ∗ 100.
$G(\ldots,F) \vee const(\ldots,F) \vee \neg max(\ldots,F)$ / *Country* ## *oversea*, *Value* #≥ *Max* ∗ 100.
$G(\ldots,F) \vee const(\ldots,F) \vee min(\ldots,F) \vee max(\ldots,F)$ / *Value* #= *G* ∗ 100.

const(…,0) / *Country* #= *ch*, *Customer* #= *office*.
const(…,30) / *Country* ## *ch*, *Customer* #= *office*.
const(…,50) / *Customer* #= *personal*.

max(*Min*, 10000,…,*F*) / *Country* #= *ch*, *Currency* #= *ch*.
max(*Min*, 20000,…,*F*) / *Country* #= *ch*, *Value* #> 1000000.

min(10,…,*F*) / *Country* #= *ch*, *Type* #= *share*, *Value* #< 100.
min(10,…,*F*) / *Country* #= *ch*, *Type* #= *obligation*, *Value* #< 100.
min(20,…,*F*) / *Country* ## *ch*, *Value* #< 200.
min(50,…,*F*) / *Country* ## *ch*, *Value* #≥ 200, *Value* #< 500.
min(80,…,*F*) / *Value* #> 500.

Figure 4. Transformed rule set.

(a) ?- *Country* :: [*ch*, *d*, *usa*, *gb*, *nl*, *oversea*],
Customer :: [*office*, *personal*, *other*],
Type :: [*share*, *obligation*],
Currency :: [*ch*, *d*, *usa*],
Value :: 0..1000000,

*F*1##*F*2,
G(…,*F*1),
G(…,*F*2),

print_answer([…,*F*1,*F*2]).

(b)

Following rates are possible	: [0] or [10]
Value-Range	: [0..99]
Country	: [ch]
Transaction Type	: [share]
Customer	: [office]
Currency	: [d,ch,usa]

Figure 5. A query and its answer.

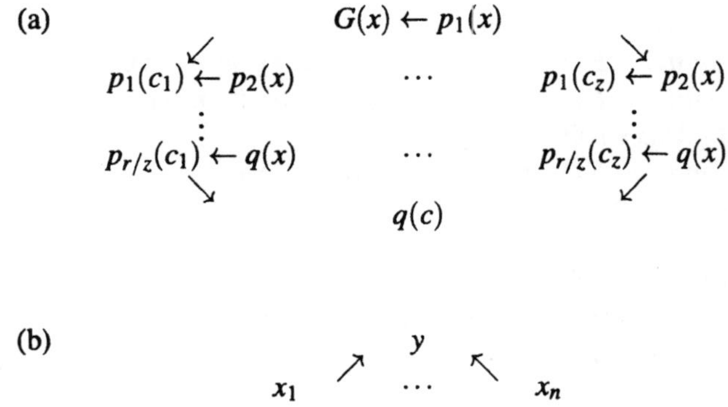

Figure 6. Rule set and constraint graph.

In order to analyze determinism of a rule set, we can combine our approach (which is a strict search) with methods from (approximate) determinism analysis. These methods give an abstract interpretation of (constraint) logic programs. If the latter methods fail—and this will often be the case—, then our algorithm has to be applied. However our procedure does not perform brute-force search since the search is guided by the constraints, and computations are bundled. We also tried to tackle the problem without using finite (integer) domain constraints, but there was no chance of solving the problem within reasonable time even for small examples.

4.4. *Complexity Examinations*

We will now try to estimate the complexity of the procedures to detect determinism and ambiguity. We split the considerations into two parts: (a) the analysis of the number of possible (top-down) computations with non-recursive rules, and (b) the effort that must be spent to propagate the constraints provided that there are no interdependencies among the input parameters.

Remember, for the analysis of determinism we have to explore all answers to the query $G(x,y)$ in the worst case. Now, can the cost of this computation be exponential in the number of rules? The answer to this question is yes, even if we forbid recursive loops. To see this look at the rule set in figure 6 (a) where the output value is left out for the ease of notation. For each p_k we have z alternative rules with $1 \leq k \leq r/z$. Thus, we have $(z \cdot r/z) = r$ rules in total (ignoring both outermost rules). For each k we can choose among z alternatives, such that there are $z^{r/z} = (z^{1/z})^r$ possible computations. The expression becomes maximal at $z = e \approx 2.71$, or $z = 3$ for integer numbers. Hence the worst-case complexity is $(\sqrt[3]{3})^r \approx 1.44^r$, i.e. exponential in the number of rules. In real rule sets this worst-case complexity should not be approached too often. Anyway, ambiguity will be detected after at most d^n computations where n is the number of parameters and d the domain size (assumed to be constant for all input variables), since there cannot be more output values than that number.

Let us now draw our attention to the constraint processing. Before discussing its complexity in more detail, we have to introduce some new notions.

DEFINITION 6. An instance of the *constraint satisfaction problem* requires us to instantiate a set of variables $x_1, \ldots, x_n$ subject to a set of constraints, i.e. relations involving these variables. We define *k-consistency* as follows: choose any set of $k-1$ variables along with values for each that satisfy all the constraints among them and any k-th variable, then there exists a value for the k-th variable such that the k values taken together satisfy all constraints among the k variables. Note that, the notion *arc-consistency* is synonymous with 2-consistency. Strong k-consistency means i-consistency for all $i \leq k$.

The finite domain solver of ECLiPSe (ECRC, 1995), used for our implementation, only applies rules similar to (6), mentioned in section 3.2. However in general, this does not always achieve global consistency. The constraints may not even be arc-consistent. But in our examples, the constraints are so simple that arc-consistency is always assured by the solver.

DEFINITION 7. A *constraint graph* is a graph where the nodes represent variables and two nodes are linked to represent the existence of a constraint that involve these variables. An ordered constraint graph arranges the nodes in a linear order. The *width* of such a graph is the maximal number of links, i.e. the number of constraints, that lead back from a node to previous ones.

Because of the restriction in our application that there are no interdependencies among the input variables, the constraint graph forms a simple tree with root y, i.e. the output value, with binary constraints only. This is illustrated by figure 6 (b). We want to order the variables such that we avoid as much backtracking as possible. We distinguish two different kinds of orderings: the vertical and the horizontal search order. An ordered constraint graph determines the vertical search order, i.e. the order in which variables are chosen for instantiation, whereas the horizontal order, i.e., the order in which values are tested for the given variables, is not determined. Backtracking is necessary, if an instantiation chosen during the search, which is consistent with all previous choices, must be discarded later in the search when no consistent instantiation can be made for a variable at a lower level in the backtrack tree. A search sometimes is accidently backtrack-free. But we only say that a given vertical search order is *backtrack-free* iff it guarantees a backtrack-free search regardless of the horizontal search order. The following theorem can be stated:

THEOREM 8. *For a given constraint satisfaction problem, a vertical search order is* backtrack-free, *if the level of strong consistency is greater than the width of the corresponding ordered constraint graph. There always is an ordering such that tree-structured constraint graphs have width* 1.

The theorem stated above—adapted from Freuder (1982)—allows us to compute solutions without backtracking during the constraint-processing phase, because the constraint graph has width 1, while 2-consistency is achieved. This means, we can do it without the built-in constraint predicate *indomain* (van Hentenryck, 1989) that tests possible values for variables by instantiating them, provided that we instantiate the output value y first. This means, $y, x_1, \ldots, x_n$ is an optimal vertical search ordering for the constraint graph. This justifies the bundling of answers as we do in figure 5 (b). Therefore, we can avoid the expensive labeling phase, i.e. guessing values for the variables, in our application. We are only concerned with binary constraints. But of course, in principle we are not bound to this restriction. It simply speeds up our procedure.

5. Final Remarks

5.1. *Results*

It has been shown how theorem proving, logic programming and constraint processing can be combined in a straightforward and easy manner for the specification and analysis of (logic) programs. The rapid implementation of a theorem proving system using finite domain constraints is possible. We applied it to the calculation of banking fees and succeeded in testing some real examples for ambiguity within seconds. We investigated a small example (sketched in this chapter) with 15 rules (14 clauses plus 1 query) and 8 parameters sketched in figure 1, and a bigger one with 169 rules and 20 parameters including 10 disjunctive rules. The latter one was a rule set of the Swiss credit institution SKA.

Example	rules	parameters	time
small	15	8	0.1s
big	169	20	3.5s

The given answers concerning ambiguity were calculated by the Protein system (Baumgartner and Furbach, 1994) with the support of definite answer calculation (Baumgartner et al., 1997) on a Sparc station 4. These proof time results are very encouraging, but unfortunately we have not yet been able to tackle the problem whether a clause set is total or not in a really satisfactory way.

5.2. *Related Work*

There are other approaches treating similar problems, e.g. Colomb and Chung (1995) and Hoover and Chen (1995). We will discuss one of them in greater detail, namely the *Tablewise* system, a decision table tool (Hoover and Chen, 1995). This tool tests decision tables for consistency and completeness, i.e. determinism and totality according to our terminology. It also supports a special form of structural analysis that localizes flaws causing decision tables to be inconsistent or incomplete. Tablewise can generate Ada code implementing the function defined by the decision table. In our system we get the implementation for free, simply as a by-product. We just have to run the transformed Prolog code we produced with the query $G(x,y)$ and an instantiated input parameter set x. This means, a rule set can be directly considered as an executable specification (i.e., a restricted kind of automatic programming is possible).

Tablewise generates English language documentation describing that function. This means, the rules are first specified by means of decision tables and then translated automatically into a natural language form. This idea, namely to use formalisms like decision tables for specifying complex rules sets, seems to be very promising. The procedure may be transferred to logic programming as formalism. One of the largest tables Tablewise dealt with so far had 17 input variables, mostly boolean, with 13 operational procedures, comprising 20 to 25 different output values in all. Consistency and completeness analyses took 1 to 2 minutes depending on the system, structural analysis longer depending on the number of variables. All details are according to e-mail correspondence with Douglas Hoover.

However the domain where Tablewise is used slightly differs from ours: Tablewise was developed in a joint project with an aircraft company and to our knowledge has not yet been used for banking applications. We have a bigger number of different output values. For example, there may be a rule saying that the banking fee y amounts to 1% of the value x of the deal in question. This is difficult to express with decision tables whose semantics is naturally defined in terms of a (generalized) propositional logic, not in terms of full first-order predicate logic as in our approach, since in propositional logic it is impossible to express binary constraints such as

(7) $y = 0.01 \cdot x$.

One solution would be to introduce propositional variables with the meaning $x = d$ or $y = d$, respectively, where d stands for a concrete value in the respective domains. Then, (7) can be expressed by the set of all instances of the rule

(8) $y = d_1 \leftrightarrow x = d_2$ where $d_1 = 0.01 \cdot d_2$.

But this is ponderous, if the domain sizes are large. Although it is possible—and also applied in practice—to make use of propositional variables meaning $x \geq d$ or a similar unary constraint, this again does not help us to encode (7), because it is a binary constraint.

Colomb and Chung (1995) emphasize that expert systems, i.e. propositional Horn clause systems can be mechanically transformed into a decision table. The transformed systems execute faster and in bounded time. The compact decision table presentation simplifies the problem of checking rule sets for totality and ambiguity, and it makes a systematic analysis of an expert system quite convenient. But since the underlying calculus is propositional the argument stated above holds here too. In addition, our approach can also cope with disjunctive, i.e. non-Horn rules.

5.3. *Summary*

Let us now return to the questions set up in section 2.2 and try to give some answers:

1. *What are the real-world applications of these ideas?*
   The application of calculating banking fees seems to be realistic. But the effort for coding the natural language rule set into logic with constraints may be too expensive, unless the rules change frequently so that repeated automatic tests of the rule sets are appropriate.
2. *Which parts of the formalisms are needed in order to express realistic problems?*
   There is only sparse use of disjunctive rules in our examples—less than 6%. But they do occur. It is interesting that we can dispense with recursive loops and even interdependencies among variables in practice.
3. *Are the approaches efficient enough to handle large data sets?*
   We are happy that we could solve the examples we had at hand. But it can be shown that the worst-case complexity is exponential. Thus, for

sufficient large examples, our procedure might not work. In this case, it is necessary to develop more efficient procedures, which e.g. are able to identify redundant computations or which make use of incomplete heuristics.

Last but not least the question remains whether, despite the costs involved, credit institutes or other companies will accept this approach. Many people do not want computers to solve tasks usually done by humans, because they are afraid of passing the decision-control to machines.

Nevertheless, we are able to solve some problems by means of deduction techniques exploiting constraint technology with a Prolog-based theorem proving system within reasonable time. At the moment we do not yet have more than a prototype of a system. We are currently planning to contact other partners in industry interested in our system. In summary, analyzing rule sets is a field with great potential for applying tableau-based deduction techniques and constraint solving.

References

Baumgartner, P. and U. Furbach: 1994, 'PROTEIN: A PROver with a Theory Extension INterface'. In: A. Bundy (ed.): *Proceedings of the 12th International Conference on Automated Deduction*. Nancy, pp. 769–773, Springer, Berlin, Heidelberg, New York.

Baumgartner, P., U. Furbach, and F. Stolzenburg: 1997, 'Computing answers with model elimination'. *Artificial Intelligence* **90**(1-2), 135–176.

Baumgartner, P. and F. Stolzenburg: 1995, 'Constraint Model Elimination and a PTTP-Implementation'. In: P. Baumgartner, R. Hähnle, and J. Posegga (eds.): *Proceedings of the 4th Workshop on Theorem Proving with Analytic Tableaux and Related Methods*. pp. 201–216, Springer, Berlin, Heidelberg, New York.

Bibel, W.: 1982, *Automated Theorem Proving*. Braunschweig, Wiesbaden: Vieweg.

Bürckert, H.-J.: 1991, *A Resolution Principle for a Logic with Restricted Quantifiers*, LNAI 568. Berlin, Heidelberg, New York: Springer.

Colomb, R. M. and C. Y. C. Chung: 1995, 'Strategies for Building Propositional Expert Systems'. *Journal of Intelligent Systems* **10**, 295–328.

Dahn, B. I., J. Gehne, T. Honigmann, L. Walther, and A. Wolf: 1995, 'Integrating Logical Functions with ILF'. Humbold-Universität, Berlin. Manual.

ECRC: 1995, 'ECLiPSe 3.5: User Manual – Extensions User Manual'. ECRC GmbH, München. Two volumes.

Freuder, E. C.: 1982, 'A Sufficient Condition for Backtrack-Free Search'. *Journal of the ACM* **29**(1), 24–32.

Hoover, D. N. and Z. Chen: 1995, 'Tablewise, a Decision Table Tool'. In: *Proceedings of the 10th Annual Conference on Computer Assurance*. Washington, pp. 97–108, IEEE.

Jaffar, J. and M. J. Maher: 1994, 'Constraint Logic Programming: a Survey'. *Journal of Logic Programming* **19/20**, 503–581.

Lloyd, J. W.: 1987, *Foundations of Logic Programming*. Berlin, Heidelberg, New York: Springer.

Loveland, D. W.: 1968, 'Mechanical Theorem Proving by Model Elimination'. *Journal of the ACM* **15**(2).

Maher, M. J.: 1993, 'A Logic Programming View of CLP'. In: D. S. Warren (ed.): *Proceedings of the 10th International Conference on Logic Programming*. Budapest, pp. 737–753, MIT Press, Cambridge, MA, London.

Rich, C. and R. C. Waters (eds.): 1986, *Artificial Intelligence and Software Engineering*. Los Altos, CA: Morgan Kaufmann.

Schmid, K.-M.: 1990, 'Eindeutigkeit und Vollständigkeit zweistelliger Relationen – Entscheidungsverfahren für eingeschränkte Fälle'. Diplomarbeit, Universität Kaiserslautern. Supervised by Hans-Jürgen Ohlbach.

Stickel, M. E.: 1985, 'Automated Deduction by Theory Resolution'. *Journal of Automated Reasoning* **1**(3), 333–355.

Stickel, M. E.: 1988, 'A Prolog Technology Theorem Prover: Implementation by an Extended Prolog Compiler'. *Journal of Automated Reasoning* **4**, 353–380.

Stolzenburg, F.: 1998, *Disjunctive Logic Programming with Constraints and its Applications*, Koblenzer Schriften zur Informatik. Koblenz: Fölbach. Dissertation.

Stolzenburg, F. and B. Thomas: 1996, 'Analysing Rule Sets for the Calculation of Banking Fees by a Theorem Prover with Constraints'. In: *Proceedings of the 2nd International Conference on Practical Application of Constraint Technology*. London, pp. 269–282, Practical Application Company.

van Hentenryck, P.: 1989, *Constraint Satisfaction in Logic Programming*. Cambridge, MA, London: MIT Press.

BERND FISCHER, JOHANN SCHUMANN, GREGOR SNELTING

CHAPTER 11

DEDUCTION-BASED SOFTWARE COMPONENT RETRIEVAL

1. REUSE BY CONTRACT

Reuse of approved components and concepts is a major characteristic of mature engineering disciplines. In software engineering it has even been considered from the very beginning (cf. McIlroy's (1968) contribution to the seminal NATO conference) but despite ongoing interest and research it is by no means an established practice.

The most basic technique, reusing code pieces, still prevails. Unfortunately, it is complicated because of the amount of information which must be considered as even some very fine details of the code must be inspected and checked against the new assumptions. And while experience tells us that this is possible, it has also taught us that it is difficult. Code reuse remains inherently risky and the spectacular stories are no success stories. The investigation of the recent Ariane 5 disaster for example revealed that it was caused by the reuse of an unmodified Ariane 4 software component which led to an uncaught exception crashing the software and hence the spacecraft (Lions et al., 1996).

It has been argued that the ultimate reason for the crash was the component's failure to make its assumptions explicit. Jézéquel and Meyer (1997) thus propose to use *contracts*, i.e., formal models of the components' behavior. The best-known contracts are axiomatic specifications with pre- and postconditions but other formats are also possible, e.g., type and class invariants or abstract statements. Jézéquel and Meyer conclude that

> "*reuse without a contract is a sheer folly.* From CORBA to C++ to VisualBasic to ActiveX to Java, the hype is on software components. The Ariane 5 blunder shows clearly that naïve hopes are doomed to produce results *far worse* than a traditional, reuse-less software process. To attempt to reuse software without assertions is to invite failures of potentially disastrous consequences."

Reuse by contract embodies this premise: the components are associated with contracts and organized, retrieved, and reused by these. This approach has three major benefits:

- it supports safe reuse and safe software composition because it can retrieve the components which provably fulfill the contracts,

W. Bibel, P. H. Schmitt (eds.), Automated Deduction. A basis for applications. Vol. III

- it promotes abstraction because the contracts become the actual medium by which components are retrieved, and
- it constitutes a formal framework in which every single reuse step can be formally justified.

In essence, reuse by contract is the application of formal methods to software reuse. Within our project NORA/HAMMR[1] we investigate not only technical problems as for example scalable and efficient architectures or contract-based library indexing and retrieval techniques but also organizational aspects such as reuse-friendly specification techniques, the integration into formal software development processes, or the interaction with conventional reuse mechanisms. Our long-term goal is to build a system which smoothly combines component design, implementation, and verification with the systematic reuse of fully specified and verified component libraries.

In this chapter we will focus on *deduction-based software component retrieval* because it is the key problem in reuse by contract: "You must find it before you can reuse it!"[2] Its purpose is to check whether the contracts of the client and a candidate component are compatible in some sense and hence the candidate is suitable for reuse. It thus builds a proof task from the contracts which formally captures the intended compatibility and uses automated deduction techniques to solve this task. If the proof obligation can be solved the candidate is retrieved, if not it is discarded—retrieval becomes a deduction problem. The crucial point is to retrieve in a short time as many relevant components as possible (high recall) while maintaining a high fraction of "right" components (high precision).

The basic idea of deduction-based retrieval is very simple and has already been proposed several times before (Katz et al., 1987; Rollins and Wing, 1991; Manhart and Meggendorfer, 1991) but a practical implementation is very difficult. It is not sufficient to concentrate on the deduction problems because only a carefully engineered complete retrieval system generates tractable problems. But one must also take care of other aspects: integration into the software development process, contract language, user interface, response times, or degree of automatization come immediately to mind. It is the consideration of such non-deductive aspects which sets NORA/HAMMR apart from other approaches.

The remainder of this chapter is organized as follows. The next section discusses various forms of proof tasks and their effects on software reuse.

[1] NORA is no real acronym; HAMMR is a highly-adaptive multi-method retrieval tool. This work has been supported by the DFG under grants Sn11/2-1, Sn11/2-2, Sn11/2-3 and Schu908-5/1 and within the SFB 342, subproject A5.

[2] *The First Golden Rule of Software Reuse*, attributed to W. Tracz.

Section 3 then reviews the requirements for a practical tool and develops the architecture of NORA/HAMMR. Its main feature is a pipeline of filters of increasing deductive strength which maintains a balance between fast response and high precision. Sections 4 to 6 are each devoted to a class of such filters: signature matching filters, rejection filters and confirmation filters with automated theorem provers (ATPs). We describe the respective steps which must be taken to translate the proof tasks from the application domain and to prepare them for the different provers. Section 7 details on the experimental evaluation of this process. We conclude the chapter with a survey of related work and with discussions of the lessons learnt in this application of automated theorem proving to software engineering and future work on NORA/HAMMR.

2. Contracts, Retrieval, and Reuse

2.1. *Contract Languages*

Reuse by contract requires that each component is associated with a contract of its own which captures the essentials of the components behavior. In practice, components are usually functions and axiomatic specifications with pre- and postconditions are the most common form of contracts. Pre- and postconditions of a component c (which we will denote by pre_c and $post_c$, respectively) could in principle be formulated in pure FOL but this is hardly acceptable by the user. Instead, a more convenient specification language with a rich set of predefined operations must be chosen as contract language or *custom logics*. In NORA/HAMMR, we apply VDM-SL (Dawes, 1991). VDM-SL allows a concise formulation and even the specification of side effects. Figure 1 shows a typical component which has been taken from our experimental library (cf. Section 7.1).[3]

Using custom logics, however, usually requires either a specialized deductive component or a translation into FOL to retain soundness. VDM-SL is based on the three-valued *logics of partial functions* or LPF (Barringer et al., 1984) for which some dedicated provers exist (e.g., *mural* (Jones et al., 1991) and Isabelle/VDM-LPF (Agerholm and Frost, 1997)). But these are not applicable in our case because they work interactively. We decided not to tailor a fully automatic proof procedure for LPF (e.g., based on a many-valued prover as 3TAP (Beckert et al., 1991)). Rather, we chose the second approach and translate the proof tasks from VDM-SL via LPF to FOL. Hence, we can re-use

[3] In VDM-SL, ^ means concatenation of lists, [] the empty list, [*i*] a singleton list with item *i*, and hd and tl the functions head and tail, respectively.

existing high-performance theorem provers and model checkers for FOL. We apply an optimized variant of the translation of Jones and Middelburg (1994) which will be described in Section 3.4.

```
module rotate
    exports functions rotate : list --p--> list
    types item = token; list = seq of item
    functions
        empty : list ---> bool
        empty(l) == l = []
        rotate(l : list) l' : list
        pre     ¬ empty(l)
        post    l' = (tl l) ^ [hd l]
end rotate
```

Figure 1. Example component

2.2. *Deduction-Based Retrieval and Proof Tasks*

The purpose of deduction-based component retrieval is to identify components with compatible contracts. However, there are different notions of compatibility modeling different approaches to reuse by contract. Match relations formalize these notions via their associated *proof tasks* formulated in the contract language.

Generally, a component c can be considered as compatible if it bridges the gap between the pre- and postconditions the client requires. Ideally, c fits completely and can thus be plugged into the place of the query q because it has a weaker pre- and a stronger postcondition than required by q. This relation is sometimes (Zaremski and Wing, 1995a; Penix et al., 1995) formalized by proof tasks of the form

$$(11.1) \quad (pre_q \Rightarrow pre_c) \wedge (post_c \Rightarrow post_q)$$

but this is not entirely adequate.[4] Reusable components are usually implemented very defensive yet general. Hence, they may work on wider domains than expected but their results on the domain extensions need not to fit the

[4] Actually, the proof tasks are universally closed with respect to the formal input and output parameters of the component and the query. However, to improve readability, we use these traditionally abbreviated formulations.

user's expectations. If we want to retrieve such components anyway we must restrict the postconditions to the domain specified by pre_q:

(11.2) $(pre_q \Rightarrow pre_c) \wedge (pre_q \wedge post_c \Rightarrow post_q)$

This variant of *plug-in compatibility* increases recall compared to the standard variant but due to the domain restrictions it still supports safe reuse. The retrieved components can be considered as black boxes and may be reused "as is", without further proviso or modification.

Under a weaker notion of compatibility, c is not required to close the gap completely but must bridge only a part of it. Then, c must have a stronger postcondition than desired, at least on its own domain, but this domain is not determined by the query:

(11.3) $pre_c \wedge post_c \Rightarrow post_q$

In contrast to plug-in compatibility, code reuse based on this *conditional compatibility* is potentially unsafe because the client still has to satisfy the open obligation pre_c.

Deduction-based retrieval is also applicable for traditional white-box code reuse. The *partial compatibility* match relation

(11.4) $pre_c \wedge pre_q \wedge post_c \Rightarrow post_q$

retrieves all components which do "the right thing" at least on the domain restricted by $pre_c \wedge pre_q$. By varying pre_q, clients can control recall and granularity of reuse. The weaker it is, the more components are retrieved but the smaller is their respective benefit. However, in the absence of a formal process, manual checks or code modifications are required in order to match the components precondition.

2.3. *Deduction-Based Retrieval and Formal Software Development*

Deduction-based retrieval works best in connection with formal software development processes. Here, the contracts arise naturally and do not impose extra work on the developers. Since retrieving components is an alternative to any single refinement step, reuse should be built into the process from the very beginning. This does not only increase productivity as the development steps become larger but also enables vertical prototyping and thus increases the confidence into the validity of the system being built.

3. The System NORA/HAMMR

3.1. *Requirements*

The goal of the project NORA/HAMMR is to build a practical deduction-based retrieval tool for the working software engineer. Since "practical" has no precise definition we will discuss some of its facets in the following. For practical purposes, *retrieval* can be measured in terms of response times, *recall* ("do we get all matching components?") and *precision* ("do we get the right components?").[5] In the case of reuse by contract, however, their relative importance depends on the applied software development process. The speed of the tool is of course also important due to the *Fourth Reuse Truism*: "You must find it faster than you can rebuild it!" (Krueger, 1992). It is, however, dominated by the time required to discard non-matches.

In a formal process, the first proven match already saves significant time which is otherwise spent in refinement steps and their justifications. Hence, users are not very much interested in the actual recall as long as it is not zero.

In a non-formal process the picture is quite different. Here, most time is spent in analyzing and modifying the retrieved components manually, in particular when partial compatibility is used. Hence, sets of promising candidates should be available for inspection immediately and at any time. Furthermore, a high recall is more important than high precision.

For a *practical tool*, "look and feel" is extremely important. Here, the most obvious aspect is the presentation. The bare theorem provers are unsuitable for non-experts and must thus be hidden by a dedicated tool interface. It must also offer facilities to control the retrieval process and to monitor its progress.

The number and structure of proof tasks pose severe problems. In contrast to other applications, e.g., in mathematics, where only a few tasks over small signatures and with few axioms have to be solved, we face an "inverse economics of scale." Each single query already induces a number of complex proof tasks which is in the order of the size of the component library. Consequently, their construction is itself a major task which must run fully automatically. It also involves steps which are usually done by "experienced experimentators", e.g., addition of key lemmas or setting of prover control parameters. Moreover, since reasonable libraries contain only few matches for any query, most of the tasks describe non-theorems. This renders the naïve generate-and-proof approach impossible. Instead, it becomes vital to rule out non-theorems using dedicated disproving tools to prevent the actual prover from drowning. For the remaining tasks, aggressive simplification is required

---

[5] Cf. Section 7.2 for exact definitions of recall and precision.

before the proof can actually be attempted. But even then it may remain necessary to use different provers, axiomatizations, or tactics in competition to solve enough tasks.

3.2. *System Architecture*

NORA/HAMMR's system architecture reflects these diverse and sometimes even conflicting requirements. NORA/HAMMR is built as a pipeline of independent filters through which the candidates are fed. This design supports an incremental retrieval approach. Each component which passes a filter can immediately be inspected by the user. This early feedback is essential for an effective control of the retrieval process as queries which are too weak (i.e., in-discriminative) or too strong (i.e., over-specified) become obvious before the entire library is checked by the first filter.

Typically, a filter pipeline starts with a *signature matching* filter which checks for compatible calling conventions. Then, *rejection* filters try to efficiently discard as many non-matches as possible. Finally, *confirmation* filters are invoked to guarantee a high precision of the retrieval results. The next sections of this chapter are devoted to these filters. The pipeline itself can be freely customized although any configuration should maintain a fair balance between initial fast responses and a final high precision. However, it still allows the integration of arbitrary non-deductive methods, e.g., full text retrieval (see e.g., (Maarek and Smadja, 1989; Maarek et al., 1991; Frakes and Nejmeh, 1987)). This is especially important in a non-formal process where the established methods need to be augmented and not to be replaced in order to gain acceptance.

The prototypical graphical user interface (cf. Figure 2) reflects the system architecture. The filter pipeline may be pre-selected via a pull-down menu and may easily be customized through the central icon pad. Each icon also hides a specialized control window which allows some fine-tuning of the filter. Additional inspectors display the intermediate results and grant easy access to the components. They also allow to save retrieved components into new library files which may be used for subsequent retrieval runs. A simple, VCR-like control panel provides single-button control of the entire system. The objective of the GUI is to hide evidence of automated deduction techniques as far as possible but to grant control over these techniques as far as necessary. Hence, using NORA/HAMMR as a *retrieval tool* requires only knowledge of the contract and target languages.

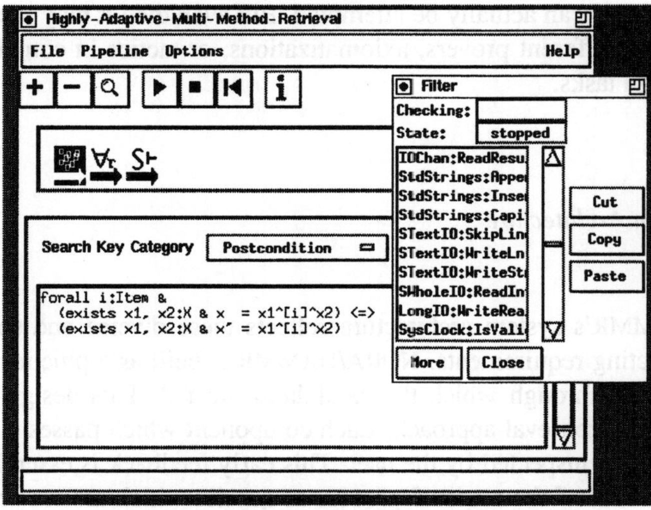

Figure 2. Graphical user interface

3.3. *Reuse Administration*

Any reuse scheme requires an initial library ("seed") and some continuous maintenance but this demands more detailed knowledge of the contract language and the applied deduction methods than the average re-user has. We thus assume that a dedicated *reuse administrator* is responsible for these tasks.

Construction of an initial library is straightforward but tedious. The components (e.g., functions, procedures, or methods) need to be extracted and indexed: the contracts must be written or checked and completed by all necessary context information (e.g., type definitions or auxiliary functions). Since we use VDM-SL as contract language, we represent a component as a VDM-SL module which contains the implicitly specified function and all referred constructs.

The components may be organized hierarchically using plug-in compatibility to reduce the number of proof tasks (Mili et al., 1994; Jeng and Cheng, 1994) but for simplicity we currently work with a flat library. We also assume without a formal justification that the contract is a faithful representation of the component and that it contains all information of interest to the clients. Hence, the actual source code is not required for retrieval purposes.

The reuse administrator must also support the deductive infrastructure. This in particular includes the construction and maintenance of a *lemma library* which contains the axiomatization of the contract language. In NORA-HAMMR, the library is structured hierarchically and also contains meta-information as for example operator rules, data type generators, and minimal axiomatizations. This supports simplification and axiom selection using similar techniques as described in Chapter III.2.9.

3.4. *Translation of Proof Tasks*

Regardless of the chosen compatibility criterion, NORA/HAMMR first generates the proof tasks in form of VDM-SL-expressions which are then translated in two phases into FOL. The first phase transforms the expressions into formulas of the underlying LPF-calculus. Following (Middelburg, 1993), it eliminates all extra-logical custom constructs. E.g., custom-bindings as *let*-bindings or pattern matching are replaced by appropriately scoped quantifiers, and local definitions of auxiliary predicates and functions are transformed into equivalence and equality clauses, respectively. But it also adds axioms to represent properties of locally defined data types, e.g., constructor properties for tagged unions. Finally, it simplifies the imposed sort structure and replaces all type invariants by explicit predicates using standard relativation techniques (Oberschelp, 1962).

The second phase takes the resulting LPF-formula φ and translates it into FOL in such a way that provability is preserved, i.e., essentially that $\vdash_{LPF} \varphi \Longleftrightarrow \vdash_{FOL} \varphi'$ holds. Following (Jones and Middelburg, 1994), we use signed translation functions to map any LPF-formula which contains an undefined subterm to an unprovable FOL-formula. E.g., the LPF-formula $\forall l : List \cdot hd\ l = hd\ l$ which has the truth value *undefined* becomes $\forall l : List \cdot l \neq [\,] \wedge hd\ l = hd\ l$. Since the quantifiers in LPF range only over proper (i.e., defined) values, we can optimize the handling of formulas and terms which contain no occurrences of partial functions. Of course, the translation into pure FOL is provability-preserving for the non-inductive part of LPF only. For the proper handling of recursive data types an inductive ATP would be required but in practice our induction approximation described in Section 6.2 works sufficiently well.

The necessity or even usefulness of a three-valued logic for software specification purposes has often been questioned and we refer to Jones (1991, 1995) for its defense. However, in our experience, the application of LPF proved ultimatively to be a matter of taste and not an insurmountable problem. The optimized translation avoids much of the usual overhead and after application of simplifications described in Section 5.1, the translated formulas are very similar in size and structure to respective formulas originally written in classical two-valued logics.

4. Signature Matching Filters

The main purpose of signature matching is to identify components which have compatible calling conventions in the target language. It thus has to abstract from possible implementation choices, e.g., order of parameters or currying. It can even work across language boundaries to increase interoperability. Signature matching filters do not use full contracts for retrieval but only type information. They can thus be implemented using simpler deduction techniques than full first-order theorem proving which makes them also suitable as fast pre-filters.

Signature matching is a disguised type inference problem: basically, a component can be retrieved if the desired type can be derived for it, i.e., for a library L the set $\{c \in L \mid \vdash_T c : \tau_q\}$ formally defines the answer for a query τ_q using the type system T. Generally, this type-theoretic interpretation of signature matching gives us sound semantic foundations which are also applicable if the semantics of the target language itself offers no suitable interpretation of types.

In order to abstract from the implementation choices, the type system T used for retrieval must of course contain additional rules beyond those used for typing the language. E.g., to take care of currying, the inference rule

$$\frac{\Gamma \vdash c : (\alpha \times \beta) \rightarrow \gamma}{\Gamma \vdash c : \alpha \rightarrow (\beta \rightarrow \gamma)}$$

which allows c to have the curried type as well is required.

The type derived for a component, however, does not need to be its *principle type* (Damas and Milner, 1982). Any query for components of type $list(int) \rightarrow int$ should also retrieve the polymorphic *hd*-function, whose principle type is usually $\forall \alpha \cdot list(\alpha) \rightarrow \alpha$. Hence, retrieving components with more general types is the equivalent to plug-in compatibility. Similarly, to support partial compatibility, the query does not need to be completely specific. It may contain free (i.e., not polymorphically bound) type variables which may be instantiated by the derivation process.

For efficient implementations, an actual type inference (which is in fact a syntax-directed proof search) is still much too expensive. Fortunately, this it not necessary because it suffices in most cases to work on the type terms only. The key idea is to replace type inference by E-unification . Here, the equational theory E captures the intended abstractions of T, matching from the library type to the query cares for more general types and matching from the query to the library handles the free variables. To maintain the semantic integrity, it is crucial to show that E is an adequate representation of T, i.e., that $\tau_c \doteq_E \tau_q$ for the most general type τ_c of any retrieved component c.

Signature matching has two important side effects. It identifies the respective formal parameters of the two components and it maps the respective domains (i.e., types) onto each other. Both effects are required to "join" the different scopes of two contracts together. This makes signature matching also a necessary prerequisite for the subsequent filters.

5. Rejection Filters

Detecting and rejecting non-theorems is an essential step in deduction-based retrieval because most of the tasks result from non-matching candidates. Unfortunately, most ATPs for full first-order logic are not suited for this. They exhaustively search for a proof (or refutation) of a conjecture (or its negation) but they usually fail to conclude that it is *not* valid (or contradictory). The basic idea of rejection filters is to use efficient and terminating proof procedures for logics which are decidable but weaker than FOL, e.g., quantifier-free or monadic fragments. In order to apply these procedures, the filter must either identify suitable subtasks or translate (i.e., abstract) the entire task into the weaker fragment.

Such filters are necessarily incomplete in a sense that non-theorems may still pass through. However, from a practical point of view, this is not too severe as it only reduces the effectiveness of the filters. Unfortunately, some rejection filters may also be unsound and valid tasks may be rejected improperly, e.g., because the applied abstraction is not not sufficiently precise. Candidates belonging to such tasks cannot be recovered again which reduces the recall of the entire pipeline. The engineering problem is to balance this loss against the achieved reduction.

Rejection filters can be based on different principles. In the sequel we will focus on two approaches: using logic simplification to detect non-matches and applying model generators to find counter-examples.

5.1. *Simplification*

Aggressive simplification of automatically generated proof tasks is always necessary because they contain much redundancy, e.g., too large quantifier scopes or redundant equations. But simplification can also be used as a rejection filter. If a task can be simplified to *false*, the candidate may obviously be rejected.

In NORA/HAMMR, we have experimented with different levels of simplification procedures. The lowest level works with the FOL connectives and quantifiers. Here we eliminate the propositional constants *true* and *false* which typically stem from the preconditions of the components, and transform the proof tasks into a (conjunctive or disjunctive) normal form and then further into anti-prenex form to minimize the quantifier scopes. The next level handles equalities. For simplicity, we do not rewrite with all equations but restrict ourselves to equations where at least one side is a single variable. Due to the structure of the proof tasks it is necessary to use inequalities for conditional rewriting, too. Both levels involve only the usual symbols of FOL and are thus hard-wired into our system. Because the preceding translation of the proof tasks already handled the LPF-interpretation of the symbols, we can apply the usual rewrite rules for simplification.

However, to achieve a significant filtering effect more semantic information must be utilized. In NORA/HAMMR, we extract this information automatically from the lemma library maintained by the reuse administrator. The lowest semantic level just identifies all axioms and lemmas suitable as rewrite rules, extracts a terminating subset and rewrites the task into the respective normal form. For operators which are identified as associative and commutative, we also use AC-rewriting. In contrast to rewrite-based *theorem proving*, rejection filters need no confluent rewrite system and hence no expensive Knuth-Bendix completion. On this level, we also consider quantifiers as "ordinary" operators and rewrite them modulo α-conversion. We can thus use the axiom $\exists x : T \cdot p(x)$ to simplify $Q \wedge \exists y : T \cdot p(y)$ to Q.

The next semantic level is concerned with properties of sorts. If a sort is marked as freely generated by some set of constructor functions, we can use this information to simplify equations between constructor terms. E.g., for lists we simplify $nil = cons(x, xs)$ to *false* and $cons(x, xs) = cons(y, nil)$ to $x = y \wedge xs = nil$. But we use the generator information even more aggressively. From the type information of the generators we can determine whether the domain of a sort contains at least two different elements or not. If this is the case, we rewrite universally quantified equations as for example $\forall x : T \cdot x = t$ to *false* provided that x does not occur free in t. Similarly, existentially quantified equations are rewritten to ***true***. Both simplifications are in general unsound when partial functions are involved but the preceding LPF-translation guarantees that they do not change the provability of the original formulas.

The final semantic level again uses the generator information to unfold sorted quantifiers by a single step. This is similar to the induction approximation described in Section 6.2. The difference is that we do not generate a proper step case. For example, we rewrite $\forall l : List \cdot F(l)$ only to $\forall i : Item, l' : List \cdot F(nil) \wedge F(cons(i, l'))$.

5.2. *Counterexample Generation*

A component c can always be rejected, if we find a "counter-example" for its associated proof task because it then cannot be valid. Model generators for FOL like Finder (Slaney, 1994) or MACE (McCune, 1994) try to find such counter-examples (which are just interpretations under which the task evaluates to *false*) by systematically checking all possible interpretations. Their highly efficient implementation (usually using BDD-based Davis-Putnam decision procedures) would make them ideal candidates for fast rejection filters.

However, the exhaustive search terminates only if all involved domains are finite. But most domains in our application are unbounded, e.g., numbers or lists. If we want to use model generation techniques for our purpose, we must map these infinite domains onto finite representations, either by *abstraction* or by *approximation.*

Domain abstraction uses techniques from abstract interpretation (Cousot and Cousot, 1977) where the infinite domain is partitioned into a small finite number of sets which are called abstract domains. Abstract model checking (Jackson, 1994) then represents the abstract domains by single model elements and tries to find an abstract counter model, using an axiomatization of the abstract functions and predicates with a standard FOL model generator. The drawback of abstract model checking is that some domains in combination with certain predicates (e.g., equality) cannot be abstracted properly.

The experiments we describe in Section 7.4 are based on a second approach which uses domain approximations. From the infinite domain, we select a number of values which seem to be "crucial" for the component's behavior. E.g., for lists, we pick the empty list *nil* and some small lists with one or two elements. Then, we search for models or a counter-example. This approach mimics the manual checking for matches: if one looks for matching component, one first makes checks with the empty list and one or two small lists. If this does not succeed, the component cannot be selected. Otherwise, additional checks have to be applied. This approach, however, is neither sound nor complete. For some invalid formulas, a counter-example can be found in the finitely approximated domain, e.g., $\exists l : List \cdot \exists i : Item \cdot l = [i]^\wedge l$. Vice versa, for some valid formulas the approximation may be no longer a model, e.g., $\exists x, y, z : List \cdot x \neq y \wedge y \neq z \wedge z \neq x$ which has a model only in domains with at least three distinct elements. While the first case is not too harmful for our application—the performance of the filter just decreases (i.e., more proof tasks can pass), the second one is dangerous: proof tasks describing valid matches might be lost.

6. Confirmation Filters

Confirmation filters are in charge of rejecting any task for which they cannot find a proof. If a proof is found, the retrieved component actually matches the query with respect to the selected compatibility criterion (see Section 2.2). Thus, the last filter in any pipeline should be such a confirmation filter.

In contrast to rejection filters, the applied deduction must be sound. Otherwise, the guaranty becomes void and the precision may become imperfect. In NORA/HAMMR, we use first-order automated theorem provers as confirmation filters. The experiments described in this chapter have been carried out with PROTEIN, (p-)SETHEO, and SPASS. These systems are sound, i.e., they always guarantee full precision of the retrieval process. However, the challenge in this application is to retain a high recall with short answer times in the range of up to a few seconds.

The proof tasks must be pre-processed by a variety of steps to generate suitable input for the different provers. While some of these steps are specially tailored for each prover (e.g., handling of sorts and equations, control of the prover), others address common problems: logic simplification of the formulas (using techniques described in Section 5.1), selection of axioms, and handling of inductive problems.

6.1. *Selection of Axioms*

Each proof task must contain—besides the theorem and the hypotheses—the properties of each data type (e.g., *List, Nat*) as a set of axioms. Further lemmata can be added to the formula to facilitate and abbreviate complex proofs. Automated theorem provers, however, are extremely sensitive with respect to the number and structure of the axioms added to the formula. Adding a single (unnecessary) axiom can increase the run-time of the prover by orders of magnitude, thus decreasing recall in an unacceptable way. In general, selecting the optimal subset of axioms is a very hard problem and has not been solved in a satisfactory way yet. Our strong time-constraints furthermore do not allow us to use time-consuming selection techniques. In our prototype, we therefore use a simple strategy similar to the one described in Chapter III.2.9 in this volume. The lemma library contains hierarchically structured sub-theories for all involved data types. By selecting all sub-theories which share function symbols with the theorem and the hypotheses and the sub-theories they rely on, we essentially obtain all axioms sufficient to find a proof of the given obligation.

6.2. *Inductive Problems*

Whenever recursive specifications are given or recursively defined data structures are used (e.g., lists) many of the proof tasks can be solved by *induction* only. The provers under consideration cannot handle induction by themselves and our strong time-constraints do not allow us to use an inductive theorem prover (e.g., (Bundy et al., 1990)). Therefore, we approximate induction by splitting up the problem into several cases. For example, for a query and a candidate with the signature $l : List$, and the corresponding proof task of the form $\forall l : List \cdot F(l)$ we obtain the following cases:[6]

$$\begin{aligned}
&\forall l : List \cdot l = [] \Rightarrow F(l) \\
&\forall l : List \cdot \forall i : Item \cdot l = [i] \Rightarrow F(l) \\
&\forall l : List \cdot \forall i : Item, l_0 : List \cdot F(l_0) \wedge l = [i]^\wedge l_0 \Rightarrow F(l)
\end{aligned}$$

After rewriting the formula accordingly, we obtain three independent first order proof tasks which then must be processed by the prover. Only if all three proof tasks can be proven the entire proof task is considered to be solved. However, we cannot solve every inductive problem. Nevertheless, in combination with the simplification filter (Section 5.1, unfolding), this approximation is highly effective. For first experiments with this approximation see (Schumann and Fischer, 1997).

6.3. *The Provers*

All provers which are being used for the experiments (Protein, SETHEO, Spass) have been developed or enhanced within the Schwerpunkt Deduktion. They all accept formulas in first order logic and try to find a refutation for it. They receive proof obligations in sorted first order logic. Each prover has its own modules to convert the formulas into clausal normal form. The sorts of the proof tasks are imposed from the VDM-SL contracts and are structured hierarchically. VDM-SL allows only very limited overloading and polymorphism such that it can actually be resolved by the signature matching filter (cf. Section 4). It also allows, however, dynamic sorts with arbitrary invariants but these have been eliminated during the translation into FOL.

Details of the proof procedures of the provers are largely irrelevant for NORA/HAMMR. We can consider the provers as black boxes which return "proof found" or "failed to find a proof" after the given time-limit. Therefore, we will not give an extended description of each prover.

---

[6] Although it would be sufficient to have cases 1 and 3 only, we also generate case 2, since many specifications are valid for non-empty lists only. For those specifications, case 1 would be a trivial proof task which does not contribute to filtering.

PROTEIN (*PRO*ver with a *T*heory *E*xtension *IN*terface, (Baumgartner and Furbach, 1994)) is a PTTP-based first order theorem prover over built-in theories for theory reasoning (e.g., handling of equations) using a linearizing completion technique. PROTEIN is based on the Model Elimination Calculus (Loveland, 1978) and refinements thereof.

SETHEO (Letz et al., 1992; Goller et al., 1994) is also based on the Model Elimination Calculus. However, the proof procedure is implemented as an extension of the Warren Abstract Machine. Equations are handled by the naïve approach of adding the corresponding equality axioms, or by the compilation approach used within E-SETHEO (Moser et al., 1997). SETHEO provides a variety of parameters to control its search. In order to optimize SETHEO's answer times, *parallel competition* over parameters is used. The parallel prover p-SETHEO (see Chapter II.3.10 for details) allows us to obtain optimal efficiency combined with short answer times.

SPASS (Weidenbach et al., 1996; Ganzinger et al., 1997) is based on the superposition calculus developed by Bachmair and Ganzinger (1994). This calculus which is able to handle equality is extended by specific sort constraints (cf. Chapter I.2.9). This means that sort-information about the variables is represented as monadic predicates which are handled specifically within SPASS.

7. EXPERIMENTS

7.1. *The Experimental Data Base*

All experiments were carried out over a database of 119 list specifications which were modified to have the type *list* → *list* because the integration of the signature matching filter is not yet completed. Approximately 75 of these specifications describe actual list processing functions (e.g., *tail*, *rotate*, or *delete_minimal*) while the rest simulates queries. We thus included underdetermined specifications (e.g., the result is an arbitrary front segment of the argument list) as well as specifications which don't refer to the arguments (e.g., the result is not empty). For simplicity, we formulated the specifications such that the postconditions only use VDM-SL's built-in sequences and comparisons.

In order to simulate a realistic number of queries we then cross-matched each specification against the entire library, using plug-in compatibility as match relation. This yielded a total of 14161 proof tasks where 1860 or 13.1% were valid. Although the domain of our experimental library is rather limited, it serves as a good test-bed, because it allows to generate a realistically large number of proof tasks with many non-trivial matches.

7.2. *Evaluation of Filters*

Information retrieval methods are evaluated by the two criteria precision and recall (Salton and McGill, 1983). Both are calculated from the set *REL* of *relevant* components which satisfy the given match relation with respect to the query and *RET*, the set of *retrieved* components which actually pass the filter. The *precision p* is defined as the relative number of hits in the response while the *recall r* measures the system's relative ability to retrieve relevant components:

$$p = \frac{|REL \cap RET|}{|RET|} \qquad r = \frac{|REL \cap RET|}{|REL|}$$

Ideally, both numbers would be 1 (i.e., the system retrieves all and only matching components) but in practice they are antagonistic: a higher precision is usually paid for with a lower recall. We also need some metrics to evaluate the filtering effect. To this end we define the *fallout*

$$f = \frac{|RET \setminus REL|}{|L \setminus REL|}$$

(where L is the entire library) as the fraction of non-matching components which pass the filter as well as the *reduction* which is just the relative number of rejected components. Finally, we define the *relative defect ratio* by

$$d = \frac{|REL \setminus RET|}{|L \setminus REL|} \cdot \frac{|L|}{|REL|}$$

as the relative number of rejected matching components in relation to the precision of the filter's input. Thus, a relative defect ratio greater than 1 indicates that the filter's ability to reject only irrelevant components is even worse than a purely random choice.

7.3. *Rejecting Tasks with Simplification*

Table I summarizes the results of the simplification filter. The columns labeled with FOL and Equality represent the respective syntactic levels, Sorts the full set of semantic simplifications except quantifier unfolding which is included in the last column. Run-times (measured on a Sun SparcStation20) are well below 2 seconds. The results justify the application of simplification as first rejection filter. All four levels are sound and thus yield a total recall

Table I. Results of simplification

Level	FOL	Equality	Sorts	Unfolding
recall r	100.0%	100.0%	100.0%	100.0%
precision p	13.1%	13.1%	13.8%	22.1%
σ_p	0.19	0.19	0.2	0.25
increase	1.0	1.0	1.1	1.7
fallout	100.0%	100.0%	94.4%	53.2%
reduction	0.0%	0.0%	4.8%	40.7%
reduction$^+$	2.8%	3.5%	9.3%	45.2%
defect ratio d	0.0	0.0	0.0	0.0

and a zero defect ratio. It is, however, somewhat disappointing that even aggressive syntactic simplification has absolutely no filtering effect and hence dumps the full fallout on the next filter. But both syntactic variants are already able to reduce a significant fraction (FOL: 21.0%, Equality: 26.7%) of the valid tasks to *true*. These can be added to the original rejections which gives the small total reductive effect shown in the reduction$^+$ entry.

Adding semantic information improves the situation slightly but only with unfolding we are able to reject more than 40% of the tasks because they are rewritten to *false*. The set of surviving candidates still contains more than 50% of the original non-matches (fallout) but its precision already increased by a factor of 1.7.

7.4. *Rejecting Tasks with Model Generation*

For the experiments with counter-example generation, we used different approximations of the domains *List* and *Item*. Due to the large number of variables in the proof tasks, however, we are confined to small approximations with at most three elements. Table II shows the results obtained with the model generator MACE (with a run-time limit of 20 seconds). Regardless of the domain approximation, this filter is able to recover at least 75% of the relevant matches. This figure is at the lower end of our expectations, since almost 25% of the valuable matches are lost in this filter. Furthermore, the large standard deviation indicates that the filter's behavior is far from uniform and that it will perform poor for some queries.

Table II. Results of model checking

Model Size: $\|List\| + \|Item\|$	2+1	3+1	3+2
recall r	74.7%	76.5%	81.3%
σ_r	0.25	0.26	0.25
precision p	18.5%	19.6%	16.5%
σ_p	0.21	0.19	0.16
increase	1.4	1.5	1.3
fallout	42.8%	41.0%	55.5%
reduction	50.1%	51.7%	39.0%
defect ratio d	0.51	0.45	0.48

On the other hand, between 41% and 55% of the non-matches (fallout) can pass this filter. Together with an increase of precision of approximately 1.5, the filtering effectiveness of the counter-example generation filter is in the range of the simplification filters. Due to the comparatively low recall, however, the defect ratios are quite high. Although this filter acts at least twice as good as blind guessing, the results are not satisfactory. The main reasons for this poor behavior are the large number of variables in the proof tasks which cause many timeouts and the deeply nested terms which would require larger domain approximations.

7.5. *Using ATPs as Confirmation Filters*

For the evaluation of the performance of the filters we only used the 1860 valid proof tasks of our library. Since the provers are not able to detect most non-matches, this selection is justified. Table III shows an overview of the results of all experiments. Proof tasks are most simple if one tries to retrieve identical components. In these cases, they are of the structure $A \wedge F \rightarrow F$. The first row in Table III shows that the provers (PROTEIN, p-SETHEO with 2 processors, and SPASS) were able to detect most of these matches. However, in these cases the unnecessary axioms A decreased the recall substantially. E.g., p-SETHEO was able to prove 86% of these cases when the axioms were dropped. The next series of experiment considered the general case, retrieval of arbitrary matching components. For unsimplified proof tasks, the recall which was obtained with a run-time limit of 60 seconds was relatively low. No prover was able to find significantly more than 60% of the matches. Table III shows the detailed results for each prover in the columns labeled with *basic*. There are two major reasons for this result. First, these experiments were not carried out with the approximation of induction as described in Section 6.2. Previous experiments with SETHEO (Schumann and Fischer, 1997) and a smaller library showed that the overall recall could be increased by about 19% even for unsimplified tasks. Second, the pre-selection of axioms and lemmas is still too coarse. In particular, it is not independent of the other preprocessing and simplification steps. Table IV shows results obtained by SPASS on four variants of the proof tasks. Two versions only contain a minimal set of axioms; for the others useful lemmas have been added. Furthermore, the variants differ in the way the predicates have been pre-processed: in the "expanded" case, all occurrences of the predicates *not_equal* and $<, >, \geq$ have been replaced by appropriate definitions which use $=$ and $\leq$ only. Table IV reveals that the recall can be increased when the definitions are expanded, i.e., the number of different predicate symbols is reduced. In this case, however, the additional lemmas are counter-productive, i.e., they increase the search space without improving the overall recall.

Table III. Results of ATP-based confirmation filters

	PROTEIN		p-SETHEO		SPASS		competition
Task type	Basic	Best	Basic	Best	Basic	Best	Best
r_{id}	50 %	65 %	71 %	66 %	89 %	82 %	94%
recall r	42.2%	56.6%	47,1%	55.9%	61.1%	71.2%	80.1%
σ_r	0.35	0.36	0.37	0.38	0.36	0.41	0.30
precision p	100 %	100 %	100 %	100 %	100 %	100 %	100 %

Both tables show that different variants of proof tasks and different provers exhibit quite a different behavior. Since our main interest is to obtain a high recall while keeping the answer-times low, parallel competition is of interest. When all provers are running different variants in a parallel competitive mode (similar to p-SETHEO), significantly better results can be obtained, as the simulated results in the last columns of both tables show. With such a technique, we were able to get an overall recall r of 80% which is high enough for practical purposes.

Table IV. Detailed results of the SPASS-based filter

	minimal axioms		axioms + lemmas		competition	
expansion	yes	no	yes	no	yes	no
r_{id}	91%	89%	76%	76%	91%	89%
recall r	66.4%	61.1%	64.8%	61.7%	71.1%	66.5%
σ_r	0.35	0.36	0.41	0.40	0.35	0.36
precision p	100 %	100 %	100 %	100 %	100 %	100 %

8. Related Work

Reuse by contract is an attempt to lift the axiomatic programming style of Hoare (1969) into the large. It is thus a combination and extension of the "Systematic Software Development" (Jones, 1990) and the "Design by contract" (Meyer, 1992) approaches. Similar work (Liskov and Wing, 1994) is done in component-oriented programming which has recently emerged as a new subfield in software engineering (Weck et al., 1997; Leavens and Sitaraman, 1997).

Signature matching has originally been developed as a library retrieval method for functional programming languages. Runciman and Toyn (1991) define a partial order on polymorphic types and then use unification to retrieve components with equivalent or more general types. Rittri (1991, 1990, 1993) explored the strong connections between typed polymorphic lambda calculi and Cartesion closed categories (CCC). His system uses equational matching with respect to the theory *CCC* and is thus semantically stronger motivated than the one of Runciman and Toyn. However, since *CCC*-unification is undecidable (Narendran et al., 1993) he has to resort to the decidable sub-theory of *linear isomorphisms*, thus losing some of his foundations. Matthews (1992) extended Rittri's work by sorts to handle Haskell's type classes. DiCosmo (1995) introduced the notion of *definable isomorphisms* which can be used to reconstruct the necessary mapping between the identified types. Moorman and Wing (1995a) work with *match predicates* which are essentially Prolog-implementations of simple inference rules used in theory unification. The predicates can be combined to obtain different filters but Moorman and Wing do not discuss any decidability questions which may result from this combination (cf. (Schmidt-Schauß, 1990) for the combination of unification algorithms). Their work also lacks any semantic foundations but is pragmatically motivated. Stringer-Calvert (1994) has applied this framework to Ada.

Most early publications on deduction-based retrieval ignored the usability and scaling problems. We will thus discuss only more recent work; for a more general overview see (Mili et al., 1998). Moorman and Wing (1995b) have investigated specification matching in a slightly more general framework than ours but their main application area is also software reuse. They use the Larch/ML specification language for component description and the associated interactive Larch prover for retrieval. Mili et al. (1994) describe a system in which specifications are given as binary relations of legal (input, output)-pairs. They then define a subsumption relation on such relations and use this for retrieval, relying on Otter as the deductive component. However, their system is still in a prototypical stage, so no relevant statistical evaluation is presented. The work of Jeng and Cheng (1994) also uses a subsumption test but to construct a hierarchical library. Scaling problems have been addressed by various authors. The Inscape/Inquire-system (Perry and Popovitch, 1993) limits the specification language to make retrieval more efficient. Similarly, AMPHION (Lowry et al., 1994) uses a GUI to foster a more uniform specification style which in turn allows an appropriate fine-tuning of the prover. Additional speed-up is achieved by automatically "compiling" axioms into decision theories (Lowry and Van Baalen, 1995). These techniques have successfully been applied to assemble over 100 FORTRAN programs from a scientific component library for solar system kinematics. Penix et al. (1995) use "semantic features" (i.e., user-defined abstract predicates which follow from the components' specifications) to classify the components and perform case-based reasoning along this classification to identify the most promising candidates. This classification process uses forward reasoning with an ATP. However, the authors give no evidence of how successful their approach is.

9. Lessons Learnt

In this chapter, we have presented the deduction-based software component retrieval tool NORA/HAMMR. From the very beginning we have considered it as an application project with strong practical emphasis. Our goal was to show that such a tool is not only theoretically possible but practical with state-of-the-art theorem provers. We can summarize our experience as follows:

1. Deduction-based software component retrieval is accepted only if any ATP evidence is hidden behind a user-friendly interface with a problem-oriented specification language; and if strong run-time constraints ("results while-u-wait") are met.

2. Only a pipeline of different signature, rejection and confirmation filters prevents the theorem prover from drowning.

3. Parallel competition between different provers and prover parameters is essential.

4. The application side (user interface, transformation of tasks into first order formulas) is at least as hard and as important as the deduction side; automated theorem proving alone is no silver bullet.

5. ATP researchers must pay more attention to the customers' point of view and provide convenient input formalisms as well as preprocessors for simplification.

6. We need fast methods to discover that a formula is not valid. Model checking is of limited value for infinite domains.

7. Experiments show that today's ATP technology makes deduction-based software component retrieval an almost realistic option in a formal development process.

In developing NORA/HAMMR, we had to deal with much low-level details. Much time was spent to transform syntactical representations of the formulas between the systems. A major problem was to generate the proof tasks from the VDM-SL specifications, due to the enormous syntactic variety. Much to our surprise, the translation from LPF to FOL was no major impediment after the optimizations. The naïve translation, however, increased the formula sizes considerably and brought retrieval to a grinding halt.

The integration of the model generators and theorem provers was easy. They can be considered as black boxes which receive a proof task in sorted first-order logic together with a run-time limit, after which they they return either *valid, not valid*, or *time-out*.

The implementation of necessary simplifications took more effort than one can usually afford in an application. Similarly, the administration of lemma libraries is an expensive task. Unfortunately, we were not able to reuse the implementation of Reif and Schellhorn (cf. Chapter III.2.9) because it is tailored towards equational logics.

A practical system must be evaluated in depth. Most other approaches in this area (cf. Section 8) are tried out with a handful of toy examples only. For the evaluation of NORA/HAMMR we spent much effort to develop libraries of reasonable size and complexity, resulting in a statistically significant body of experimental data. The evaluation of the system and its fine-tuning consumed large amounts of computation resources.

10. Future Work

Although the results of the experiments carried out so far are very encouraging, many improvements are required before NORA/HAMMR can really be used in industry. We are currently preparing experiments with a library of commercial date and time handling functions as used e.g. in stock trading software.[7]

Due to the hard time-constraints, the reduction of proof tasks, both in complexity and number, is of central importance. Powerful rejection filters must ensure that only a few proof tasks remain to be processed by the automated theorem prover. However, our current model-checking filter rejects too much valid matches due to the necessary approximate abstractions. Future work will thus include experiments with even more aggressive simplification techniques than those described in Section 5.1. We will also develop rejection filters based on decision procedures. Additionally, heuristics and knowledge-based filters similar to (Penix and Alexander, 1997) can help to reduce the number of proof tasks.

As our experiments showed, current high-performance ATPs are certainly usable as confirmation filters. Much work, however, is still necessary to adapt them for such kinds of proof tasks. In particular, the requirement of full automatization and the strong time-limits must be obeyed carefully. Further important issues are the handling of inductive proofs, and the selection of appropriate axioms. Here, more powerful heuristics must be developed. A further reduction of the search space could be achieved by axiom compilation techniques similar to those of Meta-AMPHION (Lowry and Van Baalen, 1995). However, the integration of decision procedures into the provers at hand is still an open research topic.

Acknowledgments

We are indebted to our colleagues in the DFG-Schwerpunkt for support with the experiments, particularly to Christoph Weidenbach, Peter Baumgartner, Michael Kühn, Ingo Dahn, and the other members of the ILF, PROTEIN, and SETHEO groups. Ralf Haselhorst of the DG Bank, Frankfurt, helped with the date library. Several people in Braunschweig contributed to the implementation of NORA/HAMMR, particularly Christian Lindig and Torsten Neumann, who worked on the GUI.

---

[7] This work is done in cooperation with the German DG Bank.

REFERENCES

Agerholm, S. and J. Frost: 1997, 'An Isabelle-based Theorem Prover for VDM-SL'. In: *Proceedings of the 10th International Conference on Theorem Proving in Higher Order Logics (TPHOLs'97)*. Springer.

Bachmair, L. and H. Ganzinger: 1994, 'Rewrite-based Equational Theorem Proving with Selection and Simplification'. *Joural of Logic and Computation* **4**(3), 217–247.

Barringer, H., J. H. Cheng, and C. B. Jones: 1984, 'A Logic Covering Undefinedness in Program Proofs'. *Acta Informatica* **21**(3), 251–269.

Baumgartner, P. and U. Furbach: 1994, 'PROTEIN: A PROver with a Theory Extension INterface'. in (Bundy, 1994), pp. 769–773, Springer.

Beckert, B., S. Gerberding, R. Hähnle, and W. Kernig: 1991, 'The Tableau-Based Theorem Prover $_3T^AP$ for Multiple-Valued Logics'. In: D. Kapur (ed.): *Proc. 11th Intl. Conf. Automated Deduction*, Vol. 607 of *Lecture Notes in Artifical Intelligence*. pp. 758–760, Springer.

Bundy, A. (ed.): 1994, 'Proc. 12th Intl. Conf. Automated Deduction', Vol. 814 of *Lecture Notes in Artifical Intelligence*. Springer.

Bundy, A., F. van Harmelen, A. Smaill, and A. Ireland: 1990, 'Extensions to the Rippling-Out Tactic for Guiding Inductive Proofs'. in (Stickel, 1990), pp. 132–146, Springer.

Cheng, J. H. and C. B. Jones: 1991, 'On the usability of logics which handle partial functions'. In: C. Morgan and J. C. P. Woodcock (eds.): *Proceedings of the Third refinement Workshop*. Berlin, pp. 51–69, Springer-Verlag.

Cousot, P. M. and R. Cousot: 1977, 'Abstract Interpretation: A Unified Lattice Model for Static Analysis of Programs by Construction or Approximation of Fixpoints'. In: *Proc. 4th ACM Symp. Principles of Programming Languages*. Los Angeles, California, pp. 238–252, ACM Press.

Damas, L. and R. Milner: 1982, 'Principle Type-Schemes for Functional Programs'. In: *Proc. 9th ACM Symp. Principles of Programming Languages*. Albuquerque, New Mexico, pp. 207–212, ACM Press.

Dawes, J.: 1991, *The VDM-SL Reference Guide*. London: Pitman.

DiCosmo, R.: 1995, *Isomorphisms of Types: from λ-calculus to information retrieval and language design*, Vol. 14 of *Progress in Theoretical Computer Science*. Boston: Birkäuser.

Frakes, W. B. and B. A. Nejmeh: 1987, 'Software Reuse Through Information Retrieval'. In: *Proc. 20th Annual Hawaii Intl. Conf. on Systems Sciences*. pp. 530–535.

Ganzinger, H., C. Meyer, and C. Weidenbach: 1997, 'Soft Typing for Ordered Resolution'. In: W. McCune (ed.): *Proc. 14th Intl. Conf. Automated Deduction*, Vol. 1249 of *Lecture Notes in Artifical Intelligence*. pp. 321–335, Springer.

Goller, C., R. Letz, K. Mayr, and J. M. P. Schumann: 1994, 'SETHEO V3.2: Recent Developments—System Abstract'. in (Bundy, 1994), pp. 778–782, Springer.

Hoare, C. A. R.: 1969, 'An Axiomatic Basis for Computer Programming'. *Communications of the ACM* **12**(10), 576–580.

Jackson, D.: 1994, 'Abstract Model Checking of Infinite Specifications'. In: M. Naftalin, T. Denvir, and M. Bertran (eds.): *Proc. 2nd Intl. Symp. Formal Methods Europe*, Vol. 873 of *Lecture Notes in Computer Science*. Barcelona, pp. 519–531, Springer.

Jeng, J.-J. and B. H. C. Cheng: 1994, 'A Formal Approach to Using More General Components'. In: *Proc. 9th Knowledge-Based Software Engineering Conf.* Monterey, CA, pp. 90–97, IEEE Computer Society Press.

Jézéquel, J.-M. and B. Meyer: 1997, 'Design by Contract: The Lessons of Ariane'. *IEEE Computer* **30**(1), 129–130.

Jones, C. B.: 1990, *Systematic Software Development Using VDM*. Englewood Cliffs, New Jersey: Prentice-Hall, second edition.

Jones, C. B.: 1995, 'Partial functions and logics: A warning'. *Information Processing Letters* **54**(2), 65–67.

Jones, C. B., K. D. Jones, P. A. Lindsay, and R. Moore: 1991, *mural: A Formal Development Support System*. Springer.

Jones, C. B. and K. Middelburg: 1994, 'A Typed Logic of Partial Functions Reconstructed Classically'. *Acta Informatica* **31**(5), 399–430.

Katz, S., C. A. Richter, and K. S. The: 1987, 'PARIS: A System for Reusing Partially Interpreted Schemas'. In: *Proc. 9th Intl. Conf. Software Engineering*. Montery, CA, pp. 377–385, IEEE Computer Society Press.

KBSE-10: 1995, 'Proc. 10th Knowledge-Based Software Engineering Conf. ' Boston, MA:, IEEE Computer Society Press.

Krueger, C. W.: 1992, 'Software Reuse'. *ACM Computing Surveys* **24**(2), 131–183.

Leavens, G. T. and M. Sitaraman (eds.): 1997, 'Proc. ESEC/FSE Workshop on Foundations of Component-Based Sytems'.

Letz, R., J. Schumann, S. Bayerl, and W. Bibel: 1992, 'SETHEO: A High-Performance Theorem Prover'. *Journal of Automated Reasoning* **8**(2), 183–212.

Lions et al., J. L.: 1996, 'Ariane 5 Flight 501 Failure Report'.

Liskov, B. and J. M. Wing: 1994, 'A Behavioral Notion of Subtyping'. *ACM Transactions on Programming Languages and Systems* **16**(6), 1811–1841.

Loveland, D. W.: 1978, *Automated Theorem Proving: a Logical Basis*. North–Holland.

Lowry, M., A. Philpot, T. Pressburger, and I. Underwood: 1994, 'AMPHION: automatic programming for scientific subroutine libraries'. In: Z. W. Raś and M. Zemankova (eds.): *Proc. 8th Intl. Symp. on Methodologies for Intelligent Systems*, Vol. 869 of *Lecture Notes in Artifical Intelligence*. pp. 326–335, Springer.

Lowry, M. and J. Van Baalen: 1995, 'Meta-Amphion: Synthesis of Efficient Domain-Specific Program Synthesis Systems'. in (KBSE-10, 1995), pp. 2–10, IEEE Computer Society Press.

Maarek, Y. S., D. M. Berry, and G. E. Kaiser: 1991, 'An Information Retrieval Approach For Automatically Constructing Software Libraries'. *IEEE Transactions on Software Engineering* **SE-17**(8), 800–813.

Maarek, Y. S. and F. A. Smadja: 1989, 'Full Text Indexing Based on Lexical Relations An Application: Software Libraries'. In: *Proc. 12th Annual Intl. ACM SIGIR Conf. on Research and Development in Information Retrieval*. pp. 198–206.

Manhart, P. and S. Meggendorfer: 1991, 'A knowledge and deduction based software retrieval tool'. In: *Proc. 4th Intl. Symp. on Artificial Intelligence*. pp. 29–36.

Matthews, B.: 1992, 'Reusing Functional Code using Type Classes for Library Search'. In: *Proc. ERCIM Workshop on Methods and Tools for Software Reuse*. Heraklion.

McCune, W. W.: 1994, 'A Davis-Putnam Program and its Application to Finite First-Order Model Search: Quasigroup Existence Problems'. Technical report, Argonne National Laboratory, Argonne, IL, USA.

McIlroy, M. D.: 1968, 'Mass Produced Software Components'. In: P. Naur and B. Randell (eds.): *Software Engineering: Report on a Conference*. Brussels, pp. 138–150, NATO Scientific Affairs Division.

Meyer, B.: 1992, 'Applying "Design by Contract"'. *IEEE Computer* **25**(10), 40–51.

Middelburg, K.: 1993, *Logic and Specification — Extending VDM-SL for advanced formal specification*, Computer Science: Research and Practice. Chapman & Hall.

Mili, A., R. Mili, and R. Mittermeir: 1994, 'Storing and Retrieving Software Components: A Refinement-Based System'. In: B. Fadini (ed.): *Proc. 16th Intl. Conf. Software Engineering*. Sorrento, Italy, pp. 91–102, IEEE Computer Society Press.

Mili, A., R. Mili, and R. Mittermeir: 1998, 'A Survey of Software Reuse Libraries'. *Annals of Software Engineering*. To appear.

Moser, M., O. Ibens, R. Letz, J. Steinbach, C. Goller, J. Schumann, and K. Mayr: 1997, 'The Model Elimination Provers SETHEO and E-SETHEO'. *Journal of Automated Reasoning* **18**, 237–246.

Narendran, P., F. Pfenning, and R. Statman: 1993, 'On the Unification Problem for Cartesian Closed Categories'. In: R. L. Constable (ed.): *Proc. 8th Annual IEEE Symp. Logic in Computer Science*. Montreal, Canada, pp. 57–63, IEEE Computer Society Press.

Oberschelp, A.: 1962, 'Untersuchungen zur mehrsortigen Quantorenlogik'. *Mathematische Annalen* **145**, 297–333.

Penix, J. and P. Alexander: 1997, 'Toward Automated Component Adaptation'. In: N. Juristo (ed.): *Proc. 9th Intl. Conf. on Software Engineering and Knowledge Engineering*. pp. 535–542, Knowledge Systems Institute.

Penix, J., P. Baraona, and P. Alexander: 1995, 'Classification and Retrieval of Reusable Components Using Semantic Features'. in (KBSE-10, 1995), pp. 131–138, IEEE Computer Society Press.

Perry, D. E. and S. S. Popovitch: 1993, 'Inquire: Predicate-Based Use and Reuse'. In: *Proc. 8th Knowledge-Based Software Engineering Conf.* Chicago, IL, pp. 144–151, IEEE Computer Society Press.

Rittri, M.: 1990, 'Retrieving Library Identifiers via Equational Matching of Types'. in (Stickel, 1990), pp. 603–617, Springer.

Rittri, M.: 1991, 'Using types as search keys in function libraries'. *J. Functional Programming* **1**(1), 71–89.

Rittri, M.: 1993, 'Retrieving Library Functions by Unifying Types Modulo Linear Isomorphism'. *Theoretical Informatics and Applications* **27**(6), 523–540.

Rollins, E. J. and J. M. Wing: 1991, 'Specifications as Search Keys for Software Libraries'. In: K. Furukawa (ed.): *Proc. 8th Intl. Conf. Symp. Logic Programming*. Paris, pp. 173–187, MIT Press.

Runciman, C. and I. Toyn: 1991, 'Retrieving re-usable software components by polymorphic type'. *J. Functional Programming* **1**(2), 191–211.

Salton, G. and M. J. McGill: 1983, *Introduction to Modern Information Retrieval*. New York: McGraw-Hill.

Schmidt-Schauß, M.: 1990, 'Unification in a Combination of Arbitrary Disjoint Equational Theories'. In: C. Kirchner (ed.): *Unification*. London: Academic Press, pp. 217–266.

Schumann, J. M. P. and B. Fischer: 1997, 'NORA/HAMMR: Making Deduction-Based Software Component Retrieval Practical'. In: *Proc. Conf. Automated Software Engineering '97*. pp. 246–254.

Slaney, J.: 1994, 'FINDER: Finite domain enumerator'. in (Bundy, 1994), pp. 798–801, Springer.

Stickel, M. E. (ed.): 1990, 'Proc. 10th Intl. Conf. Automated Deduction', Vol. 449 of *Lecture Notes in Computer Science*. Kaiserslautern:, Springer.

Stringer-Calvert, D. W. J.: 1994, 'Signature matching for Ada software reuse'. Master's thesis, University of York.

Weck, W., J. Bosch, and C. Szyperski (eds.): 1997, 'Proc. Second Intl. Workshop on Component-Oriented Programming'. Jyväskylä:, Turku Centre for Computer Science. TUCS General Publication No 5.

Weidenbach, C., B. Gaede, and G. Rock: 1996, 'Spass and Flotter version 0.42'. In: M. A. McRobbie and J. K. Slaney (eds.): *Proc. 13th Intl. Conf. Automated Deduction*, Vol. 1104 of *Lecture Notes in Artifical Intelligence*. pp. 141–145, Springer.

Zaremski, A. M. and J. M. Wing: 1995a, 'Signature Matching: A Tool for Using Software Libraries'. *ACM Transactions on Software Engineering and Methodology* pp. 146–170.

Zaremski, A. M. and J. M. Wing: 1995b, 'Specification Matching of Software Components'. In: G. E. Kaiser (ed.): *Proc. 3rd ACM SIGSOFT Symp. Foundations of Software Engineering*. Washington, DC, pp. 6–17, ACM Press.

CHAPTER 12

REWRITE BASED HARDWARE VERIFICATION WITH REDUX

1. INTRODUCTION

This chapter reports on experiences in verifying synchronous circuit designs using the ReDuX term rewriting laboratory (Bündgen, 1993; Bündgen et al., 1996c; Bündgen and Lauterbach, 1996a). For this purpose a technique is presented that transforms the proof requirements for the verification of a processor design at gate level to purely equational problems. This technique has been successfully applied to the verification of two processors of the Sparrow family.

The Sparrow processor is a RISC CPU that has been designed in the computer engineering laboratory course at the University of Tübingen (Thole et al., 1993), and it has been implemented on a Xilinx field programmable gate array (FPGA). Sparrow processors exists in two versions: a 4-bit CPU and an 8-bit CPU. The hardware specification was available in form of a BLIF (Berkeley, 1992) net list that can be used to program an FPGA. The requirements specification against which the hardware specification was verified was drawn from the informal specification given in the class assignment of the laboratory course.

The hardware specification of synchronous circuits in BLIF format can automatically be transformed into a complete term rewriting system. This enforces that the hardware specification used in the verification really describes the actual implementation of the circuit. Since we assume that the tool to program the FPGA from a net list is correct our approach to verifying a BLIF gate level circuit description automatically extends to verifying all levels of the circuit design.

Given a rewrite specification of the hardware design, we show that all formulae needed to describe the correctness of the design can be transformed to equations that must be proven to hold modulo a complete term rewriting system. Such complete systems can be constructed without explicitely running a completion procedure. Therefore normalization is the crucial operation within this kind of verification. Standard normalization techniques turned out to be much to inefficient for this task. Therefore we have experimented with three orthogonal methods to speed-up the normalization process: mem-

W. Bibel, P. H. Schmitt (eds.), Automated Deduction. A basis for applications. Vol. III

oization of critical intermediate results, building-in Boolean rings by partial evaluation domains, and parallel or distributed normalizations.

The use of term rewriting systems for Boolean rings (Hsiang, 1985) in hardware verification was proposed for the first time by Chandrasekhar et al. (1987). The verification of some aspects of isolated circuits has been investigated by several researchers (e. g., Staunstrup et al., 1992 or Berregeb et al., 1996). Typically those circuits are rather small. So far a few articles have been published on verifications of complete processors. The work of Hunt (1989) that verifies the FM8502 processor using the Boyer-Moore prover is probably the most prominent of them. Yet the FM8502 was designed only for the purpose of being verified. It never existed as hardware. Even less work has been reported on rewrite based verifications of complete processors (e. g., Narendran and Stillman, 1989 or Sekar and Srivas, 1989). Some recent work in rewrite based hardware verification uses induction methods to verify state of the art arithmetical circuits that have been parametrized by the word size (Kapur and Subramaniam, 1996).

In contrast to our approach the hardware for all those systems has been specified independently from the hardware realization. A first successful verification of the 4-bit Sparrow has been reported in (Bündgen et al., 1996b). Here we describe the verification of both a 4-bit and an 8-bit design using improved techniques. Since both designs are (functionally) very similar the comparison of the two verifications gives an idea of the scalability of our approach. A key result of this work is the importance of efficient normalization techniques including built-in data types for the Boolean ring and memoization of intermediate normalization results. We also made a first step towards software-hardware co-verification when verifying an optional instruction of the sparrow processor against an equivalent assembler program. Very recently the method described here has been adapted by Geser and Küchlin (1997) to verify commercial hardware.

For the remainder of this chapter we assume the reader is familiar with the basic concepts and results of term rewriting systems.

2. The Sparrow Processor

The Sparrow processor is a RISC CPU that has been designed in the computer engineering laboratory course at the University of Tübingen. Its design is part of a larger project that also comprises the development of an assembler and a small operating system (Bubeck et al., 1993; Thole et al., 1993). Sparrow designs are implemented on a Xilinx XC3195 FPGA (XILINX, 1991) and can

> Let n be the width of a word (i.e. $n \in \{4, 8\}$). Then there are n input lines for the data read bus and two additional input lines for the clock and a reset signal. In addition there are $3n+6$ output lines: $2n$ lines for the address bus, n lines for the data write bus which directly connects to the accumulator, a carry-out line connecting to the carry bit of the ALU and a line for the RAM read signal that controls whether data is read or written to the data bus and four lines displaying the current opcode. Note that since the address bus has twice the width of the data bus two cycles are needed to read an address (as in STA, ADD, NOR, JNC, SUB) whereas only one cycle is needed to read data (as in LDA).

Figure 1. The requirement specification of the Sparrow (Part I)

be tested on hardware that in addition to the CPU comprises main memory, displays and dials for I/O, and serial and parallel interfaces.

Over the time two versions of Sparrow designs evolved: a first version with a 4-bit data bus and an 8-bit address bus and second version with an 8-bit data bus and a 16-bit address bus. Both versions support five one-address instructions: LDA *load accumulator immediate*, STA *store accumulator*, ADD *add*, NOR *nor* and JNC *jump if carry is not set*. So far we analyzed both a 4-bit design that we call *Sparrow-4*[1] and an 8-bit design that we call *Sparrow-8*. The particular 8-bit design we analyzed supports an additional SUB *subtract* instruction.

The specification of the Sparrow was given informally in the course material (Rosenstiel and Schubert, 1995). Figures 1 and 2 rephrase this specification in English.

No such specification was given for the SUB instruction. For the verification of *Sparrow-8* the following assumptions were used: Instruction and operand fetches are performed in the same way as in the ADD instruction. The function to be computed and stored in the accumulator and the carry bit is specified by a SPASS[2] assembler program provided by the designer of the *Sparrow-8*. The SPASS assembler language allows to introduce new instructions using macro techniques. The SUB macro (cf. Figure 4 in Appendix A) used as a base for the requirements specification of the SUB instruction was intended to supply a subtraction instruction to those Sparrow designs that do not provide a hardware SUB instruction. This means that when verifying the SUB instruction the equivalence of a software and hardware implementation of an instruction was shown. This is actually a task that falls in the realm of software-hardware co-verification. Note we did *not* verify the SPASS assem-

[1] It was called *Sparrow-0* in (Bündgen et al., 1995; Bündgen et al., 1996b).

[2] *SPASS* is the Sparrow assembler described in (Bubeck et al., 1993).

LDA (encoding: 0…000) <data>:
Load the accu immediately with data from the data read bus. Do not change the carry flag. The LDA instruction takes 2 clock cycles.
cycle 1: load opcode
cycle 2: load accu with data

STA (0…010) <address, low word> <address, high word> :
Write the contents of the accu to the address of the RAM. Do not change the carry flag and the accu. The STA instruction takes 4 cycles.
cycle 1: load opcode
cycle 2: load <address, low word>
cycle 3: load <address, high word>
cycle 4: write the contents of the accu to the loaded RAM address

ADD (0…101) <address, low word> <address, high word> :
Add the contents of the accu and the contents of the address of the RAM into the accu. Set the carry flag if an overflow is signaled. The ADD instruction takes 4 clock cycles.
cycle 1: load opcode
cycle 2: load <address, low word>
cycle 3: load <address, high word>
cycle 4: add the data in the RAM at the loaded address

NOR (0…100) <address, low word> <address, high word> :
Perform a bitwise NOR on the contents of the accu and the contents of the address of the RAM. Store the result in the accu. Do not change the carry flag. The NOR instruction takes 4 clock cycles.
cycle 1: load opcode
cycle 2: load <address, low word>
cycle 3: load <address, high word>
cycle 4: nor bitwise the data in the RAM at the loaded address onto the accu

JNC (0…011) <address, low word> <address, high word> :
Jump if not carry. Do neither change the carry flag nor the accu. The JNC instruction takes 3 clock cycles.
cycle 1: load opcode
cycle 2: if carry then load <address, low word>
cycle 3: if carry then load <address, high word>
cycle 4: if carry was not set in cycle 1 continue computation at loaded RAM address

All other instruction codes are invalid and shall be ignored.

Figure 2. The requirement specification of the Sparrow (Part II)

bler program for SUB. As a matter of fact the SPASS implementation does not coincide with the author's understanding of an integer subtraction.

3. Rewrite Specification for Synchronous Circuits

The first step of a rewrite based verification is the construction of a complete rewrite specification R_{HW} for the hardware specification of the processor. The hardware specification is given in form of a net list (e. g., in BLIF format) which is a machine readable circuit specification produced by some hardware design tool. For a purely synchronous circuit such a net list contains definitions of several kinds of hardware variables:

1. hardware variables describing input lines of the circuit,
2. hardware variables describing output lines of the circuit,
3. internal hardware variables defining Boolean functions depending on a set of other hardware variables and
4. hardware variables describing the output of latches (i. e., D-flip-flops, registers) which in turn depend on a reset state and the latch input at the preceding clock cycle.

Hardware variables do not correspond to first order variables. Instead each hardware variable reflects a function of time because its value, *Low* or *High* changes over time. Hence a hardware variable V is represented by a function symbol $V : TIME \rightarrow SIGNAL$ where the domain sort *TIME* represents a data type isomorphic to the natural numbers and the result sort *SIGNAL* consists of the two values *Low* and *High*. For simplicity we assume the sort *TIME* is freely constructed by a constant 0 and unary successor operator $s : TIME \rightarrow TIME$.

Hardware variables defining Boolean functions translate to rules of the form $V_{out}(t) \rightarrow f(V_{in_1}(t), \ldots, V_{in_k}(t))$ where t is a variable of sort *TIME*, the V_{out} and V_{in_j} denote hardware variables, and f is a Boolean function constructed from Boolean operators. Each latch L with input line V is represented by two rules. A rule $L(0) \rightarrow R$ for $R \in \{Low, High\}$ defines the reset state, and a rule $L(s(t)) \rightarrow V(t)$ that defines its value in case no reset has been signalled. The meaning of the Boolean operators is defined by the term rewriting system R_{BR} for Boolean rings (Hsiang, 1985) that is canonical modulo the associativity and commutativity (AC) of the logical-and and exclusive-or operators. Our tool *blif2rdx* automatically translates BLIF net lists (Berkeley, 1992) of a hardware *HW* to an equivalent rewrite specification R_{HW} in ReDuX syntax. R_{HW} consists of R_{BR} and rules specifying hardware variables for Boolean functions and latches defined in the net list. The specification is then extended to support a more convenient formulation of the requirement specifications. E. g., sorts and constructors may be introduced to define bit vectors. Equally we may define operators specifying functions of the ALU and the control unit like adding two bit vectors or incrementing an address.

EXAMPLE 1. The BLIF specification of *Sparrow-8* contains the following definitions:

The function of a simple internal hardware variable in BLIF format is coded as a list of (negated) minterms that are to be combined by logical-or operations. The hardware variable `/U30/X_7` in the Sparrow-8 net list is defined as follows:

```
.names /U30/X1_1 1_1-000081 /U30/X_7
#XOR (/U30/U37)
01 1
10 1
```

The first line starts with the keyword `.names` and specifies two input variables `/U30/X1_1`, `1_1-000081` and the output variable `/U30/X_7`. The second line is a comment and lines three and four each contain a binary encoding of a minterm. The terminating `1` in each line specifies that the minterm is positive (a `0` would negate the whole minterm). The tool *blif2rdx* translates this net list definition to the ReDuX rule

```
[125] V122(t) == or(and(not(V120(t)),V121(t)),
                    and(V120(t),not(V121(t))));
```

The names of the hardware variables in the BLIF net list are translated to ReDuX operator names of the form V*i* for natural numbers *i*. Here `V122` represents `/U30/X_7` and `V120` and `V121` stand for `/U30/X1_1` and `1_1-000081`. The correspondence between BLIF names and ReDuX names can be looked up in a dictionary recorded by *blif2rdx*. The reasons for this decision are syntactical restrictions of the ReDuX identifiers and the fact that most BLIF names are meaningless to human readers because they were generated mechanically by some hardware design tool or they were taken from a hardware module library whose specification was not available to us. In addition this name translation prevents any "prejudice" of the verifier on the meaning of certain hardware variables.

A BLIF latch definition has the format
`.latch` input-variable output-variable control-type clock reset-value
The definition of the latch `U136/QQ1`

```
.latch U136/QQ1_D U136/QQ1 re CLK 0
```

then translates to two rules

```
[470] V32(0)    == L;
[471] V32(S(t)) == V31(t);
```

where V32 represents U136/QQ1 and V31 represents U136/QQ1_D. V31, V32, V120, V121 and V122 are declared to be unary functions from sort TIME to SIGNAL. t is a variable, 0 a constant and S is the successor function (+1) of the sort TIME. L is a constant of sort SIGNAL denoting *Low* voltage — the zero of the Boolean ring (H for *High* is its one). □

Note that the resulting term rewriting system is complete. Each hardware variable is defined only once. Therefore there are no critical pairs and the system is locally confluent . In addition, in a synchronous circuit each cycle must be 'buffered' by a latch and therefore termination can be checked by a path ordering with the successor operator as maximal symbol. Both properties can be checked automatically.

In general the resulting term rewriting system is not interreduced. Interreducing R_{HW} is an extremely complex task that leads to very large terms on the right-hand sides of rules. This is not surprising since hardware is generally designed to yield a minimal circuit in terms of the number of gates needed on a chip. Hardware variables are introduced to allow sharing Boolean functions. Interreducing the rules of R_{HW} then macro expands (i. e., copies) shared Boolean functions.

For the *Sparrow-4* processor we obtained a $R_{Sparrow\text{-}4}$ with about 250 hardware variables (including 27 latches) and 300 rules. The rewrite specification $R_{Sparrow\text{-}8}$ of the *Sparrow-8* processor contained 475 hardware variables (including 50 latches) and 571 rules. Hence *Sparrow-4* is equivalent to an automaton with more than $1 \cdot 10^8$ states and *Sparrow-8* corresponds to an automaton with approximately $1 \cdot 10^{15}$ states. However not all of these states need to be reachable from the initial state.

The trial to normalize the right-hand sides of the rules in $R_{Sparrow\text{-}8}$ w. r. t. $R_{Sparrow\text{-}8} \setminus R_{BR}$ resulted in a file containing a rewrite system that measured over two MBytes. Note that this does not yet include a potential additional exponential explosion due to the distributivity law of the Boolean ring. Therefore the verification experiments were executed with the non-interreduced specification.

4. Proof Requirements

An internal processor state at time t_0 is completely determined by the setting of all its latches at time t_0. By an *instruction fetch state* we mean an internal processor state that is valid at the beginning of an instruction fetch cycle. Typically an instruction fetch state can be characterized by a subset of all latches in a processor.

The Sparrow processors were verified by showing that for all instruction fetch states, any instruction I can be executed such that

1. after the execution of I the processor is again in an instruction fetch state,
2. the instruction and argument fetches are performed correctly and
3. the result of the instruction is correct and depends only on the arguments supplied to I and a few internal states (accu, program counter, carry bit).
4. In addition, all invalid instructions behave like a no-operation instruction (NOP) and
5. the reset signal should reset all latches to their state at time 0.

A typical formula describing the above requirements has the form

$$\phi = \forall t\ ((\bigwedge_{i=1}^{k} u_i(t) = v_i(t)) \supset x(t) = y(t))$$

where $u_i(t)$, $v_i(t)$, $x(t)$ and $y(t)$ are Boolean terms containing at most one free variable, namely the variable t of sort *TIME*. Showing that ϕ is valid w. r. t. R_{HW} is equivalent to showing that $x(t_0) = y(t_0)$ is valid in the equational theory of $R_{HW} \cup \{u_i(t_0) = v_i(t_0) \mid 1 \leq i \leq k\}$ where t_0 is a new (Skolem) constant. Consequently we can prove the implication ϕ by showing that $x(t_0)$ and $y(t_0)$ have the same normal form w. r. t. R_ϕ a complete term rewriting system equivalent to $R_{HW} \cup \{u_i(t_0) = v_i(t_0) \mid 1 \leq i \leq k\}$.

Note that we can take advantage of *don't cares* in the circuit. Instead of proving the formula $\phi = \phi_1 \wedge \phi_2$ with

$$\phi_1 = \forall t\ ((u_0(t) = Low \wedge \bigwedge_{i=1}^{k} u_i(t) = v_i(t)) \supset x(t) = y(t)),$$

$$\phi_2 = \forall t\ ((u_0(t) = High \wedge \bigwedge_{i=1}^{k} u_i(t) = v_i(t)) \supset x(t) = y(t))$$

it is sufficient to prove the formula

$$\phi_{1,2} = \forall t\ ((\bigwedge_{i=1}^{k} u_i(t) = v_i(t)) \supset x(t) = y(t)).$$

This need not yield a computational advantage (cf. Chandrasekhar et al., 1987). Yet it clearly cuts down the size of the requirement specification.

In the particular situation of processor verification, each premise equation $u_i(t) = v_i(t)$ in ϕ has one of two special forms. Either u_i is a function symbol denoting a latch or it is a term $v_{in}(t+n)$ for some input variable v_{in} and an integer constant n[3]. In both cases, we have $v_i(t) = R_i \in \{Low, High\}$. Thus $u_i(t_0) = R_i$ can always be oriented to the rule $u_i(t_0) \to R_i$, which does not superpose with any other rule in R_{HW}. Note that hardware variables denoting inputs of the processor may only occur in the right-hand sides of R_{HW}. Consequently

$$R_\phi = R_{HW} \cup \{u_i(t_0) \to R_i \mid 1 \leq i \leq k\}$$

is a complete term rewriting system and ϕ holds if and only if the equation $x(t_0) = y(t_0)$ holds modulo R_ϕ if and only if $x(t_0)$ and $y(t_0)$ have the same normal form w.r.t. R_ϕ. Figure 3 shows the anotated ReDuX specification of the formulae required to verify the STA instruction. The set of all the equations to be proved during the verification of the Sparrow-8 processor is given in the appendix.

The complete verification now consists of proving a (large) set of formulae of the above form each of which describes a small aspect of a correct processor behavior. Typically all formulae describing the behavior of the same instruction called in the same instruction fetch state use the same premises. So R_ϕ needs not to be reconstructed for every formula.

In order to verify the *Sparrow-4* processor we had to construct 45 term rewriting systems, and to prove about 1400 bit level equations (500 if data and address busses are represented as vectors as shown in Figure 3). For the *Sparrow-8* 60 rewrite systems had to be constructed and about 3400 (resp. 650) equations had to be proved.

See (Bündgen et al., 1995) for a comprehensive description of all verification requirements for *Sparrow-4*.

[3] $t+n$ may of course be denoted as the n-fold successor of t.

To prove the correctness of STA given that the processor is in the standard instruction fetch state we must add the following rules to $R_{Sparrow\text{-}8}$:

```
%% Reset: no reset signalled
[9110] RESET(t) == H;
%% standard instruction fetch state
[9501] V32(t0) == H;              [9502] V34(t0) == L;
%% Inputs for Instruction STA, Opcode: 0010
%% BLIF ids        DR7  DR6  DR5  DR4  DR3  DR2  DR1  DR0
%% ReDuX ids       In10 In6  In9  In5  In8  In4  In7  In3
[9120] IN(t0) == [I07, I06, I05, I04, L,   L,   H,   L];
```

where `t0` is a constant of sort `TIME`, `I04 ...`, `I07` are constants of sort `SIGNAL`, `[ ...]` constructs an 8-bit vector. `IN` is a function from sort `TIME` to 8-bit vectors. The processors inputs `DR`*i* correspond to the *i*-th component of `IN`. Using this augmented rewrite system we must prove the following:

```
rr(t0)== H;      rr(S(t0)) == H;      rr(S(S(t0))) == H;
rr(S(S(S(t0)))) == L;          rr(S(S(S(S(t0))))) == H;
```

meaning that during the first three clock cycles, data is read, in the fourth cycle data is written and after termination of the STA instruction the processor again expects to read an opcode. `rr` is a renaming of the output line for the RAM read signal. Next the address bus must be set correctly:

```
address(S(t0))             == Sv16(address(t0));
address(S(S(t0)))          == Sv16(Sv16(address(t0)));
address(S(S(S(t0))))       == dr(S(S(t0))) | dr(S(t0));
address(S(S(S(S(t0)))))== Sv16(Sv16(Sv16(address(t0))));
```

After reading the STA instruction the address must be incremented twice. In the subsequent cycle the address read in before must be on the address bus and after that the address must be set to read the next instruction. `address` is a function from the sort `TIME` to a 16-bit vector comprising the output data lines for the address bus. `Sv16` is specified as a function incrementing a 16-bit vector and | denotes the concatenation of two 8-bit vectors to a 16-bit vector. In addition both the accumulator (dw denotes an 8-bit vector describing the data write lines directly connected to the accumulator) and the carry bit must be left unchanged by STA.

```
dw(S(S(S(t0))))            == dw(t0);
carry(S(S(S(S(t0))))) == carry(t0);
```

Finally we show that after termination the processor is again in the standard instruction fetch state.

```
V32(S(S(S(S(t0))))) == H;    V34(S(S(S(S(t0))))) == L;
```

Figure 3. Proof requirements to verify the STA instruction

5. Verification and Disproof

5.1. *Symbolic Simulation*

A complete rewrite specification R_{HW} as described in Section 3 can be used to simulate the processor behaviour at clock cycles t_0 and later. Therefore the setting of all latches at time t_0 and all inputs at times $t_0 \leq t \leq t_0 + n$ must be known. This information must be added to R_{HW} as follows. Let

$$R_{ini}(t_0) = \{L_i(t_0) \to R_i \mid R_i \in \{Low, High\}, L_i \text{ a latch}\}$$

and

$$R_{inp}(t_0) = \{I_i(t_0 + k) \to R_i \mid R_i \in \{Low, High\}, 1 \leq k \leq n, I_i \text{ an input line}\}.$$

Then $R_{sim}(t_0) = R_{HW} \cup R_{ini}(t_0) \cup R_{inp}(t_0)$ is a complete term rewriting system and for output line O_i the $R_{sim}(t_0)$-normal form of $O_i(t_0 + k)$ describes the processor output at time $t_0 + k \leq t_0 + n + 1$.

Note that t_0 may either be a new constant or a ground term denoting a numerical value. For a simulation to make sense, care must be taken that $R_{ini}(t_0)$ describes a reachable processor state.

A simulation can be parameterized by leaving out some information from $R_{ini}(t_0)$ and/or $R_{inp}(t_0)$. The missing information will be dealt with symbolically and instead of truth values the simulation will result in Boolean expressions in those values for which no information was provided. This is why we call this technique *symbolic simulation*. Due to ω-completeness of $R_{BR} \subset R_{HW}$, symbolic simulation computes constant values whenever Boolean simplification results in constant values.

5.2. *Instruction Fetch States*

The control unit of a processor determines whether an instruction is fetched, an operand is read, the ALU performs a calculation, or whether a result is written. The control unit is normally realized as a small finite automaton. One of the states of this finite automaton is the *standard instruction fetch state*. However additional processor states must be treated as instruction fetch states too:

1. the starting state of the processor and
2. any state following an instruction fetch state if during the instruction fetch an invalid opcode has been read.

Both Sparrow designs used a finite automaton with four states (coded in two latches) one of which was intended to be the standard instruction fetch state.

In the course of the verification we found out that in both designs the JNC instruction (if the carry bit was set) terminated in a state different from the standard instruction fetch state. This state turned out to be an instruction fetch state too, provided that a JNC instruction was performed right before this state. A discussion with the designer of the chip showed that this additional instruction fetch state was not intended. It may have been introduced by an automated optimization step.

In a typical hardware verification situation the verifiers and the designers will cooperate and therefore the designers may inform the verifiers concerning the intended instruction fetch states. Yet this information may be incomplete. If there are additional instruction fetch states, they can be found using symbolic simulation.

5.3. *Partial Verification*

In contrast to other approaches to hardware verification (e. g., Chandrasekhar et al., 1987 and Narendran and Stillman, 1989) our verification method uses direct proofs. Therefore all equations proved during the verification are valid for the processor. This allows for a partial verification of the processor. E. g., we can prove the correctness of a single instruction w. r. t. to a set of instruction fetch states. Similarly we can restrict ourselves to the verification of the I/O-control of the processor, therefore omitting the proof of the functional correctness of the ALU. The latter case can be very important for the verification of very complex arithmetic instructions (like multiplication or division) that require special purpose verification methods. Then we may combine a rewrite based verification of the processor control with a different verification method for ALU functions.

5.4. *Fault Detection*

Given a complete term rewriting system R and a term t, let $t\downarrow_R$ denote the normal form of t w. r. t. R. Each equation $x(t_0) = y(t_0)$ to be proved consists of a description showing how to compute an output value $x(t_0)\downarrow_{R_\Phi}$ that is compared with a description $y(t_0)\downarrow_{R_\Phi}$ of how this value is computed using the requirement specification. If $x(t_0)$ and $y(t_0)$ have the same normal form the equation is proved. In any case the normal forms of $x(t_0)$ and $y(t_0)$ must be Boolean expressions in hardware variables that denote either latches or processor inputs. Otherwise the specification is incomplete.

If $x(t_0)$ and $y(t_0)$ have different normal forms we must show that $x(t_0)$ and $y(t_0)$ can actually evaluate to distinct truth values in a way that is consistent with our model (the processor design). That is to say we must show that there exists a reachable processor state $S(t_0)$ such that

$$S(t_0) \supset x(t_0)\downarrow_{R_\Phi} \neq y(t_0)\downarrow_{R_\Phi} .$$

Let $\sigma = \{v_1 \mapsto V_1, \ldots, v_n \mapsto V_n\}$ be a substitution where the v_i are first order variables and $V = \{V_1, \ldots, V_n\}$ is the set of ground terms in $x(t_0)\downarrow_{R_\Phi}$ and $y(t_0)\downarrow_{R_\Phi}$ that denote hardware variables for latches and processor inputs at times $t_0 + k_i$. Further let $\bar{x}$ and $\bar{y}$ be terms (generated by Boolean connectives, constants and first order variables only) such that $\bar{x}\sigma = x(t_0)\downarrow_{R_\Phi}$ and $\bar{y}\sigma = y(t_0)\downarrow_{R_\Phi}$. The equation $x(t_0) = y(t_0)$ can be disproved if there is a Boolean ring unifier β (Martin and Nipkow, 1988) such that $\bar{x}\beta\downarrow_{R_{BR}} = (\bar{y}\beta \; xor \; High)\downarrow_{R_{BR}}$ and the set

$$\{V_i = v_i\beta \mid 1 \leq i \leq n, V_i \text{ denotes a latch}\}$$

describes a reachable state of the hardware.

Hence using our method, disproving an equation is rather complicated because it involves an additional Boolean ring unification and an analysis of the reachable states of the processor. However in many cases a disproof is much less involved. Namely if V contains only inputs or latches that may take any value (like latches in program counters, accumulators etc.) then $x(t_0) = y(t_0)$ is disproved if

$$x(t_0)\downarrow_{R_\Phi} \neq y(t_0)\downarrow_{R_\Phi} .$$

In many other cases analyzing $x(t_0)\downarrow_{R_\Phi}$ and $y(t_0)\downarrow_{R_\Phi}$ will supply valuable hints to possible flaws which then may be detected by an appropriate simulation. As a matter of fact all flaws we detected[4] during our verification efforts did not require an explicit Boolean unification.

6. Efficient Normalization of Equations

As argued in Section 4, the verification mainly consists of normalizing a (large) set of equations. Using naïve rewriting methods this task becomes easily intractable. Our first trials verifying *Sparrow-4* ran for days on a SPARC 10-40 without producing any result. Here we propose three methods to improve over naïve normalization.

---

[4] both in the hardware and in the requirement specification

6.1. *Intelligent Memoization*

One observes that the normal form of some term $V(t_0+n)$ mostly depends on the normal form of terms at a previous cycle (say $W(t_0+n-1)$) and so on. Therefore it should be awarding to compute normal forms of hardware variables at earlier points in time first and store the results (as "lemmas") for later computations. This yields a normalization technique similar to reductions with memoization in OBJ3 (Goguen et al., 1993). Yet we cannot afford to (pre-)compute the values of all possible variables at times earlier than those needed in the actual equation to be proved. We decided to precompute only the normal forms of ground terms denoting the values of latches, hardware variables that expand to large terms and hardware variables that occur often in right-hand sides of R_{HW}.

Using this kind of improvement alone allowed for the verification of the *Sparrow-4* in less than four hours.

6.2. *Building-in Evaluation Domains*

A further improvement is possible if we use built-in data structures to model the Boolean ring properties of the Boolean operators. We therefore used partial evaluation domains (Bündgen and Lauterbach, 1996b). Let R be a complete rewrite specification with a complete subspecification R'. Then a model D of R' is a *partial evaluation domain* for R if D is a free algebra in the models of R' over a countable set of generators. In addition we need an efficiently computable mapping from an implementation of D to terms in R'-normal form. This mapping is fixed by a bijection from the generators of D to variables and terms outside of the signature of R'. The R'-normal form of any term t can be computed by retranslating the D-interpretation of t back to a term. Typical examples of evaluation domains are shown in Table I.

Given that R' is a proper subset of R, we must alternate evaluation based normalizations w.r.t. R' and reductions w.r.t. $R \setminus R'$ until a fixed point is reached in order to compute R normal forms. Experiments have shown that for many important rewrite specifications evaluation support drastically improves normalization times.

Normalization with support from built-in evaluation domains has been implemented in ReDuX (version 1.6). So far we have experimented with two different partial evaluation domains modeling the subspecification describing Boolean rings. The first is an implementation of functional decision diagrams (FDDs) (Kebschull et al., 1992) and the second is an implementation of Boolean polynomials (Stone, 1936) based on the SAC-2 computer algebra

Table I. Evaluation domains

canonical TRS for	evaluation domains
monoid	lists over the natural numbers
group	reduced words over the integers
Abelian groups	the functions from the naturals to the integers
commutative rings	multivariate integral polynomials
Boolean rings	Stone polynomials or functional decision diagrams (FDDs)

system. Note that for the special kind of term rewriting systems R_ϕ, the use of these evaluation domains totally eliminates the need for AC-matching.

Combining memoization techniques and building in partial evaluation domains we observed similar speed-ups for FDDs and Boolean polynomials for the verification of *Sparrow-4*. For the verification of *Sparrow-8* the FDDs outperformed Boolean polynomials. The same observation has been made for formulae with under-specified premises. Such formulae have very large normal forms in general.

The main difference between Boolean polynomials and FDDs is that the latter exploit structure sharing of common subformulae in an eager way. The probability of this feature playing a significant rôle increases with the size of the Boolean terms to be normalized.

6.3. *Parallelization*

The sheer amount of independent formulae to be proved suggests that these formulae should be proved in parallel. Yet by virtue of efficient memoization techniques not all formulae are completely independent. The proofs of all formulae with a common premise depend on the precomputation of hardware variables at earlier points in time than those occurring in the conclusions. Clearly precomputed lemmas should be shared for all these formulae. However the normal forms of hardware variables of the same generation (viewed at the same clock cycle) can be computed in parallel. Parallel normalization of equations has already been successfully employed in fine grained parallelization of completion procedures implemented on a shared memory multiprocessor (Bündgen et al., 1996a). For the verification of *Sparrow-8* a simple process level parallelization resulted already in a speed-up of 3.9 on a four processor SPARCstation.

7. Experiences and Results

Both the *Sparrow-4* and the *Sparrow-8* could be verified. More precisely, we found sensible requirement specifications w. r. t. which both designs are correct. Particularly for the 8-bit version it turned that the design did not fulfil the requirements for invalid opcodes. According to the requirements specification in (Rosenstiel and Schubert, 1995) any word where one of the upper four bits is set must be interpreted as invalid opcode. However using symbolic simulation the author found out that *Sparrow-8* was designed so that it ignores the upper four bits of a word to be interpreted as opcode. The latter specification corresponds to the intention of the designer of *Sparrow-8* which was designed before the latest version of (Rosenstiel and Schubert, 1995) was released. In both designs the synchronous reset did not work properly. Yet the Sparrow has an additional asynchronous reset that is not modelled by our specifications. This shows that the method presented in this chapter does not only perform well in verifying a correct circuit. It is also helpful in detecting errors in the design (or the specification).

Table II summarizes the results for verifications using ReDuX with FDDs as partial evaluation domains. It shows the time needed to verify each instruction measured on a SPARCstation with a 90 MHz hyperSPARC processor[5]. For most instructions the verification of the 8-bit variant was one order of magnitude slower than the corresponding verification of the 4-bit variant. Only the conditional jump was two orders of magnitude slower. The most striking figure of Table II is the enormous amount of time needed to verify the SUB instruction that dominates the whole verification. Remember that this part of the verification was of a completely different nature than the verification of the remaining instructions because it consisted of verifying hardware against software given by a non-trivial assembler program. Within the verification of SUB most time was spent verifying the computation of the ALU and in particular in computing the value of the carry bit. A reason for this may be that the specification of the subtraction is rather complicated (cf. Figure 4). As a matter of fact the normalized term describing the resulting carry bit takes more than 3000 lines to print out.

Table III shows the impact of built-in data structures for the Boolean ring. It shows that their importance increases with the size of the design. For a system of the size of *Sparrow-8* pure rewriting must be considered as intolerable and for larger designs FDDs perform better than Boolean polynomials.

We also parallelized the verification using process level parallelism on a four processor workstation. This resulted in speed-ups of over 3 for Sparrow-

[5] roughly equivalent to a 50 MHz superSPARC processor

Table II. Verification of *Sparrow-4* and *Sparrow-8*

Instruction	*Sparrow-4* proof time (CPU time)	*Sparrow-8* proof time (CPU time)
LDA	7.95 s	74.61 s
STA	13.98 s	135.02 s
JNC	23.43 s	2737.23 s
ADD	17.18 s	385.26 s
NOR	16.53 s	218.89 s
SUB	–	39588.88 s
invalid opcodes	36.71	326.83 s
Σ	1:56 min	12:24:57 h
	total time (wall clock time)	total time (wall clock time)
Σ	10:50 min	13:13:58 h

Table III. Verification of LDA using different evaluation domains

built-in evaluation domain	*Sparrow-4* proof time (CPU)	*Sparrow-8* proof time (CPU)
none	139.60 s	> 20 h
Boolean polynomials	11.63 s	323.30 s
FDDs	7.95 s	74.61 s

4 and of almost 4 for Sparrow-8 w. r. t. the wall clock times measured.

8. Concluding Remarks

Let us conclude with a few general remarks on rewrite based hardware verification.

For the verification we tried to keep things as simple as possible. Thus we could use theoretically well established techniques. These techniques were then tailored to the particular needs of the application. Exploiting some special properties of the application and powerful optimization methods for the most important operations form the basis for successfully for verifying specifications of non-trivial complexity.

The semantics of complete term rewriting systems is relatively simple and the way we specify the processor is rather straightforward and can be explained to hardware designers without a strong background in logic. The author considers this to be an important precondition for a verification technique to become eventually accepted by practitioners.

Another important aspect w. r. t. the acceptance of a verification method is its degree of automation. Our method is not completely automated yet. The functions needed for the requirement specification and the equations to be proved were specified by the verifier. This holds also for the tuning of the normalization by memoization. These are interesting open problems to be tackled in the future.

ACKNOWLEDGEMENTS

The author wants to thank Till Bubeck, the designer of *Sparrow-8* for answering many questions concerning his design.

REFERENCES

Berkeley: 1992, 'Berkeley Logic Interchange Format (BLIF)'. University of California, Berkeley, CA.

Berregeb, N., A. Bouhoula, and M. Rusinowitch: 1996, 'Automated Verification by Induction with Associative-Commutative Operators'. In: R. Alur and T. A. Henzinger (eds.): *Computer Aided Verification (LNCS 1102)*. pp. 220–231, Springer-Verlag. (Proc. CAV'96, New Brunswick, NJ, USA, July/August 1996).

Bubeck, T., U. Kebschull, and W. Rosenstiel: 1993, '— Sparrow — Ein Mikroprozessor-Entwurfs- und Ausbildungssystem'. Technical Report 93–07, Wilhelm-Schickard-Institut, Universität Tübingen, D-72076 Tübingen. (in German).

Bündgen, R.: 1993, 'Reduce the Redex → ReDuX'. In: C. Kirchner (ed.): *Rewriting Techniques and Applications (LNCS 690)*. pp. 446–450, Springer-Verlag. (Proc. RTA'93, Montreal, Canada, June 1993).

Bündgen, R., M. Göbel, and W. Küchlin: 1996a, 'Strategy Compliant Multi-Threaded Term Completion'. *Journal of Symbolic Computation* **21**(4–6), 475–505.

Bündgen, R., W. Küchlin, and W. Lauterbach: 1996b, 'Verification of the Sparrow Processor'. In: *IEEE Symposium and Workshop on Engineering of Computer-Based Systems*. IEEE Press, pp. 86–93. (Proc. ECBS'96, Friedrichshafen, Germany, March 1996).

Bündgen, R. and W. Lauterbach: 1996a, 'Combining Reductions and Computations in ReDuX'. In: M. Wirsing and M. Nivat (eds.): *Algebraic Methodology and Software Technology (LNCS 1101)*. pp. 633–636, Springer Verlag. (Proc. AMAST'96, Munich, Germany, July 1996).

Bündgen, R. and W. Lauterbach: 1996b, 'Experiments with Partial Evaluation Domains for Rewrite Specifications'. In: M. Haveraaen, O. Owe, and O.-J. Dahl (eds.): *Recent Trends in Data Type Specifications (LNCS 1130)*. pp. 125–142, Springer Verlag. (WADT'95, Oslo, Norway, September 1995, selected papers).

Bündgen, R., W. Lauterbach, and W. Küchlin: 1995, 'A License to Fly: Verifying the Sparrow-0 Processor'. Technical Report 95–04, Wilhelm-Schickard-Institut, Universität Tübingen, D-72076 Tübingen.

Bündgen, R., C. Sinz, and J. Walter: 1996c, 'ReDuX 1.5: New Facets of Rewriting'. In: H. Ganzinger (ed.): *Rewriting Techniques and Applications (LNCS 1103)*. pp. 412–415, Springer-Verlag. (Proc. RTA'96, New Brunswick, NJ, USA, July 1996).

Chandrasekhar, M. S., J. P. Privitera, and C. K. W.: 1987, 'Application of Term Rewriting Techniques to Hardware Verification'. In: *24th ACM/IEEE Design Automation Conference*. pp. 277–282.

Geser, A. and W. Küchlin: 1997, 'Structured Formal Verification of a Fragment of the IBM 390 Clock Chip'. In: *Technical Report 97-50, RISC Linz Report Series*.

Goguen, J. A., T. Winkler, J. Meseguer, K. Futatsugi, and J.-P. Jouannaud: 1993, 'Introducing OBJ'.

Hsiang, J.: 1985, 'Refutational Theorem Proving using Term-Rewriting Systems'. *Artificial Intelligence* **25**, 255–300.

Hunt, Jr, W. A.: 1989, 'Microprocessor Design Verification'. *Journal of Automated Reasoning* **5**(4), 429–460.

Kapur, D. and M. Subramaniam: 1996, 'Mechanically Verifying a Family of Multiplier Circuits'. In: R. Alur and T. A. Henzinger (eds.): *Computer Aided Verification (LNCS 1102)*. pp. 135–146, Springer-Verlag. (Proc. CAV'96, New Brunswick, NJ, USA, July/August 1996).

Kebschull, U., E. Schubert, and W. Rosenstiel: 1992, 'Multilevel Logic Synthesis Based on Functional Decision Diagrams'. In: *Proc. EDAC 1992*.

Martin, U. and T. Nipkow: 1988, 'Unification in Boolean Rings'. *Journal of Automated Reasoning* **4**, 381–396.

Narendran, P. and J. Stillman: 1989, 'Formal Verification of the Sobel Image Processing Chip'. In: G. Birtwistle and P. A. Subrahmanyam (eds.): *Current Trends in Hardware Verification and Automated Theorem Proving*. New York, NY: Springer-Verlag, pp. 92–127.

Rosenstiel, W. and E. Schubert: 1995, 'Praktikum VLSI-Design mit VHDL, WS 95/96'. Course notes for the VLSI desing laboratoy at the University of Tübingen (in German).

Sekar, R. C. and M. K. Srivas: 1989, 'Formal Verification of a Microprocessor Using Equational Techniques'. In: G. Birtwistle and P. A. Subrahmanyam (eds.): *Current Trends in Hardware Verification and Automated Theorem Provin*. New York, NY: Springer-Verlag, pp. 171–217.

Staunstrup, J., S. J. Garland, and J. V. Guttag: 1992, 'Mechanized Verification of Circuit Descriptions using the Larch Prover'. In: V. Stavridou, T. F. Melham, and R. T. Boute (eds.): *Theorem Provers in Circuit Design*. pp. 277–299, North-Holland.
Stone, M. H.: 1936, 'The Theory of Representations of Boolean Algebras'. *Trans. American Math. Society* **40**, 37–111.
Thole, P., U. Kebschull, E. Schubert, and W. Rosenstiel: 1993, 'The design and implementation of an educational computer system based on FPGAs'. In: *Proc. 4th Eurochip Workshop on VLSI Design Training*.
XILINX: 1991, 'XILINX The Programmable Gate Array Data Book'. XILINX Inc., 2100 Logic Drive, San Jose, CA 95124.

A. The Specification of the SUB Instruction

The SPASS macro for the SUB instruction is shown in Figure 4. It uses the

```
;"subtract" from accu
;This macro replaces the internal hardware command 'sub
;address'. This is sometimes necessary, because not all
;CPUs have a SUB-Command.
.macro          sub     address
                sta     ~normal_sub+1
                sta     ~special_sub+1
                ldam    ?1
                not
                add     _one
                sta     _tmp_sub
                jnc     ~normal_sub
~special_sub:   lda     #0
                jmp     ~clear_carry
~normal_sub:    lda     #0
                add     _tmp_sub
                jnc     ~set_carry
~clear_carry:   clc
                jnc     ~end
~set_carry:     sec
~end:
.mend
```

Figure 4. The SPASS macro for the SUB instruction

SPASS macros load accumulator from memory (`ldam` *address*), invert all bits of accumulator (`not`), jump to address (`jmp` *address*), clear carry bit (`clc`) and set carry bit (`sec`). Symbolic addresses start with an underscore (_one should contain the value 1) constant values are preceeded by a hash sign and labels (relative addresses) are preceded by a tilde.

The `sub` macro first computes the two's complements of the argument. If this operation raises an overflow the accumulator is restored and the carry is set otherwise the result is added to the original contents of the accumulator. If the carry is set after the SUB instruction no (intermediate) overflow occurred.

B. Equations to be Proved

Here we want to present all equations that have been proved for the verification of *Sparrow-8*. The equations are given in ReDuX syntax.

B.1. *Instruction Fetch States*

The following sets of equations describe the different intruction fetch states. The variables `V32` and `V34` store the base state of a finite automaton. The vector `[V46, V42, V44, V56, V234, V238, V242, V246]` stores the last instruction code read in.

The regular state:

```
[9501] V32(t0)    == H;  [9502] V34(t0)    == L;
```

The starting state for $t_0 = 0$:

```
[9501] V32(t0)    == L;  [9502] V34(t0)    == L;
[9503] V46(t0)    == L;  [9504] V42(t0)    == L;
[9505] V44(t0)    == L;  [9506] V56(t0)    == L;
[9506] V234(t0)   == L;  [9506] V238(t0)   == L;
[9506] V242(t0)   == L;  [9506] V246(t0)   == L;
```

The state after a successful JNC instruction with carry bit set :

```
[9501] V32(t0)    == L;  [9502] V34(t0)    == L;
[9503] V46(t0)    == H;  [9504] V42(t0)    == H;
[9505] V44(t0)    == L;  [9506] V56(t0)    == L;
[9507] V74(t0)    == H;
```

The next six equation sets describe states that may be reached after an invalid instruction:

1. `[9501] V32(t0)    == L;  [9502] V34(t0)    == H;`

```
     [9503] V46(t0)    == H;  [9504] V42(t0)    == L;
     [9505] V44(t0)    == L;  [9506] V56(t0)    == L;
     [9507] V242(t0)   == L;  [9508] V238(t0)   == L;
     [9509] V234(t0)   == L;  [9510] V246(t0)   == L;
2.   [9501] V32(t0)    == L;  [9502] V34(t0)    == H;
     [9503] V46(t0)    == H;  [9504] V42(t0)    == H;
     [9505] V44(t0)    == H;  [9506] V56(t0)    == L;
     [9507] V242(t0)   == L;  [9508] V238(t0)   == L;
     [9509] V234(t0)   == L;  [9510] V246(t0)   == L;
3.   [9501] V32(t0)    == L;  [9502] V34(t0)    == H;
     [9506] V56(t0)    == H;  %% OPC3
4.   [9501] V32(t0)    == L;  [9502] V34(t0)    == H;
     [9508] V238(t0)   == H;  %% OPC5
5.   [9501] V32(t0)    == L;  [9502] V34(t0)    == H;
     [9509] V234(t0)   == H;  %% OPC6
6.   [9501] V32(t0)    == L;  [9502] V34(t0)    == H;
     [9510] V246(t0)   == H;  %% OPC7
```

B.2. *ADD, NOR, SUB*

For each instruction fetch state the following equations must be proved in order to verify the ADD, NOR and SUB instructions:

```
rr(t0)                        == H;
rr(S(t0))                     == H;
rr(S(S(t0)))                  == H;
rr(S(S(S(t0))))               == H;
rr(S(S(S(S(t0)))))            == H;
address(S(t0))                == Sv16(address(t0));
address(S(S(t0)))             == Sv16(Sv16(address(t0)));
address(S(S(S(t0))))          == dr(S(S(t0))) | dr(S(t0));
address(S(S(S(S(t0)))))       == Sv16(Sv16(Sv16(address(t0))));
V32(S(S(S(S(t0)))))           == H;
V34(S(S(S(S(t0)))))           == L;
```

In addition the following equations must be proved to verify the ADD instruction

```
dw(S(S(S(S(t0)))))            == addv8(dw(t0),dr(S(S(S(t0)))));
carry(S(S(S(S(t0)))))         == add7c(dw(t0),dr(S(S(S(t0)))));
```

the NOR instruction

```
dw(S(S(S(S(t0)))))            == norv8(dw(t0),dr(S(S(S(t0)))));
```

```
carry(S(S(S(S(t0)))))     == carry(t0);
```

and the SUB instruction

```
dw(S(S(S(S(t0)))))        == subv8(dw(t0),dr(S(S(S(t0)))));
carry(S(S(S(S(t0)))))     == subcarry(dw(t0),
                                 dr(S(S(S(t0)))),carry(t0));
```

B.3. *JNC*

For each instruction fetch state the following equations must be proved in order to verify the JNC instruction:

```
rr(t0)                  == H;
rr(S(t0))               == H;
rr(S(S(t0)))            == H;
rr(S(S(S(t0))))         == H;
address(S(t0))          == Sv16(address(t0));
address(S(S(t0)))       == Sv16(Sv16(address(t0)));
dw(S(S(S(t0))))         == dw(t0);
carry(S(S(S(t0))))      == carry(t0);
V34(S(S(S(t0))))        == L;
```

If the carry bit is set the following equations must be proved additionally:

```
address(S(S(S(t0)))) == Sv16(Sv16(Sv16(address(t0))));
V32(S(S(S(t0))))        == L;
Out13(S(S(S(t0))))      == L; %% OPC3
Out12(S(S(S(t0))))      == L; %% OPC2
Out11(S(S(S(t0))))      == H; %% OPC1
Out10(S(S(S(t0))))      == H; %% OPC0
```

If the carry bit is not set the next two equations must be proved instead:

```
address(S(S(S(t0)))) == dr(S(S(t0))) | dr(S(t0));
V32(S(S(S(t0))))        == H;
```

B.4. *LDA*

For each instruction fetch state the following equations must be proved in order to verify the LDA instruction:

```
rr(t0)               == H;
```

```
rr(S(t0))            == H;
rr(S(S(t0)))         == H;
address(S(t0))       == Sv16(address(t0));
address(S(S(t0)))    == Sv16(Sv16(address(t0)));
dw(S(S(t0)))         == dr(S(t0));
carry(S(S(t0)))      == carry(t0);
V32(S(S(t0)))        == H;
V34(S(S(t0)))        == L;
```

B.5. *STA*

See Figure 3.

B.6. *Invalid Opcodes*

For each instruction fetch state the following equations must be proved in order to verify that invalid instruction codes are handeled correctly:

```
rr(t0)            == H;
rr(S(t0))         == H;
address(S(t0))    == Sv16(address(t0));
dw(S(t0))         == dw(t0);
carry(S(t0))      == carry(t0);
V32(S(t0))        == L;
V34(S(t0))        == H;
```

INDEX

APPLIED LOGIC SERIES

1. D. Walton: *Fallacies Arising from Ambiguity*. 1996 ISBN 0-7923-4100-7
2. H. Wansing (ed.): *Proof Theory of Modal Logic*. 1996 ISBN 0-7923-4120-1
3. F. Baader and K.U. Schulz (eds.): *Frontiers of Combining Systems*. First International Workshop, Munich, March 1996. 1996 ISBN 0-7923-4271-2
4. M. Marx and Y. Venema: *Multi-Dimensional Modal Logic*. 1996 ISBN 0-7923-4345-X
5. S. Akama (ed.): *Logic, Language and Computation*. 1997 ISBN 0-7923-4376-X
6. J. Goubault-Larrecq and I. Mackie: *Proof Theory and Automated Deduction*. 1997 ISBN 0-7923-4593-2
7. M. de Rijke (ed.): *Advances in Intensional Logic*. 1997 ISBN 0-7923-4711-0
8. W. Bibel and P.H. Schmitt (eds.): *Automated Deduction - A Basis for Applications*. Volume I. Foundations - Calculi and Methods. 1998 ISBN 0-7923-5129-0
9. W. Bibel and P.H. Schmitt (eds.): *Automated Deduction - A Basis for Applications*. Volume II. Systems and Implementation Techniques. 1998 ISBN 0-7923-5130-4
10. W. Bibel and P.H. Schmitt (eds.): *Automated Deduction - A Basis for Applications*. Volume III. Applications. 1998 ISBN 0-7923-5131-2

(Set vols. I-III: ISBN 0-7923-5132-0)

KLUWER ACADEMIC PUBLISHERS – DORDRECHT / BOSTON / LONDON

MIX
Papier aus verantwortungsvollen Quellen
Paper from responsible sources
FSC® C105338

If you have any concerns about our products,
you can contact us on
ProductSafety@springernature.com

In case Publisher is established outside the EU,
the EU authorized representative is:
Springer Nature Customer Service Center GmbH
Europaplatz 3, 69115 Heidelberg, Germany

Printed by Libri Plureos GmbH
in Hamburg, Germany